Business Experiments with R

Business Experiments with R

B. D. McCullough
Drexel University

Published by John Wiley & Sons, Inc., Hoboken, New Jersey.
Published simultaneously in Canada.

Registered Office
John Wiley & Sons, Inc., 111 River Street, Hoboken, NJ 07030, USA

Editorial Office
The Atrium, Southern Gate, Chichester, West Sussex, PO19 8SQ, UK

For details of our global editorial offices, customer services, and more information about Wiley products, visit us at www.wiley.com.

Wiley also publishes its books in a variety of electronic formats and by print-on-demand. Some content that appears in standard print versions of this book may not be available in other formats.

Library of Congress Cataloging-in-Publication Data

Names: McCullough, Bruce D., author.
Title: Business experiments with R / Bruce. D. McCullough, Drexel University.
Description: Second edition. | Hoboken : Wiley, 2021. | Includes index.
Identifiers: LCCN 2020029253 (print) | LCCN 2020029254 (ebook) | ISBN 9781119689706 (hardback) | ISBN 9781119689904 (adobe pdf) | ISBN 9781119689881 (epub)
Subjects: LCSH: Commercial statistics.
Classification: LCC HF1017 .M366 2021 (print) | LCC HF1017 (ebook) | DDC 658.4/034–dc23
LC record available at https://lccn.loc.gov/2020029253
LC ebook record available at https://lccn.loc.gov/2020029254

Cover Design: Wiley
Cover Image: © MR.Cole_Photographer/Getty Images

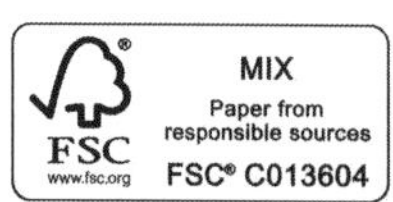

To my lovely wife and wonderful children

Contents

Preface

Rigorous experiments are rapidly becoming a critical tool for businesses. While experiments have been a popular tool in engineering for decades, driven largely by the quality movement of the 1980s, experiments are rapidly coming out of the engineering lab to be deployed more broadly across business operations. Companies today are using experiments to solve a broad range of problems from determining whether improved training results in better employee productivity and retention to gauging whether store renovations result in higher profitability, to determining which subject lines in a promotional email attract customer attention, to assessing which behavioral nudges are most likely to get utility users to shift energy consumption to nonpeak times, and to deciding which color to use for the checkout buttons on a website. Articles in the business press such as those by Davenport (2009), Anderson and Simester (2011), and Thomke and Manzi (2014) have made the case that experiments are a key component of "data-driven" business and companies are taking note and incorporate experiments into their day-to-day operations. With this rapid adoption of experimentation, there is a need for a comprehensive resource for business leaders and analysts who want to use experiments to understand how business systems work and solve tactical business problems.

Yet, material on business experiments remains siloed within specific business functions. For example, there are dozens of books and articles on website testing that focus intently on experimentation with websites and never mention that the same basic ideas – randomization, measurement, and comparison – can be applied in other business domains. Our goal for this book is to step back from this highly tactical approach and give business students a broad, theoretically grounded understanding of the role of experiments in business, providing them with a practical foundation for designing and analyzing experiments that will be useful to them throughout their careers.

Of course, experiments have long been a foundational tool in engineering, physical sciences, psychology, medicine, and statistics, and excellent books on how to design and analyze experiments are available within each of these fields, yet none of these books provides the right coverage for business students. The literature on experiments in the physical sciences and engineering provides great depth on multivariable test design, yet often focuses on situations where the experimenter is studying a physical process in a lab setting, where everything can be measured and controlled and error can be reduced nearly to zero – conditions that seldom hold in business. The classic textbook on experimental design for engineers by Montgomery (2017) naturally assumes a level mathematical skill that is not common among business students. Psychologists like Campbell and Stanley (1963) focus more on experiments with human subjects, raising many of the same issues that come up in marketing and human resources experiments, yet they focus intently on using experiments to develop theory and so cover issues like construct validity that are less relevant in the types of tactical experiments that business uses to drive day-to-day decisions. Useful discussions of the practicalities of running experiments in the field have appeared recently in the economics literature championed by Harrison and List (2004) and Gerber and Green (2012), yet the experiments reported tend to focus on testing economic theory. In many ways, the types of tactical experiments that are valuable in business are much more similar to clinical trials and other biomedical experiments, where there is a tradition of tactical experiments designed to determine, "Should we give this treatment to patients?" and a well-established culture of randomized controlled trials as the gold standard of evidence. While much of this work from other disciplines is relevant to business experiments, it doesn't provide students with concrete examples of how to apply these methods to answer important business questions.

This book has drawn from all of these literatures selecting the key ideas that *business* students need to know about experiments. This book approaches experimentation as a skill that one becomes better at over time, like playing a sport or an instrument, which requires equal measures of creativity and technical skill. To help students practice, the book is structured around a series of examples, focusing first on a specific business question, then turning to the design or analysis of an experiment that then will answer that question, and then building up the appropriate statistical methods. This way, the technical material is presented in context. This approach provides an ideal "second course in statistics," where students can review material they have seen before (specifically two-sample tests, power, ANOVA, confidence intervals, and regression) in the context of business problems. My hope is that students who work through this book will move beyond the mechanics and develop their abilities to frame ill-defined problems, determine what data and analysis would provide information about that problem, and examine the evidence for or against a particular business decision.

For clarity in the use of pronouns, I have adopted the convention that the analyst is always female. Other persons are male.

Suggested Courses Using This Book

This book can be used both with general business students, such as MBA or executive education students, and with the growing number of master's programs in business analytics and marketing analytics.

The first four or five chapters (depending on instructor preference) provides students with the tools they need to design and analyze two-treatment experiments (i.e. A/B tests) to answer business questions, focusing on the strategic and technical issues involved in designing experiments that will truly affect organizations. The book begins with a discussion of the importance of causal analysis in business and the risks of observational data analysis in Chapter 1. By design, the next chapter begins with the analysis of A/B tests, with a review of the statistical methods required for analyzing experiments, focusing on visualization and hypothesis testing. Once students understand how experiments are analyzed, in Chapter 3, they can move on to understanding issues involved in designing experiments including selecting treatments, measuring responses, planning sample sizes, and the importance of randomization for causal interpretation. Chapter 4 returns to the analysis of A/B tests, covering several advanced analysis techniques including analyzing tests with matched pairs, analyzing tests with more than two treatment groups, analyzing subgroups within an A/B test, and determining the minimal detectable effect for an experiment that has been completed. Chapter 5, which covers issues involved in the design of A/B tests when the number of units available for tests is small, as is the case when doing geo-testing or testing across store locations. This coverage makes Part I ideal for a half-semester graduate course for second-year MBA students or strong undergraduate students. These chapters can be covered over five weeks, skipping some of the more technical material and allowing time for the students to apply the material to a project where they plan, execute, and analyze an A/B test.

Part II builds on the foundation in Part I, expanding to multivariable testing. Chapter 6 begins Part II with a review of the primary tool used to analyze multivariate tests: regression analysis. Next comes full and fractional factorial designs in Chapters 7 and 8,which builds students' understanding of the core issues in multivariate design: interactions and confounding. Fractional factorial designs are useful for building understanding. Next comes custom design, which is ideal for designing experiments in constrained business contexts. The development of software for optimal design has made this sophisticated tool accessible to users, and my goal for Chapter 9 is to provide students with a basic introduction to optimal design, while avoiding the algorithmic details that would be more appropriate to a "technical manual." Parts I and II together form the core of a full-semester course that is appropriate for more analytically oriented graduate students including MBA students concentrating in analytics and master's of business analytics students.

Not all experiments are technically complicated. There is a great deal of "low-hanging fruit" in the business world that can be profitably plucked by persons whose knowledge of experimental design does not exceed 2^k designs. The goal of this book is to enable students who have taken business statistics to pluck that fruit. In particular, after finishing the first half of this book, the student should be able to conduct an A/B test. After finishing the second half, the student should be able to design, execute, and analyze a 2^k experiment. To this end, instructors should require students to complete a project where they design, execute, analyze, and report on an experiment. Students benefit greatly from the experience, which helps them see the challenges involved in designing a conclusive experiment and cement the ideas in the book by applying them in practice. A/B tests are easy to dream up, multivariable tests not so much. Hence, Box's helicopter experiment is very important. It is perhaps the only easily executed multivariable experiment available to students that is amenable to screening designs, full factorial designs, and even custom designs. By the end of this book, the reader should be able to execute Box's helicopter experiment Box (1992a) even if the professor does not make it part of the class.

Philadelphia, PA

B. D. McCullough

Acknowledgments

Without constant support and encouragement over the past three years from my esteemed colleague, Dr. Elea Feit, this book would never have been started, let alone finished. She allowed me to sit in on the business experiments course that she developed at Wharton and brought to Drexel. Despite the publish-or-perish demands of an assistant professor's life, she took the time to read and comment on some chapters, suggested many books and articles for me to read, and always had time for me when I ran into difficulties. Her influence can be found in every chapter, but any errors or omissions are mine and mine alone.

I thank Russell Lavery and three sections of STAT 335 students for reading some of the early chapters, and I thank Malcolm Hazel for reading all the chapters.

B. D. McCullough

Bruce McCullough

Bruce McCullough passed away in Fall 2020, just before this book was published. Bruce was an extraordinary scholar and friend, who will be greatly missed. He was fair, direct, true-to-his-word, and expected the highest level of rigor and integrity from himself and those around him. He was also unusually protective of those he took into his care including junior colleagues like me, as well as his wife and children. I'm sure it troubles him greatly that he can't be here in-person to see us all through.

Based on many discussions we had about the book, I know he felt just as protective towards his readers. He devoted the last few years of his professional life to providing you with a book that would open up the potential of experimentation in business and prevent you from making mistakes when you design and analyze your own business experiments. I hope you enjoy learning from Bruce as much as I did.

Elea McDonnell Feit
Philadelphia, PA

About the Companion Website

This book is accompanied by a companion website:

www.wiley.com/go/mccullough/businessexperimentswithr

The website includes datasets.

1

Why Experiment?

We can learn from data, but what we can learn depends on the way the data were generated. When nature generates the data and all variables are free to operate as they will, it is very easy to learn about correlations. Very often though, we don't need to know that price decreases are correlated with increases in quantity sold, but we need to know by how much the quantity sold will increase when price is decreased by a specific amount, and correlations are not up to this task. To learn about causation, we need to restrict the way some variables can act, and this can only be done with an experiment. We don't need sophisticated statistical methods to conduct experiments; the statistics learned in a first non-calculus-based statistics course often is more than sufficient to conduct a wide variety of useful experiments. Before conducting any experiments, we need to be very precise about the reasons that observational data are not up to the task of yielding causal insights.

After reading this chapter, students should:

- Distinguish between observational and experimental data.
- Understand that observational data analysis identifies correlation, but cannot prove causality.
- Know why it is difficult to establish causality with observational data.
- Understand that an experiment is a systematic effort to collect exactly the data you need to inform a decision.
- Explain the four key steps in any experiment.
- State the "Big Three" criteria for causality.
- Identify the conditions that make experiments feasible and cost effective.
- Give examples of how experiments can be used to inform specific business decisions.
- Understand the difference between a tactical experiment designed to inform a business decision and an experiment designed to test a scientific theory.

Business Experiments with R, First Edition. B. D. McCullough.

Companion Website: www.wiley.com/go/mccullough/businessexperimentswithr

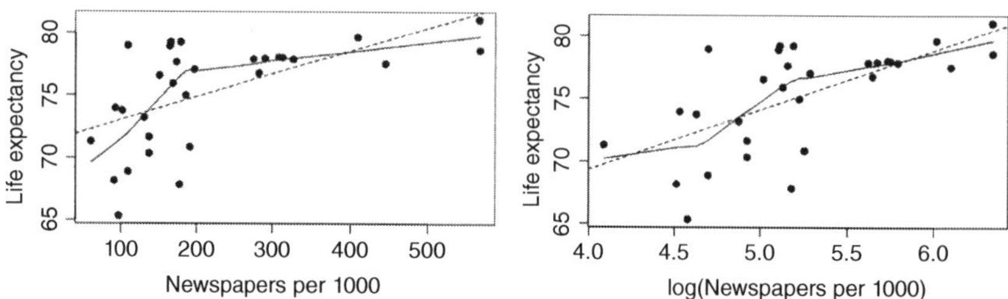

Figure 1.1 Life expectancy vs. newspapers per 1000 (left) and log(newspapers per 1000) (right) for several countries.

1.1 Case: Life Expectancy and Newspapers

Suppose we are interested in determining the reasons that some countries have long life expectancies while others do not. We might begin by examining the relationship between life expectancy and other variables for various countries. The left panel of Figure 1.1 shows a scatterplot of the average life expectancy for several countries versus the number of newspapers per 1000 persons in each country. The data are in `WorldBankData.csv`. The fitted linear regression line compared to the data shows the obvious curvature, and linear regression is not appropriate. In the usual fashion, linearity is induced by applying the natural logarithm transformation to the independent variable, as shown in the right panel. The log data are still not completely linear, but are much more linear than the original data.

Software Details

Reproduce the above graphs using the data file `WorldBankData.csv`...

Below is code for the first graph. To create the next graph, you will have to create a new variable, the natural logarithm of newspapersper1000.

```
df <- read.csv("WorldBankData.csv")  # "df" is the data frame.

plot(df$newspapersper1000,df$lifeexp,xlab="Newspapers per1000",
     ylab="Life Expectancy",pch=19,cex.axis=1.5,cex.lab=1.15)

abline(lm(lifeexp~newspapersper1000,data=df),lty=2)
lines(lowess(df$newspapersper1000,df$lifeexp))

plot(log(df$newspapersper1000),df$lifeexp,xlab="log(Newspapers
     per1000)",ylab="Life Expectancy",pch=19,cex.axis=1.5,
     cex.lab=1.15)

abline(lm(lifeexp~log(newspapersper1000),data=df),lty=2)
lines(lowess(log(df$newspapersper1000),df$lifeexp))
```

To analyze these data, we can run a regression of life expectancy (LE) in years against the natural logarithm of the number of newspapers per 1000 persons (LN) for a large number of countries in a given year. The results are

$$\text{LE} = \underset{(5.6)}{50.4} + \underset{(1.1)}{4.7}\text{LN} \tag{1.1}$$

where standard errors are in parentheses, so both the coefficients have very high *t*-statistics and are significant. This means that there is a relationship between life expectancy and the number of newspapers per 1000 people. But does this show that a country having more newspapers *leads* to longer lives for its citizens? Common sense says probably not. The natural logarithm of the number of newspapers is probably a proxy for other variables that drive life expectancy; countries that can afford newspapers can probably also afford better food, housing, and medical services. What we are observing is most likely a mere correlation, and, unfortunately, this sort of observational analysis should not be interpreted as *causal*.

Try it!

Run the above simple regression. You should get the same coefficients and standard errors.

A better analysis would add more variables to the regression to "control" for other factors. So, let's try adding other variables that we expect to drive life expectancy: LHB (natural logarithm of the number of hospital beds per 1000 in the country), LP (natural logarithm of the number of physicians per 1000 in the country), IS (an index of improvements in sanitation), and IW (an index of improvements in water supply). Since we don't believe that newspapers cause longer life expectancy, we would expect that once we include these variables in the regression, the coefficient on LN will be reduced. The results are

$$\text{LE} = \underset{(24.7)}{33.5} + \underset{(1.3)}{3.7}\text{LN} - \underset{(1.2)}{3.2}\text{LHB} - \underset{(1.6)}{0.31}\text{LP} + \underset{(0.14)}{0.06}\text{IS} + \underset{(0.37)}{0.23}\text{IW} \tag{1.2}$$

The coefficient on LN has not gone to zero; in fact, it hasn't changed much. The coefficients on all but one of the other variables that we know affect life expectancy are insignificant. What are we to make of this?

Try it!

Run the above multiple regression. You should get the same coefficients and standard errors. Be sure you understand why the variables LN and LHB are "significant" while the others are not.

In reality, life expectancy is affected by a large set of variables in a complex way, and the natural logarithm of newspapers is a good proxy for these other variables. If we have some beliefs about which variables are more likely to be the true causes of an increase in life expectancy, we might be able to build a model that we think represents the cause and effect relationships. If we really want to find the causal effect of newspapers, we might also try more sophisticated methods that involve trying to find and compare countries that are similar in all respects except the number of newspapers per 1000 people. This sort of analysis leads into "the garden of forking paths," a phrase used by the statistician Andrew Gelman (who specializes in causal inference) to describe the many decisions a researcher may take that can lead the researcher to unknowingly reaching spurious statistical conclusions.

For example, if an analyst analyzing the above data actually added variables and dropped variables until LN was insignificant and some other health-related variables all were significant, she would have taken a trip through the garden of forking paths and would come up with a useless model. The model would be useless because she tested many hypotheses on the same set of data and her "results" almost assuredly are contaminated by false positives (i.e. type I errors): she thinks the coefficients are significant when they're really not.

To better illustrate this idea, let us have 10 covariates (independent variables) to use in building a model to describe a particular dependent variable, and we will be allowed to include anywhere from 1 to all 10 of the variables. For each variable there is a decision (fork) to include or exclude the variable. Then there are 2^{10} possible models to choose from. A researcher just tries a sufficient number of models, dropping and including variables, until $p < 0.05$ at which point she freezes the model, and the choice of variables to include is justified after the variables have been included. Even a researcher who does not deliberately try all possible models still will make choices about including and excluding variables ("I thought X1 would be significant, but it wasn't, so I dropped it and tried X2"), which implies that her model is but one of many possible models that she just happened to select. Because of the garden of forking paths, a seemingly objective analysis is really quite subjective, and causality cannot be determined from subjective analyses.

The purpose of this example is to drive home the point that, in general, observational data simply are not up to the task of answering causal questions. In this book, we focus on an alternative approach to answering business questions, which is to conduct experiments.

We do not suggest that observational studies have no valid uses. To the contrary, there are many situations when experiments are not possible, and in such cases, there is no alternative to the use of observational data:

- Sometimes it is impossible to run an experiment. For example, it would be unethical to randomly assign people to smoke versus not smoke, so our

understanding of the causal relationship between smoking and cancer was built on observational data. (However, it took a long time to convince everyone, since observational data is easy to question.)
- If you want to build a new store, it is foolish to construct several stores in random locations to test hypotheses about where to locate stores.
- Establishing causality is not always necessary, and documenting correlations is sometimes sufficient for the purpose at hand. In fact, the whole field of "predictive analytics" focuses on prediction problems, where causality is not important. For example, if we are predicting defaults on mortgages, very often we only need to know the probability that a person will default, not the causal factors that determine the probability of default; correlation is sufficient, and causation is not necessary.
- The outcome of interest is sufficiently rare that running an experiment with a large enough number of trials is expensive. Perhaps you can only afford a sample size of 100, but the response rate is 2%; you're never going to get a good estimate of the response rate with such a comparatively small sample.
- Each trial is so expensive that even a small experiment is too expensive.
- The population is too small to support an experiment. There is no point in remodeling a sample of stores to decide whether to remodel the entire population if the population is only 10 stores.
- Hypothesis generation is an excellent use of observational data. The researcher explores the observational data looking for interesting ideas that might merit a follow-up experiment.

Observational data can be used to shed light on causal questions, but the statistical machinery necessary to do so is very sophisticated and still not as good as actually running an experiment. These types of sophisticated analyses are usually performed when an experiment is impossible, not in lieu of an experiment. A standard reference for this type of analysis is Rosenbaum (2010). Using observational data to answer causal questions is hard; so we try to use experimental data, which is comparatively easy.

Notwithstanding the above, it is *very* common for people to use observational data to (mistakenly) make causal assertions outside of the narrow range in which it can be done. It is very common, therefore, for "causal" results from observational data to be later contradicted by experiments. Many instances can be found in the article by Young and Karr (2011) and the references therein. With respect to establishing causality, unless you are a very skilled statistician, analyzing observational data can only provide what researchers call hypothesis-generating evidence. In other words, analyzing observational data can give you good ideas for experiments to conduct, but it can't tell you anything about causality.

It is important for the analyst to know whether she has an "umbrella problem" or a "rain dance problem." If all she wants to know is whether or not she should carry an umbrella, then she has a pure prediction problem and causal questions are of secondary importance; she only needs to know whether the probability of rain is high or low. On the other hand, if there has been a long drought and she wants to end it, prediction is of little value: causal questions are of primary importance. If she wants to induce rainfall, she needs to know what variables cause rain and then try to manipulate those variables. We recall the experience of one analyst, trained in the field of predictive analytics, who had been given some observational data on donations from specific individuals and asked for insight as to whether or not attending fundraisers increased donations. She dropped the variable "number of fundraisers the person attended" because it had no predictive value. She just couldn't understand that her boss had given her a rain dance problem, not an umbrella problem.

Exercises

1.1.1 Find an example of an observational study. Answer the following questions: (i) What makes this an observational rather than an experimental study? (be specific) (ii) What is the purpose of the study? (iii) What is the primary variable of interest?

1.1.2 For each of the following, indicate whether the data are observational or experimental, and defend your answer.

(a) A broadcaster moves a popular television show from Tuesday to Thursday, and its viewership increases.

(b) A psychologist wants to know how often students take a break from studying. To do this, he installs cameras in the library's reading room for one day. He notes that students take more breaks in the evening.

(c) A sports manufacturer wonders if his quick-dry exercise shirts dry more quickly than regular shirts. He finds some basketball players playing a game in the gym. He gives one team quick-dry shirts, and the other team gets regular 100% cotton shirts.

(d) A child psychologist asks several parents whether their children play violent video games. He also asks how many times a week their children display violent behavior. He finds that children who play violent video games display more violent behavior.

1.1.3 Give two situations where experiments can't be conducted.

1.2 Case: Credit Card Defaults

You work for a credit card company, and you want to figure out which customers might default. In the `credit.csv` dataset are 30 000 observations on six variables: credit limit (how much can be charged on the credit card), sex of the cardholder, education level of the cardholder (high school, undergrad, grad, other), whether the cardholder is married (single, married, other), the age of the cardholder in years, and whether or not the cardholder defaulted (1 = default, 0 = non-default).

In this problem we are confronted with the ultimate questions confronting all credit issuers: whether to grant credit to each potential customer and, if so, how much? Generally, we don't want to give credit to people who are likely to default, and if we do give credit, we don't want to give more than the person can repay.

A simple crosstab in Table 1.1 with the data shows that men are more likely to default than women. Another crosstab in Table 1.2 shows that divorced/widowed (other) persons are more likely to default.

Try it!

We encourage you to replicate the analysis in this chapter using the data in the file `credit.csv`. Computing crosstabs can be done in a spreadsheet using pivot tables. Most statistical tools also have a cross-tabulation function.

```
df <- read.csv("credit.csv",header=TRUE)
# Table 1.1
table1 <- table(df$default,df$sex) # to get the counts
table1  # to print out the table

prop.table(table1,2) # to get column proportions
prop.table(table1,1) # to get row proportions
```

Table 1.1 Credit default rates for men and women.

	Female	Male
0	14 349 (79%)	9 015 (76%)
1	3 763 (21%)	2 873 (24%)
Total	18 112	11 888

Table 1.2 Credit default rates by marital status.

	Married	Single	Other
0	10 453 (77%)	12 623 (79%)	288 (76%)
1	3 206 (23%)	3 341 (21%)	89 (24%)
Total	13 659	15 964	377

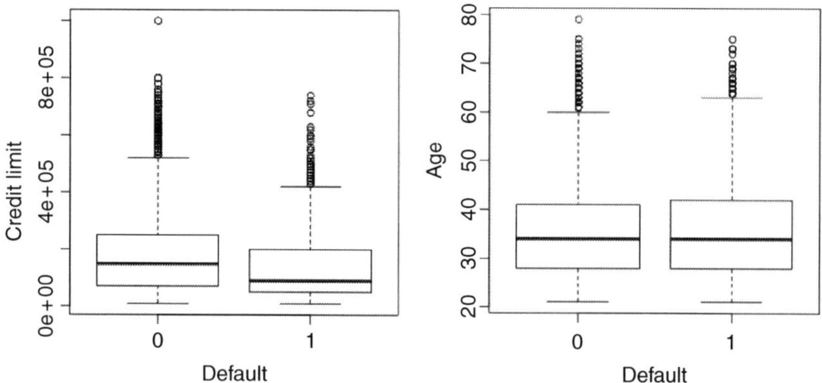

Figure 1.2 Boxplot of default vs. non-default for credit limit and ages.

In addition to the categorical variables in our data set like sex and marital status, we also have continuous variables like age. Perusing the boxplots in Figure 1.2, it appears that persons who do not default have higher credit limits than persons who default, while age appears to have no association with default status.

If it is really the case that persons with higher credit limits are less likely to default, can we decrease the default rate simply by giving everybody a higher credit limit?

Software Details

To reproduce Figure 1.2, load the data file `credit.csv`...

```
boxplot(limit~default,xlab="default",ylab="credit limit",
    data=df)
```

We have thus far looked at how the four variables are associated with default, individually. How might we examine the effects of all the variables at one time in order to answer the two fundamental questions?

The answer, of course, is to use regression to relate default to all four variables at once. Since default is a categorical variable with two levels, linear regression is not appropriate. We would have to use logistic regression instead. As for the independent variables, credit limit and age are continuous and require no special treatment before being included in the regression (though it may be advantageous to turn each into a categorical variables with, say, categories "low," "medium," and "high"). Sex and marital status are categorical variables and will have to be included as dummy variables. If you are unfamiliar with the creation of dummy variables, sex can be represented by a single dummy variable, say, S:

$$S = \begin{cases} 1, & \text{if male} \\ 0, & \text{if female} \end{cases}$$

Marital status (married, single, or divorced/widowed) will be represented by two dummy variables, M_1 and M_2:

$$M_1 = \begin{cases} 1, & \text{if married} \\ 0, & \text{if otherwise} \end{cases} \quad M_2 = \begin{cases} 1, & \text{if divorced/widowed} \\ 0, & \text{if otherwise} \end{cases}$$

For a married person, $M_1 = 1$ and $M_2 = 0$, for a person who is divorced/widowed $M_1 = 0$ and $M_2 = 1$, while for a single person $M_1 = 0$ and $M_2 = 0$.

1.2.1 Lurking Variables

It is not uncommon for an analyst to reach mistaken conclusions based on observational data that are incorrect due to lurking variables.

1. During WWII, an analysis of the accuracy of strategic bombing runs showed that Allied bombers were more accurate at lower altitudes than at higher altitudes (this makes sense). The analysis also showed that Allied bombers were more accurate when opposed by enemy fighters than when enemy fighters were not present. Explain.
2. A scatterplot shows a strong relationship between the number of firefighters at a fire and the dollar amount of the damage caused by the fire. While this relationship may be predictive, it is not causal: it is not true that if fewer firefighters are sent to a fire, the dollar amount of the damage will decrease. What is the missing causal variable?
3. On a daily basis in a coastal town, there is a positive relationship between ice cream sales and drowning deaths. What is the missing causal variable?
4. The observational data repeatedly say that persons who eat five fruits and veggies per day have a lower cancer rate than those who don't eat fruits and veggies. The experimental results find no difference in cancer rates. Explain the discrepancy.

5. A large, expensive observational study by the National Institutes of Health concluded that hormone replacement therapy (HRT) prevents heart disease in postmenopausal women. Consequently many women were placed on HRT. Later, an experiment showed that HRT does not prevent heart disease in postmenopausal women. Explain the discrepancy.

The resolutions of the above dilemmas are given below:

1. Cloud cover. Planes couldn't fly in the clouds and had to fly above the clouds. If the weather was cloudy, the enemy wouldn't bother to send up fighters, and accuracy was terrible because in that era, bombing depended on sighting landmarks on the ground.
2. There is a third variable in the background – the seriousness of the fire – that is responsible for the observed relationship. More serious fires require more firefighters and also cause more damage.
3. The lurking variable that causes both ice cream sales and an increase in drowning deaths is season of the year, i.e. summer.
4. Of course, persons who eat five fruits and veggies per day are different than those who do not. How, precisely, they are different we do not know. Just because there is a lurking variable does not mean that we can identify it.
5. The women who *chose* HRT were different from other women in ways for which the observational study could not control. Again, just because we can deduce the existence of a lurking variable does not follow that we can say what the variable is.

The above "experiments" (the word is in quotes because they really aren't experiments) are actually just observational data masquerading as experiments, and the way to see this is to perform a hypothetical thought experiment and think about *manipulating* one of the variables as it would be manipulated in a true experiment. In the fire example above, imagine there was a fire and firemen had responded, and then we ordered 100 more firemen to show up to the fire. Would we expect there to be more damage simply because more firemen were present? Of course not. As will be seen, designed experiments eliminate the effect of the lurking variables.

Here we mention that many authors conflate the concepts of "lurking variable" and "confounding variable," treating them as one and the same, but this is a mistake. Though they both make it difficult for the analyst to interpret results, they do so through different mechanisms. A lurking variable affects observational data, while a confounding variable affects experimental data. In this chapter we only encounter lurking variables. In later chapters we will encounter confounding. The "Learning More" section for this chapter describes the differences in detail.

1.2.2 Sample Selection Bias

Sample selection bias plagues nonexperimental, i.e. observational, data. Its effects are especially pernicious when selection is based on the dependent variable (the effects are not so bad when selection is based on an independent variable). To motivate this important idea, we generated some linear data with a zero intercept and a slope of unity. X takes on values from 10 to 20. If we fit a line $Y = a + bX$, we get $a = 0.76(2.49)$ and $b = 0.96(0.16)$ with standard errors in parentheses. The t-statistics to test the null hypotheses that the coefficients equal zero are the coefficients divided by the standard errors. The t-stat on the intercept is $0.76/2.49 = 0.30$, and the t-stat for the slope is $0.96/0.16 = 6$. Using 2 as a rough cutoff for a 5% significance level, we observe that the intercept is not significantly different from zero while the slope is significantly different from zero.

Try it!

Use the data in the file `SampleSelection.csv` to repeat the above analysis by running the regression for the full sample and again only for those observations for which $Y > 15$.

These results are consistent with the true intercept of zero and the true slope of unity. Suppose that we only get to observe observations when $Y > 15$. Then the results are $a = 15.35(3.05)$ and $b = 0.39(0.19)$. These results are not consistent with the truth: the intercept is significantly above 0, and the slope is significantly below unity. Figure 1.3 depicts the situation. Note that for observations close to the cutoff $Y > 15$, for any value of X, observations with positive errors will be included in the sample and observations with negative errors will be dropped from the sample. Thus, some values of observed X are correlated with the error: values of X close to but above $Y = 15$ are quite likely to have a positive error. This violates one of the assumptions for linear regression to be unbiased. Any time that selection into the sample depends on the value of the dependent variable, sample selection bias is a problem.

If we have an umbrella problem and we only want to make a prediction of Y for some value of X when $Y > 15$, then there is no difficulty. We can get good predictions of Y for values of Y larger than 15. If we have a rain dance problem and we need a good estimate of the true slope, then clearly we cannot use the regression to determine the effect that a change in X has upon Y, because the only regression we can run (the dashed line) has a biased slope (see the solid line for the true slope).

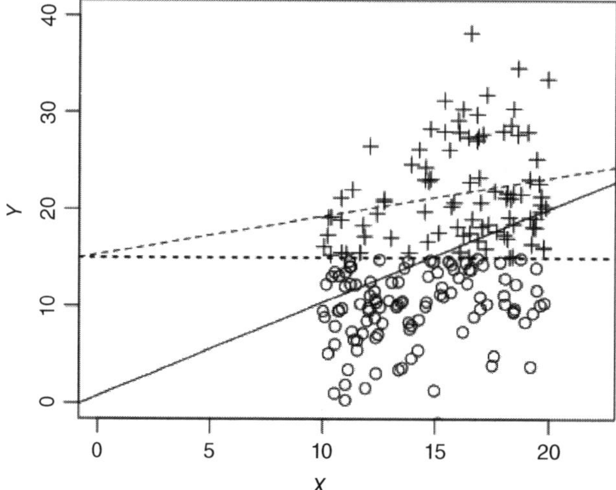

Figure 1.3 Sample selection bias. Dashed line for observations $Y > 15$ and solid line for all observations and horizontal dotted line at $Y = 15$.

Now let us return to the credit example, and suppose that we had lots of variables that included all possible lurking variables. Suppose we knew the model so that there was no garden of forking paths problem. Now, could we really get causal answers out of these data? Suppose we brought in a statistical expert on getting causal results from observational data. Could he do it? The answer is "no."

Aside from the garden of forking paths, there is a more serious problem with these data, and it is rather subtle. Let us consider where our data came from. People apply for credit. Some are granted credit, while others are not. Of those who are granted credit, some default and some do not. It should be apparent that our data do not constitute a random sample from the population, but a *selected* sample from the population. In the present case, there is no data on those who were denied credit, some of whom would have defaulted and some of whom would not have defaulted. This is a general problem called *sample selection bias*, and it plagues observational data; in such a case, the sample is not representative of the population. Let us be more explicit. The population of credit applicants includes four types of persons:

1. Non-defaulters who get credit.
2. Non-defaulters who are denied credit.
3. Defaulters who get credit.
4. Defaulters who are denied credit.

Meanwhile, since the sample consists only of persons who have been granted credit, the sample consists only of categories (1) and (3) above and does not look like the population. Therefore, inferences from the sample do not extrapolate to the population, and using such a model would likely be a mistake.

This is a very important problem that bedevils the credit industry, and this problem even has a name: "reject inference," which is how to conduct inference when there is no data on persons whose credit applications were rejected. Very sophisticated statistical machinery, far beyond the level of this book, has been unleashed on this problem, with only a modicum of success. Indeed, some credit card companies deliberately conduct designed experiments and issue credit to persons who otherwise would have been rejected in order to collect data from categories (2) and (4) so that they may extrapolate their results to the population. However, this is a very expensive solution to the problem, so these types of experiments are rarely performed. The credit card industry largely makes do with sophisticated statistical analyses to answer causal questions.

Finally, there is a modeling problem in the data to which we must draw attention. Suppose there was no reject inference problem, and we knew all the lurking variables. We might run a regression (perhaps a logistic regression, for those of you who know that method) with "default" on the loan as the dependent variable and "age" as one of the independent variables. Suppose further that the coefficient on age is positive and statistically significant. What does this mean?

Such a regression coefficient would not agree with what creditors generally know about the relationship between age and default probability. Based on years of empirical data, they know that young people tend to default more than older people. The nonlinearity in this relationship cannot be captured by a linear regression. What does this nonlinearity mean for our linear regression model? We now have a very substantial modeling problem, and we will get different answers depending on how we model the effect of age on defaults (Is it linear or nonlinear; if nonlinear, what type of nonlinearity?). Does a modeler really want results to be dependent on the choice of how the regression model is built? This is just part of what makes drawing causal inferences from observational data so fraught with danger. When we run experiments, we don't have to worry about *any* of these things.

Exercises

1.2.1 Consider the five examples of lurking variables. We already described an experiment and a manipulated variable for the fire damage example. Come up with experiments and manipulated variables to expose the falsity of the observational conclusion for the bombing and drownings examples (we don't know the lurking variable for the other two).

1.2.2 For the following cases of observational data, articulate the precise nature of the sample selection.

(a) We wish to determine the effect of education on income by running a regression. (Think of the minimum wage.)

(b) Mutual funds that have been in business longer tend to have higher returns than newer mutual funds. We "know" this because we collected observational data on mutual funds, regressed return on number of years in business, and found that funds with more years had higher returns. (What happened to mutual funds with low returns?)

1.2.3 In the text, we asked, "If it is really the case that persons with higher credit limits are less likely to default, can we decrease the default rate simply by giving everybody a higher credit limit?" Definitely not! Why not? What is the lurking variable?

1.2.4 Using the data set `credit.csv`, create the variable "percentage default by age" (y) and plot it against age (x). The relationship between y and x is definitely nonlinear. Describe the nonlinearity and a reason for it.

1.2.5 We know that regression results are biased when data have been selected due to a condition on the dependent variable, e.g. $Y > 15$. What happens when the data are selected due to a condition on X? Subset the data for those values $X < 15$ and run the regression. Is the slope estimate biased? What about the intercept? What can you conclude?

1.3 Case: Salk Polio Vaccine Trials

By the 1950s, polio had killed hundreds of thousands worldwide and infected tens of thousands per year in the United States. Many victims who did not die were condemned to spend the rest of their lives in "iron lungs" since they were unable to breathe on their own. To say nothing of the misery wrought by the disease, the effect on the community was devastating: parents kept their children indoors all summer, playgrounds were vacant, and one sick student in a class was reason for many healthy students to stay home from school. Jonas Salk developed a vaccine in 1952, and in 1954 two separate field trials were conducted, involving nearly 2 million children and 300 000 volunteers in the United States, Canada, and Finland. These constituted the largest clinical trials in history.

Table 1.3 Results of first Salk vaccine field trial.

Group	Size	Rate per 100 000
Grade 2 vaccinated	225 000	25
Grades 1 and 3	725 000	54
Grade 2 no consent	125 000	44

The first trial monitored nearly a million first, second, and third graders. The second graders were given permission slips for their parents to sign, and those children who got parental consent were vaccinated by injection. These children were compared with the non-vaccinated first and third graders. The observant reader will have noted already that the sample is not random, and generalizing the results to the population would be problematic. Summaries are presented in Table 1.3. It appears that the vaccine worked; children who got vaccinated had a lower infection rate than children who did not get vaccinated.

Notice, however, that the number of second graders who did not receive permission markedly was nearly half the number who did get permission. If families who gave consent were similar to families who did not give consent, then we can assume that the difference in polio rates between the grade 2 vaccinated students and the grades 1 and 3 unvaccinated students is a good estimate of the overall effect of the vaccine at reducing polio. But are the families similar? Further analysis of the trial data showed that, on average, families who gave consent had higher incomes than families who did not give consent. What else might we deduce from this income difference, with respect to educational differences? Can we infer something about how often the children visited the family doctor and other such variables that are correlated with income and might affect the children's susceptibility to polio? How might this affect the results of the trial?

A more subtle problem arises because everybody involved in the experiment, including the physicians who monitored the children for polio, knew who had been vaccinated and who had not. Specifically, the experiment was not *double-blinded*. In a blinded experiment (or single-blind experiment), only the experimenter knows whether the subject gets the treatment or the placebo; the subject does not know. In a double-blind experiment, neither the experimenter nor the subject knows whether the treatment or the placebo is administered. The reason that lack of blinding is a problem in this situation is that polio, when it is mild and not severe, can seem like a flu or other diseases. Physicians (who generally believe in the power of vaccines), when diagnosing a child with flu-like symptoms, may be subconsciously biased to diagnose polio more often in the unvaccinated children than in the vaccinated children.

Table 1.4 Results of second Salk vaccine field trial.

Group	Size	Rate per 100 000
Treatment	200 000	28
Control	200 000	71
No consent	350 000	46

A formal review of this trial published in the leading statistical journal characterized the design as "stupid and futile" and the results of the trial as "worthless" (Brownlee, 1955).

To remedy these deficiencies in the first trial, a second trial focused only on children whose parents gave consent for vaccination. These children were randomly assigned to treatment and control groups, both of whom received an injection. The former received the vaccine, and the latter received a placebo saline injection. The observant reader will have noted already that there is no longer a sample selection problem. The results are presented in Table 1.4.

Comparing Tables 1.3 and 1.4 is instructive. Note that the incidence for vaccinated children is roughly the same, 25 vs. 28, as is the incidence for children who did not get consent, 44 vs. 46. The incidence for the "control" groups has changed markedly, and it did not go down, but it went up! Apparently the children of parents who give permission are *more susceptible* to polio than the children of parents who do not give permission. Thus, the first experiment was biased *against* the vaccine: the treatment group that got the vaccine was more susceptible to polio, while the control group that did not get the vaccine was less susceptible to polio. No one involved in the trial foresaw this possibility, though an analyst trained in the design of experiments would have been able to guard against the possibility and eliminate it, even if she couldn't foresee it! How? By *randomization*! We will have much more to say about randomization in subsequent chapters.

The variable "parents' propensity to give consent" is a lurking variable that is correlated with the outcome (the child getting the disease). Several mechanisms have been suggested for this correlation. For example, less educated persons are less likely to give consent, less educated persons are likely to have less income, persons with less income are less likely to live in sterile environments, and children who live in less sterile environments are more likely to have robust immune systems that offer more protection against polio.

Regardless of the true mechanism, if treatment and control had been assigned randomly, then the effects of the lurking variable would have been spread out between the two groups instead of having too many consenters in the experimental group. Thus, randomization protects against the effect of lurking variables, even when we don't know what variables we need protection against! Remember, lurking variables are not to be confused with confounding variables; again, see the "Learning More" section for details on this point.

A postscript to the experiments is in order. The success of the trials was formally announced in 1955, and widespread vaccination began. However, a problem with the injectable vaccine quickly arose. There were six different manufacturers of the vaccine, and not all of them followed Salk's instructions for preparing the vaccine. Hundreds of cases of paralysis were reported, including cases when the paralysis began in the arm that received the injection! The program was suspended in 1955. In 1960 an even better oral vaccine was introduced and put into widespread use. This is why, when examining the data, you might not see a large drop after 1955.

Exercises

1.3.1 List the problems with the first Salk trial.

1.3.2 How did the second Salk trial overcome the problems in the first Salk trial?

1.3.3 Why were the sample sizes so large for the Salk trials?

1.3.4 Find an example of another failed medical experiment. Be sure to articulate why it failed.

1.3.5 Plot the number of cases of polio in the United States by year; the data are in `polio.csv`. The eradication of polio is obvious as time increases. What is noticeable about the period before the introduction of the polio vaccine?

1.4 What Is a Business Experiment?

We have seen that when using observational data to answer a business question, the answer we get can depend on which variables are included in the analysis, how relationships are modeled, and a host of other decisions. These types of analyses are difficult to rely on, easy to argue about, and hard to do well. Consequently, their value for informing business decisions is limited. None of these criticisms apply to business experiments.

Some persons are under the mistaken belief that an experiment is just "trying something" to see "whether or not it works," and this approach very often leads to unusable results. For example, a large bank wanted to improve its customer service, so it designated some of its branches as "laboratories" where many ideas were implemented. So many things were changed in haphazard fashion across so many branches that it was impossible to determine which of the ideas was responsible for improved results. This was a very expensive "experiment" by one of the major banks, and it produced nothing useful – it was a complete waste of resources.

Sometimes a simple experiment can point the direction toward millions of dollars as happened at Intuit. And this was after a formal usability study gave a decidedly negative recommendation about the idea! As recounted by Thomke (2020, chapter 2),

> [An] engineer noticed that about 50 percent of prospective customers tried the company's small business product 20 minutes before they had to make payroll. The problem was that all payroll companies took hours or even days to approve new customers before the first employee could be paid. Wouldn't potential customers be very pleased if they could make payroll before the long approval process was done? To make sure there was a genuine need, the engineer and product manager ran a usability study. The result: none of the twenty participants were interested in a fast payroll solution. But instead of shelving the idea, Intuit modified its webpage within 24 hours and ran a simple experiment that offered two versions of the software – one with the option to click on "pay employees first" and another one with "do set-up first." (When users clicked on "pay employees first" option, they got a message that the feature wasn't ready.) Contrary to the usability test results, the experiment revealed that 58 percent of new users picked the faster payroll option. Ultimately, the feature became hugely popular, lifted the software's conversion rate by 14 percent and generated millions of additional revenue.

An experiment is a statistical test by which a hypothesis is subjected to data produced according to a specific procedure in which some variable thought to affect the output is deliberately manipulated. A business experiment is merely an experiment whose purpose is to inform a business decision. By contrast, some disciplines, e.g. medicine, psychology, or biology, develop theories and then use experiments to test the theory; not so for us. Each of these disciplines has its specific theories and subjects, and therefore experimental methods need to be adapted to each discipline. For example, engineers usually have well-specified physical models, subjects that are often physical entities that respond predictably, and prediction methods that have small errors. In business, well-specified models are an exception, subjects are often human (who can respond in two different ways to the same stimulus!), and prediction methods have large errors. Therefore, it is usually the case that engineering methods for experimental design will not be applicable in a business setting, and vice versa.

Examples of Experiments

Example I Slow payment of invoices creates numerous problems for companies, not the least of which is a decrease in cash flow and the attendant need to incur short-term debt to cover the shortfall. Decreasing customers' payment times has obvious advantages, but how to achieve this goal is not always clear. How long after the customer receives the goods or services should the invoice be sent? If it isn't paid on time, should another invoice be sent? Or a formal letter? Or should would a phone call be more effective? One company had been in the habit of just sending repeat invoices every month and decided to take action. One set of late-paying customers was sent a formal letter asking for payment of the late debt. Another set of customers was called. The times of these letters and calls were varied across several trials. At the end of the experiment, the conclusion was that a phone call 10 days after the due date was most effective. The overall time from initial billing to final payment was reduced from 110 to 75 days.

Example II A large, multi-office financial services corporation wanted to improve the process by which its customers apply for credit. The immediate problem was that 60% of the applications had to be sent back for reprocessing, usually due to incomplete information (e.g. the applicant didn't write down his age or made a transposition error when writing his phone number). Not only is this expensive, but also it greatly adds to the overall processing time, which leads to many potential customers dropping out of the application process entirely. After a brainstorming session, the project team tasked with addressing this problem decided to focus on three variables: the type of application, how detailed are the instructions, and whether examples were provided. Two additional factors thought to affect the reprocessing rate are type of application (loan or lease) and region (Midwest vs. Northeast). Thus five factors with two levels each were used in the experiment (it doesn't matter which one is called level 1 or level 2) are presented in Table 1.5

Only two of the five factors were found to affect the reprocessing rate, C and D. The application form was redesigned, and the completeness rate increased to 95%, virtually eliminating the need for reprocessing. The number of completed applications increased dramatically, since customers were no longer dropping out. All the persons who had been tasked with reprocessing were reassigned to new tasks, increasing productivity without increasing the payroll.

Example III Progressive Insurance observed that when its policyholders hired a lawyer to settle a claim, settlement time went up from 90 days to 6 months, and the payout to the policyholder *went down* by $100. The costs to Progressive increased

Table 1.5 Factors and levels for financial services example.

Factor	Level 1	Level 2
(A) Application type	Loan	Lease
(B) Region	Midwest	Northeast
(C) Instructions	Current instructions	Instructions with more detail
(D) Example	Current example	Examples with more detail
(E) Negative example	None (current)	Example of what *not* to do

by $1600 due to the need to engage lawyers for these cases. Clearly, policyholders (and Progressive) would be better served if lawyers were not needlessly involved in the process. To achieve this goal, the project team focused on the dependent variable: percentage of claimants who hired an attorney within 60 days of the accident, which had been about 36%. Brainstorming produced 59 ideas for reducing this percentage; excluding ideas that were not "practical, fast, or cost-free" culled the number to 19. This number finally was reduced to 13, which were tested via designed experiments. When all was said and done, the percentage was reduced by *eight* points, with each one-point drop representing six million dollars in savings and better service to policyholders.

One of the more surprising innovations as a result of this experiment was that Progressive began paying out *more* in claims! If a person's car is totaled in an accident and the insurance company insists on paying book value rather than replacement value, what is the person likely to do? Hire a lawyer! In the experiment, districts that paid more in claims had a five-point drop in attorney involvement. The decrease in legal fees more than made up for the increase in payments to policyholders.

Example IV A company that sold telecommunications equipment to large corporations contemplated changing its customer management system to a desk-based account manager (DBAM) system. These account managers would not work from the field, but solely from the office, making use of the telephone and video calls. This would save on travel time, increase efficiency, and, hopefully, lead to greater profit. A small number of field account managers were provided with the DBAM and trained in its use. The accounts of these managers were the experimental group. A carefully matched set of accounts from other managers constituted the control group. (We will discuss *matching* in Chapter 5.)

Various measures on customer satisfaction and employee satisfaction were taken on each group before and after the experiment. Costs and revenues associated with each group were measured as well. The key performance indicator (KPI) for the experiment was cost-to-revenue ratio. The experimental group had a KPI that was 6% lower than the control group; this implies that the DBAM was more profitable than the existing method. The ancillary measures showed that employee satisfaction did not decrease and customer satisfaction actually increased. The primary reason for the increased customer satisfaction was that it was easier for customers to contact their representatives, who were no longer unavailable while traveling. The company rolled out the DBAM to all its field agents and increased its profits while increasing customer satisfaction.

Example V Anheuser-Busch, a beer company, wanted to determine how much money to spend on advertising. The sample of 15 marketing areas was divided into three groups: (i) 50% increase, (ii) no change, and (iii) 25% decrease in advertising expenditure over a 12-month period. At the end of the experiment, group i achieved a 7% increase in sales, group ii had no change, and group iii had a 14% increase! A follow-up experiment produced the same result, something that no one ever expected: decreasing advertising produced an increase in sales. This led the firm to conclude that they had supersaturated the market with advertising, and indeed the firm substantially reduced advertising without hurting sales in other markets.

1.4.1 Four Steps of an Experiment

From the previous examples, we can list the four steps of an experiment:

1. Randomly divide the subjects (e.g. customers) into groups. The researcher does not allow the subjects to pick the group – that would be a form of self-selection that makes the data observational rather than experimental. Moreover, this division is made randomly – the researcher doesn't assign the groups for that, too, would be a form of selection that would render the data observational.
2. Expose each group to a different treatment. The researcher does not allow the subjects to choose which treatment to receive; this assignment has to made randomly.
3. Measure a response for each group. The outcome of interest has to be chosen before the experiment is conducted, and the method for performing the measurement has to be decided in advance, too.
4. Compare group responses to determine which treatment is better. This is accomplished with a statistical test. It can be something as simple as a two-sample test of means. Whatever test is applied, the test will tell us whether the difference between the groups could be just random or is more likely due to some systematic effect.

We know that observational data can only give correlations and they can't give causation. When we have a well-designed experiment, we can answer causal questions: Does changing X cause a change in Y? Experiments are a systematic way to avoid lurking variables and are the gold standard for causal inference. Thomke Thomke (2020, chapter 2) describes an example of this phenomenon:

> [Yahoo did a study] to assess whether display ads for a brand, shown on Yahoo sites, can increase searches for the brand name or related keywords. The observational part of the study estimated that ads increased the number of searches by 871 percent to 1,198 percent. But when Yahoo ran a controlled experiment, the increase was only 5.4 percent. If not for the control, the company might have concluded that the ads had a huge impact and wouldn't have realized that the increases in searches was due to other variables that changed during the observation period.

1.4.2 Big Three of Causality

The "Big Three" criteria for being able to make causal inference are as follows:

1. When X changed, Y also changed. If X changes and Y doesn't change, then we cannot assert that X causes Y (sometimes this is useful information).
2. X happened before Y. If X happens after Y, then X cannot cause Y. This issue arises sometimes in marketing research, where a commercial is shown one day and sales on that same day are measured. How can we know that today's sales weren't affected by something that happened yesterday?
3. Nothing else besides X changed systematically. If variables W and Z change at the same time that X changes – not every time, but often enough – then we cannot rule out the possibility that W and Z are causing the changes in Y. Observational data cannot rule out this possibility. The random treatment assignment of an experiment *can* rule this out.

Experimentation is the art of making sure these criteria are met so that valid causal statements can be made. Much more will be said about this in the next two chapters. The problem with observational data is that at least one of three is always missing, usually the third.

You should consider performing an experiment when you have lots of items on which to experiment (i.e. "experimental units"), you have the capability to take measurements on these units and the outcomes from the experiments can be measured easily, and you have control over the treatments. In manufacturing, for example, lots of items come off the assembly line, so there is an abundance of experimental units. Measurement typically is easy: Does the item work or how well does it work? Often it is very easy to apply treatments to some units and not to others.

In digital marketing, experimental units are available in very large quantities – think website visitors. However, measurement can sometimes be problematic – what constitutes a "successful" website visit? Is success an immediate purchase or a purchase three weeks later? How can you tell the same visitor returned three weeks later? Control of treatments can be difficult. We can put an ad on a webpage that a customer visited, but how do we know he actually saw the ad? In the case of television, we can run a commercial, but how do we know who actually saw it? How do we know which of the persons who bought our product this week have seen the ad? These are some of the problems we will deal with in later chapters.

1.4.3 Most Experiments Fail

It is important to remember that the purpose of an experiment is to test some idea, not prove something and also that most experiments fail! This may sound depressing, but it is hugely effective if you can create a process that allows bad ideas to fail quickly and with minimal investment:

- "[Our company has] tested over 150 000 ideas in over 13 000 MVT [multivariate testing] projects during the past 22 years. Of all the business improvement ideas that were tested, only about 25 percent (one in four) actually produced improved results; 53 percent (about half) made no difference (and were a waste of everybody's times); and 22 percent (that would have been implemented otherwise) actually hurt the results they were intended to help" (Holland and Cochran, 2005, p. 21).
- "Netflix considers 90% of what they try to be wrong" (Moran, 2007, p. 240).
- "I have observed the results of thousands of business experiments. The two most important substantive observations across them are stark. First, innovative ideas rarely work… When a new business program is proposed, it typically fails to increase shareholder value versus the previous best alternative" (Manzi, 2012, p. 13).
- Writing of the credit card company Capital One (Goncalves, 2008, p. 27): "We run thirty thousand tests a year, and they all compete against each other on the basis of economic results. The vast majority of these experiments fail, but the ones that succeed can hit very big[.]"
- "Given a ten percent chance of a 100 times payoff, you should take that bet every time. But you're still going to be wrong nine times out of ten." Amazon CEO Jeff Bezos wrote this in his 2016 letter to shareholders.
- "Economic development builds on business experiments. Many, perhaps most experiments fail" (Eliasson, 2010, p. 117).

You are not going to get useful results from most of the experiments that you conduct. But, as Thomas Edison said of inventing the lightbulb, "I have not failed.

I've just found 10 000 ways that didn't work." Failed experiments are not worthless; they can contain much useful information: Why didn't the experiment work? Did we maintain false assumptions? Was the design faulty? Everything learned from a failed experiment can help make the next experiment better.

When dealing with human subjects, where response sizes are small and there are lots of noise, there can be a tendency toward false positives (especially when sample sizes are small!), so follow-up experiments of small sample experiments are important to document that the discovered effect really exists.

Even with large samples, it is best to make sure that a discovered effect really exists. In webpage development, an experiment to optimize a webpage might prove fruitful, yet the improvement will not immediately be rolled out to all users. Instead, it might be rolled out to 5% of users to guard against the possibility that some unforeseen development might render the improvement futile or worse, harmful. Only after it has been deemed successful with the 5% sample will it be rolled out to all users.

Exercises

1.4.1 Suppose the company in the invoice example billed quarterly and had, on average, $10 million in accounts receivable each quarter. If short-term money costs 6%, how much does the company save?

1.4.2 Give an example of a business hypothesis, e.g. we think that raising price from $2 to $2.25 won't cost us sales. Describe an experiment to test your hypothesis. What data need to be collected? How should the data be collected?

1.4.3 Find an example of a business experiment reported in the popular business literature, e.g. *Forbes* or *The Wall Street Journal.*

1.5 Improving Website Designs

One of the most popular types of experiment in business is the A/B website test. For example, Figure 1.4 shows two different versions of an offer made to website visitors of an iconic clothing retailer to induce them to sign up for the retailer's mailing list. The rationale for the test was that these visitors were already at the website and knew about the store and its products, so maybe a monetary inducement was unnecessary. If, indeed, it was unnecessary, then the $10 coupon would be just giving money away needlessly. Visitors to the website are randomly shown

Figure 1.4 A/B test for mailing list sign-ups. Source: courtesy GuessTheTest.com.

one of the two ads. The two groups are typically labeled "A" and "B," thus the name "A/B testing." Digital analytics software allows website owners to track the online behavior of visitors in each group, such as what customers click on, what files they download, and whether they make a purchase, allowing comparison between the two groups. In this case, the software tracked whether a visitor signed up for the mailing list or not. A test like this will typically run for a few days or weeks, until enough users have visited the page so that we have a good idea of which version is performing better. Once we have the results of the test, the retailer can deploy the better ad to all visitors. In this case, over a 30-day period, 400 000 visitors were randomly assigned to see one of the two ads. Do you think a $10 coupon really mattered to people who spent hundreds of dollars on clothes?

In this test, the $10 incentive really did make a difference and resulted in more sign-ups. While it may not be surprising that the version with the $10 incentive won the test, the test gives us a quantitative estimate of how much better this version image performs: it increased sign-ups by 300% compared with the version without the incentive. The reason tests like this have become so popular is that they allow us to measure the *causal impact* of the landing page version on sales. The landing pages were assigned to users *at random*, and when we average over a large number of users and see a difference between the A users and the B users, the resulting difference must be due to the landing page and not anything else. We'll discuss causality and testing more in Chapter 3.

Website A/B testing has become so popular that nearly every large website has an ongoing testing program, often conducting dozens of tests every month on every possible feature of the website: colors, images, fonts, text copy, layouts, rules governing when pop-ups or banners appear, etc. Organizations such as GuessTheTest.com regularly feature examples of tests and invite the reader to guess which version of a website performed better. (The example in Figure 1.4 was provided by GuessTheTest.com.) In Figures 1.5–1.7, we give three more example website tests where users were randomly assigned to see one of two different versions of a website. As you read through them, try to guess which test performed better or whether they were the same.

Website tests can also span across multiple pages in a site. For example, an online retailer wanted to know how best to display images of skirts on their website. Should the skirt be shown as part of a complete outfit (left image in Figure 1.5), or should the image of the skirt be shown with the model's torso and face cropped out to better show the details of the skirts? In this test, users were assigned to one of the two treatments and then shown either full or cropped images for every skirt on the product listing pages. (Doing this requires a bit more setup than the simple one-page tests but is still possible with most testing software.) The website analytics software measured the sales of skirts (total revenue in $) for the two groups. Which images do you think produced more skirt sales?

As mobile websites and apps have become more popular, website owners have also conducted tests on mobile devices. Figure 1.6 shows two different versions of a mobile webpage where users can find information about storage locations near them. The version on the left in Figure 1.6 directs the user to enter his zip code and then press a button to search for nearby locations. The version on the right lets the user employ his current GPS location to look up locations nearby. The test measured how many customers signed up to visit a location and how many customers actually rented a storage unit. Which version do you think would get more customers to visit a physical location and to rent?

Our last example shows a test to determine whether it is beneficial to include a video icon on the product listing to indicate that there is a video available for the product. The images in Figure 1.7 show a product listing without the icon (left) and with the icon (right). These images appear on the product listing page that shows all the products in a particular category (e.g. dresses, tops, shoes). In this test, users were assigned to either never see the video icons or to see the video icons for every product that had a video available. The two groups were compared based on the percentage of sessions that viewed a product detail page, which is the page the user sees when she clicks on one of the product listing images. The hypothesis was that the icons would encourage more people to click to see the product details where they can view the video. They also measured the total sales ($) per session. Do you think the icons will encourage users to click through to the product page?

Here we have shown four examples of website tests, but the options for testing websites and other digital platforms like apps or kiosks are nearly limitless. The growth in website testing has been driven largely by software that manages the randomization of users into test conditions. Popular software options include Optimizely, Maxymiser, Adobe Test&Target, Visual Website Optimizer, and Google Experiments. These tools integrate with digital analytics software like Google Analytics or Adobe Analytics that track user behavior on websites, which provides the

Figure 1.5 Skirt images test. Source: photograph by Victoria Borodinova.

Figure 1.6 Mobile landing page test for storage company.

Figure 1.7 Video icon test. Source: Elias de Carvalho/Pexel.

data to compare the two versions. Most major websites will have testing software installed and often have a testing manager whose job is to plan, conduct, analyze, and report the results of tests. Newer software tools also make it possible to run tests on email, mobile apps, and other digital interfaces like touch-screen kiosks or smart TV user interfaces. These website tests represent the ideal business experiment: we typically have a large sample of actual users, users are randomly assigned to alternative treatments, the user behavior is automatically measured by the web analytics solution, and the comparison between the two groups is a good estimate of the causal difference between the treatments. It is also relatively easy to implement the better version as soon as you get the results, and so these types of tests have a major impact on how websites are managed.

So, how good are you at guessing which version is better? Table 1.6 shows the winning treatment for each of the tests (in bold) along with the lift in performance. If you were able to guess the results to all four examples, then you are gifted web designer. Even experienced website managers frequently guess incorrectly and user behavior changes over time and from one website to another, which means that the only way to figure out which version is better is to run a test.

Notice that Table 1.6 shows a different response measure for each test. The *response measure* (KPI) in an experiment is simply the measure that is used to compare the performance of two treatments. This measure is usually directly related to the business goal of the treatment. For instance, the purpose of the Dell Landing Page is to get people to sign up to talk to a Dell representative, so the percentage of users who submit a request is a natural response measure. Dell could have selected a different response measure, such as the % of users who actually speak with a Dell representative or who sign up and pay for services. In some cases, the test will include several response measures; the video icon test used both the % of users that viewed a product detail page, which is closely related to the goal of the video icon, and the sales per session, which reflects the ultimate

Table 1.6 Summary of web test results.

Test	A treatment	B treatment	Response measures	Result
Email sign-up	No incentive	**$10 incentive**	# of sign-ups	300% lift
Skirt images	Head-to-toe	**Cropped**	Skirt sales ($/session)	7% lift
Location search	**Zip search**	GPS search	Sign-ups	40% lift
			Rentals	23% lift
Video icon	No icon	Icon	% to product detail	No significant
			Sales ($/session)	Difference

Note: Winning treatment shown in boldface.

business goal of any retail website. We will discuss the selection of response measures later, but for now it is sufficient to recognize that choosing a response measure that relates to business goals is a critical (and sometimes overlooked) part of test design.

Table 1.6 reports the test results in terms of a percentage *lift*. For example, in the Dell Landing Page test, the lift was 36%, which means that the hero image produced 36% more submissions. Table 1.6 only reports the lift numbers for test results that were found to be *significant*. Significance tests are used to determine whether there is enough data to say that there really is a difference between the two treatments. Imagine, for example, that we had test data on only five users: two who saw version A and looked at product details and three who saw version B and did not look at product details. Is this enough data to say that A is better than B? Your intuition probably tells you that it isn't, which is true, but when samples are a bit bigger, we can't rely on intuition to determine whether there is enough data to draw a conclusion. Testing for significance is one of the tools we use in analyzing A/B tests, and Chapter 2 will show you how to do it. As we will explain in the next few sections, we need more than just the lift numbers to perform the significance test.

Most website testing managers will tell you that more than half of the website tests that they run are not significant, meaning that they cannot conclude that one version is better than the other. For example, in the video icon test in Figure 1.7, there were no significant differences in the % of users who viewed the product detail pages or the average sales per session. If we looked at the raw data, there were probably some small differences, but that difference was not great enough to rise to the level of significance. The analyst has wisely chosen not to report the lift numbers, and instead simply said, "there was no significant difference." While the manager who came up with this video icon idea might not be too happy to find that it doesn't work, it is important to know that it doesn't work so that attention can be shifted to more promising improvements to the website. Smart testing managers realize that it is important to run many tests to find the features of the website that really do change user behavior.

Exercises

1.5.1 Visit a retail website and identify five opportunities for A/B tests on the website. For each test, clearly define the A and B treatments that you would test and identify a response variable to measure performance.

1.5.2 Find an article that reports the results of a medical experiment, a business experiment, or a psychological experiment. How are the results reported? Do they use a graph to display the data? Does the article indicate whether the difference between treatments was significant?

1.5.3 Visit a retail store and identify five opportunities for A/B tests. For each test, clearly define the A and B treatments that you would test and identify a response variable to measure performance.

1.6 A Brief History of Experiments

Experiments are as old as the bible. From The Book of Daniel (1, 11-16),

> Daniel then said to the guard whom the chief official had appointed over Daniel, Hananiah, Mishael, and Azariah, "Please test your servants for ten days: Give us nothing but vegetables to eat and water to drink. Then compare our appearance with that of the young men who eat the royal food, and treat your servants in accordance with what you see." So he agreed to this and tested them for ten days. At the end of the ten days they looked healthier and better nourished than any of the young men who ate the royal food. So the guard took away their choice food and the wine they were to drink and gave them vegetables instead.

The first clinical trial was conducted in 1747 by the Scottish physician James Lind, who was trying to find a cure for scurvy. Scurvy was a serious problem, since it killed more British sailors than the French and the Spanish combined. After two months at sea, when the men were afflicted with scurvy, Lind divided 12 sick sailors into six groups of 2. Each day the groups were administered cider, 25 drops of sulfuric acid, vinegar, a cup of seawater, and barley water, and the final group received two oranges and one lemon. After six days the fruit ran out, but one sailor was completely recovered and the other was almost recovered.

Randomization was introduced into experimental design in the nineteenth century by Peirce and Jastrow (1885) (many people incorrectly attribute this to R. A. Fisher in the twentieth century, but they are wrong). Designed experiments that were not randomized, but instead were *balanced*, were developed by William Gosset, pseudonymously writing as "Student" (who also invented Student's t-distribution) in the early twentieth century. Fisher later popularized

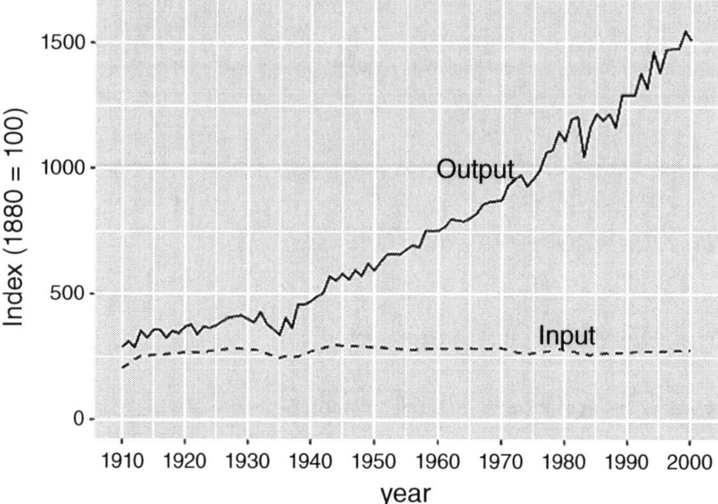

Figure 1.8 US agricultural output and input in the twentieth century.

randomized experiments in the 1920s. Much of the early work in the design of experiments in the early twentieth century focused on agriculture, and it was extremely fruitful. Moses and Mosteller (1997, p. 217) noted,

> The development of greatly increased agricultural productivity in the twentieth century has rested largely on field experiments in which new varieties of crops (and new agricultural practices) are compared to standard ones. So important is this empirical testing to agricultural progress that a large part of modern statistical design of experiments actually grew up in the context of agricultural experimentation.

In agriculture, experimental design is used to maximize output by determining the proper amount of fertilizer, growing conditions, etc. and also in the breeding of plants that have higher yields. The increase in agricultural production can be seen in Figure 1.8; note the rapid rise beginning in about 1940, after experimental methods began to spread throughout the agricultural sector.

The next field to be revolutionized by the design of experiments was manufacturing. Chemical production greatly increased as a result of the design of experiments and this spread to other process industries. Variability in successive batches of output was decreased, allowing for a more uniform, higher quality product and a concomitant decrease in waste. In the 1950s, the statistical pioneer W. Edwards Deming taught statistical methods to Japanese manufacturers at a time when “made in Japan” was synonymous with “low quality.” Deming taught

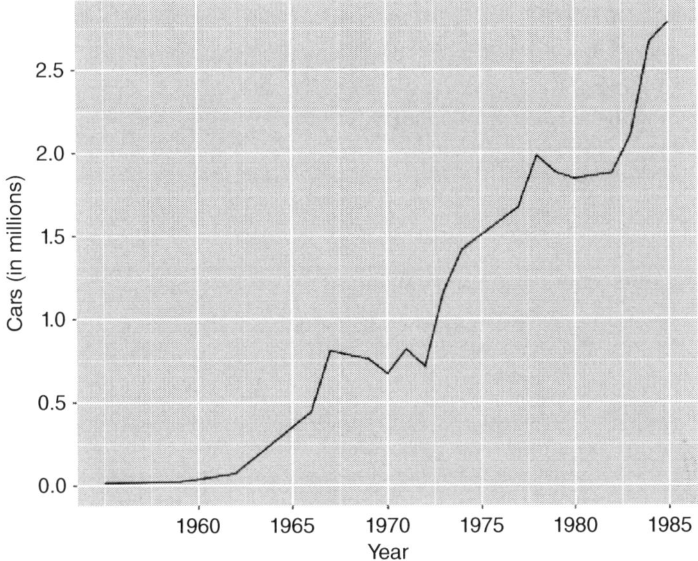

Figure 1.9 US car imports from Japan.

them to greatly increase quality and output, especially in the automotive and electronics industries.

By the 1970s, Japanese exports of autos and electronics to the United States threatened the American auto industry and decimated its electronics industry. American consumers began to realize that the Japanese cars were much better made: "The Japanese vehicle surpassed the American car in measurable, definable, observable quality" (Lightstone et al., 1993, p. 777). American cars had lifespans of 50 000–75 000 miles, while Japanese cars could be counted on to last more than 100 000 miles. Figure 1.9 shows the remarkable growth of American imports of Japanese cars from the 1960s through the 1980s. (After the 1980s, Japan started building car plants in the United States to avoid import restrictions, so it kept selling lots of cars in the United States, but was no longer exporting them from Japan to the United States.) About this time Six Sigma and Total Quality Management, both of which emphasize designed experiments, were taken up by many companies, the former's most famous adherent being General Electric. Medical research adopted experimental design and embraced clinical trials. Previously, research had been of the variety: "Dr. Smith saw six patients and this is what he thinks." Capital One came to dominate the credit card industry by relying on experimental design to optimize its postal credit card solicitations. In the present day, design of experiments continues to revolutionize various fields including economics with its field experiments, marketing with its web analytics, and even art conservation, as well as many others too numerous to mention here.

1.7 Chapter Exercises

1.1 Define the following terms:
(a) independent variable / dependent variable
(b) predictor / response
(c) sample / population
(d) treatment / response
(e) random sample
(f) observational data / experimental data

1.2 What are the four key steps in any experiment?

1.3 Ice cream sales are positively correlated with shark attacks in the Eastern United States. What is the lurking variable?

1.4 What is necessary to show causation?

1.8 Learning More

In this chapter we have twice quoted Stefan Thomke's book, *Experimentation Works: The Surprising Power of Business Experiments*, and we highly recommend it as an introduction to how experimental methods are used in business. It describes many, many business experiments and how to adopt an experimentation culture in a business. Thomke, a professor at Harvard Business School, has been conducting research in this area for decades. His last book, in 2003, was entitled *Experimentation Matters*.

Section 1.1 "Life Expectancy and Newspapers"

• The life expectancy example is based on Zaman (2010). The associated data are from the World Bank Indicator Tables for the year 2010, including only all observations that have no missing values.
• The smoother used by R in Figure 1.1 is called "lowess," which stands for LOcally WEighted Scatterplot Smoother. It's used to detect nonlinearities in a scatterplot.
• The classic reference on eliciting causal information from observational data is Rosenbaum (2010), but it is a technical book written for people who have taken advanced courses in statistics. A much more accessible book that covers both experimental and observational methods is Rosenbaum (2017), which is filled with words and has practically no equations; someone who wants to understand the topic and who hasn't taken advanced courses in statistics can do no better

than this book. It also contains a very good example of how observational data can be used to inform causal questions; see the section entitled "A Simple Example: Does a Parent's Occupation Put Children at Risk?" (pp. 119-124).

• Looking only at correlations (or lack thereof) can make it impossible to uncover causal relations. Suppose a car travels at a constant speed over hilly roads. The driver will have to accelerate on the inclines and brake on the declines to maintain a constant speed. A person who knows nothing of automobiles might observe these data and conclude that depressing the accelerator or the brake has nothing to do with what speed the car travels.

Section 1.2 "Case: Credit Card Defaults"

• The credit card data set is the "default of credit card clients Data Set" from https://archive.ics.uci.edu/ml/index.html. For the education variable, numbers 4, 5, and 6 were converted to "other," similarly for the marriage variable values 0 and 4.

• It is not a good idea to run a linear regression with the variable default on the left-hand side because default is binary (takes on only the values zero and one) and linear regression is for continuous dependent variables. There is a special method for binary dependent variables called "logistic regression," but that's something for an advanced statistics course.

• The idea of the garden of forking paths is discussed clearly and nontechnically in Gelman and Loken (2014), which article was included in *Best Math Writing of 2015*; although nontechnical, it's an excellent read for the statistically inclined person, too.

• In general, lurking variables affect observational data and confounding variables affect designed experiments. A lurking variable connects two otherwise unconnected variables, creating the appearance of a causal relation between two other variables. Consider the firefighter example, where the number of firefighters is highly correlated with the damage caused by the fire. Adding more firefighters doesn't increase the amount of damage (the variables are really unconnected). Rather, the lurking variable "intensity of the fire" connects them. A lurking variable (say, A) creates the illusion of a causal relationship between two other variables, B and C. A good article on how to detect lurking variables is Joiner (1981).

• Variables are confounded when we cannot separate their respective effects on the response. A confounding variable X has an effect on the response Y, but another variable Z also has an effect on Y, and we are unable to separate the effects of X and Z. For example, Y might be store sales and Z is a store promotion, while X is bad weather. We cannot determine the true effect of the promotion on sales because it is confounded with the weather.

Confounding can occur in a poorly designed experiment. Suppose you wish to determine the effects of fees and interest rates on credit card use. Suppose you offer a low fee and low interest rate to one group and a high fee and a high interest rate to the other group. The first group will have more credit card use and the second group less use, but you won't be able to tell whether the low fee or the low rate caused more use in the first group or whether the high fee and the high interest rate caused less use in the second group. The rate and the fee are confounded.

On the other hand, we could isolate the effect of rate by offering low fee and high rate to the first group and low fee and low rate to the second group. In more advanced designs, we sometimes will have many effects and be unable to isolate them all. In such a situation, we will deliberately confound the effects that we don't care about so much so that we can isolate the effects that we do care about. We will address this in Chapter 8.

Section 1.3 "Case: Salk Polio Vaccine"

• A layman's overview of the Salk trials is given in Meier (1989).

• The source for the polio data is http://www.post-polio.org/ir-usa.html. The source for the "under 18" US population is https://www.census.gov/data/tables/time-series/demo/popest/pre-1980-national.html. The source for the US population data is US *Current Population Reports Series P25.*

Section 1.4 "What Is a Business Experiment?"

• The financial services example is based on Watson-Hemphill and Kastle (2012).

• The Progressive Insurance example comes from Chapter 22 of Holland and Cochran (2005).

• The Anheuser-Busch example comes from Ackoff (1978).

• The number of conditions necessary to establish causality varies from discipline to discipline and even author to author. For example, the epidemiologist Hill (1965) gave nine rules. We stick with just three.

• While observational data can be useful, they are no substitute for an experiment (if an experiment can be conducted!):

> But even if done perfectly, an observational study can only approach, but never reach, the credibility of randomization in assuring that there is no missing third variable that accounts for the differences observed in the experimental outcome. (Wainer, 2016, p. 48)

• The book by Schrage (2014) is entertaining and describes numerous business experiments; most of these are small, inexpensive experiments. The book by Holland and Cochran (2005), which is written for laymen, describes several larger, more complicated experiments. The article by Ganguly and Euchner (2018) describes the approach of the Goodyear Tire Company to experimentation and gives many interesting examples, including one where they conducted an experiment to determine whether tire-pressure monitoring equipment could generate more than enough savings to pay for the monitoring equipment by reducing roadside breakdowns of tractor-trailers. It is worthwhile to read materials such as these, for it is important for the novice experimenter to develop an idea of what has been done and what is possible.

• The idea that small sample sizes, small effect sizes, and lots of noise can lead to false positives and even sign reversals (truly positive coefficients being estimated as negative) is discussed in terms about as nontechnical as possible in Gelman and Carlin (2014), but you'll have to know what "power" is to follow the argument, so maybe you should wait until after Chapter 2 to read it.

Section 1.5 "Improving Website Design"

All the web tests, in particular the results in Table 1.6, are real. Due to difficulties obtaining permissions for the original web ads, some of the web ads were mocked-up to simulate the real ads. The photograph in Figure 1.5 is from pixabay.com, and the photograph in Figure 1.7 is from pexels.com.

GuessTheTest.com is a resource for digital marketers who want objective A/B test case studies and helpful information to get split-testing ideas, insights, and best practices. There are many aspects of A/B testing on the web that are not covered in this book, and the interested reader may profitably spend some time at this website. Also, if you think you're any good at predicting the outcome of an A/B web test, to disabuse yourself of such an errant notion, try guessing at a dozen or so of the many cases presented at this website and see if you can beat 50% accuracy by a statistically significant amount.

• We barely scratched the surface of A/B testing, which, according to two recent surveys, is the most important topic in business: a survey of online marketers found "Conversion Rate Optimization" to be a top priority for the foreseeable future (SalesForce.com, 2014); a survey of businesses that engage in conversion rate optimization used A/B testing more than any other method (Econsultancy, 2015).

• An entertaining layman's article on the rise of A/B testing can be found in *Wired* magazine (Christian, 2012). On A/B testing and the Obama presidential campaigns, see the interesting article in *Bloomberg Businessweek* by Joshua Green (2012). This is of historical interest because the Obama campaign was the first to really use analytics for fundraising and get-out-the-vote activities. For those who want to learn more about the technology behind website testing and the types of tests that are possible on websites, we recommend the chapter on web testing in Waisberg and Kashuk's book titled *Web Analytics 2.0* (Waisberg and Kaushik, 2009) or the succinct book by McFarland (2012) with the catchy title *Experiment!*.

There are also numerous other books with more in-depth coverage of website experiments such as Siroker and Koomen (2013). Technology tools for website testing are rapidly evolving, and the evaluation of software tools is a critical first step in any website testing program. A good article on the mechanics of A/B testing on the web is Kohavi et al. (2009b). He also co-authored an informative article, "Trustworthy Online Controlled Experiments: Five Puzzling Outcomes Explained" (Kohavi et al., 2012), as well as "Seven Rules of Thumb for Web Site Experimenters" (Kohavi et al., 2014).

Section 1.6 "A Brief History of Experiments"

- An entertaining history of experimental design that gives proper due to Gosset is Ziliak (2014). To see how the methods that Gosset developed were used for commercial purposes at the Guinness Brewery, see Ziliak (2008).
- Through the 1980s, American cars didn't last long. In the 1960s, American automobile manufacturers offered warranties for 4000 miles or three months, whichever came first. By the 1990s, these warranties had been increased to 70000 miles or seven years (Lightstone et al., 1993, p. 774).
- W. E. Deming was a statistics professor at NYU in the late 1930s when he developed "statistical quality control." He took his idea to the Detroit automakers, but they didn't want his advice. After WWII, he then took his ideas to Japan, which embraced them wholeheartedly. Deming is largely credited for the postwar Japanese economic miracle. How would the course of history have been altered, if Detroit had embraced Deming instead of rejecting him?
- As a general rule, observational results can't be trusted until they have been verified by a well-designed experiment. Medical practices often become popular as a result of observational studies, only to be overturned years later by experiments, long after much damage has been done. Some examples are hormone replacement therapy for menopausal women, stenting for coronary disease, and a specific medicine thought to retard heart disease (fenofibrate to treat hyperlipidemia). See Huded et al. (2013) for further examples.

1.9 Statistics Refresher

If these questions utterly confuse you, you probably shouldn't be reading this book.

Which of the below questions you can answer depends on the statistics course you took. You should be able to answer many of these questions. If you can answer them using the statistical software you already know but not with R, that's okay. We give R code for everything we do in this book. If you can't answer a particular question, don't worry. The methods will be explained as needed later in the book.

1 What are the following percentiles for a z-distribution?
(a) 0.90
(b) 0.95
(c) 0.975

2 For a z-distribution, what value of z gives the following proportion in the upper tail?
(a) 0.01
(b) 0.0005
(c) 0.08

3 What are the following percentiles for a t-distribution with the given degrees of freedom (df)?
(a) 0.90, df=10
(b) 0.95, df=20
(c) 0.975, df=30

4 For a t-distribution with the given degrees of freedom (df), what value of t gives the following proportions in the upper tail?
(a) 0.01, df=10
(b) 0.0005, df=20
(c) 0.08, df=30

5 Use the data in file `SR1.csv` and perform a two-sided test of the null hypothesis that $\mu = 10$ for $\alpha = 0.05$ Specifically state the non-rejection region and the rejection region.

6 Use the data in file `SR1.csv` and produce a 95% confidence interval for the population mean.

7 For a two-sided test of the null hypothesis that $\mu = 10$ for $\alpha = 0.05$, compute power for the one-sample test of means if the true value of $\mu = 7$. Assume $\sigma = 6.0$ so that we can use the z-table, as power calculations with the t-table are problematic. The data are normal, and let $n = 36$. Compute power if $\mu = 9$. What can you infer about power? (If your first course didn't use "power," don't worry about this one.)

8 Use the data in file `SR2.csv` and perform a two-sided test of the null hypothesis that $\pi = 0.7$ for $\alpha = 0.05$ Specifically state the non-rejection region and the rejection region.

9 Use the data in file `SR2.csv` and produce a 95% confidence interval for the population proportion.

10 Use the data in file `SR3.csv` and compute a confidence interval for the difference between two means (hint: equal or unequal variances?).

11 Use the data in file `SR3.csv` and compute two confidence intervals, one for X and one for Y. Do they overlap? What does this tell you?

12 Suppose that the data in `SR3.csv` are paired data, i.e. two measurements on the same experimental unit. Let X be "before" and let Y be "after" some treatment. Perform a paired t-test. Suppose another analyst had done a two-sample test on these data, as in the previous two exercises. How do the analyses differ?

13 Use the data in file `SR4.csv` and compute a confidence interval for the difference between two proportions.

14 Use the data in file `SR1.csv` to make a histogram.

15 Use the data in file `SR3.csv` to make a scatterplot.

16 Use the data in file `SR3.csv` to run a simple regression, interpret the coefficients, and test whether the slope equals zero.

17 Use the data in file `SR5.csv` to run a multiple regression, $Y = b_0 + b_1X1 + b_2X2 + b_3X3$, interpret the coefficients, and test whether a slope equals zero. (If your first course didn't cover multiple regression, don't worry about this one.) Perform model checking (i.e. check whether the assumptions are true).

18 Use the data in file `SR5.csv` to run a multiple regression with dummy variables, $Y = b_0 + b_1X1 + b_2X2 + b_3X3 + b_4D1 + b_5D2$. (If your first course didn't cover multiple regression, don't worry about this one.)

19 Suppose there is a categorical variable with three levels. Suppose it is "type of house" and it can take on the values colonial, split level, and rancher. How can this categorical variable be represented by dummy variables?

20 Using the file `Houses.csv`, regress price (in thousands of dollars) on square footage (in square feet), age (in years), and a dummy (yes/no) for whether or not the house has a fireplace. Interpret the output.

21 Using the file `Houses.csv`, regress price (in thousands of dollars) on square footage, age (in years) and dummies for fireplace (yes/no), and type of house (colonial, rancher, or split level). Interpret the output.

22 Use the data in file `SR5.csv` to use ANOVA to test the null hypothesis that the means of X1, X2, and X3 are all equal. Perform model checking.

2

Analyzing A/B Tests: Basics

In this chapter, we will show how to analyze an A/B test using several cases. We assume that you have seen basic statistical analysis (confidence intervals and hypothesis tests, especially two-sample tests) once before, perhaps in an introductory statistics class, and so our focus here is on applying statistical methods to typical A/B test cases and drawing business conclusions. It may seem odd to begin with the analysis of the test; in any real testing project, the design of the experiment naturally comes first. However, it is difficult to plan an A/B test when you don't know how you will analyze it. So, we have intentionally started with the analysis here and will cover the design of A/B tests in Chapter 3. As you read these cases, you can imagine that the tests have been designed by an expert, and you've been assigned to analyze the test and come to a conclusion.

A/B tests have a long track record in business. The chain department store Kohl's, for example, used A/B testing to answer two important questions about the future of the entire chain: Should opening be delayed an hour on the days Monday through Saturday? Should Kohl's sell furniture? There were arguments for and against both propositions, but only an experiment could answer the questions definitively. To answer the first question, Kohl's delayed opening at 100 stores and compared sales to a matched control group of 100 stores. The answer is: Being open six fewer hours a week did not decrease sales, and it represented a great cost savings. The later opening was rolled out to the entire chain. To answer the second question, Kohl's added a furniture department to 70 stores and compared sales to a control group. The result is: the stores with furniture departments did not increase sales – in part because other departments had to be made smaller to make room for the furniture, and smaller departments have smaller sales. See the *Harvard Business Review* article by Thomke and Manzi (2014) for more details.

Contrast the above with what JC Penney did. The retailer, with 11 000 stores across the United States, had stagnating sales. Ron Johnson, who had great success at Target and Apple, became CEO in November 2011. He remade the store by

Business Experiments with R, First Edition. B. D. McCullough.

Companion Website: www.wiley.com/go/mccullough/businessexperimentswithr

changing its pricing scheme and implemented boutique shops within the larger stores, among other things. Sales crashed from $17 billion to $13 billion, and the stock price fell from $35 to $13. Johnson was replaced in April 2013. An article in *Time* magazine (Tuttle, 2013) noted that he did not test his ideas before implementing them and thus alienated JC Penney customers on a large scale. The same article reports that when he was asked about conducting experiments before implementing his ideas on a large scale, Johnson replied, "We didn't test at Apple."

After reading this chapter, students should:

- Use some of the basic vocabulary of experiments including *treatment* and *response variable.*
- Summarize and visualize the outcomes of an A/B test and present clear, concise findings from an A/B test to a decision maker.
- Use confidence intervals for comparing group averages and group proportions.
- Understand the relationship between a hypothesis test and a confidence interval.
- Relate the concepts of *power* and *confidence* to business decisions. Understand that the potential costs associated with type I and type II errors may make one or the other more important.
- Understand that classical tests are approximate and only apply to large samples.
- Be able to perform an exact finite sample comparison of proportions using the beta-binomial model.

2.1 Case: Improving Response to Sales Calls (Two-Sample Test of Means)

A company that sells printer/copier supplies to small- and medium-sized businesses regularly conducts audio-only sales calls to established customers. One of the managers suggests that using video in its sales calls will increase the amount that each customer purchases. Some of the sales staff are reluctant to use video; they feel that technical problems with the video streaming disrupt the flow of the call and may actually decrease sales.

To find out whether the manager is right or the sales staff is right, the operations director conducts an A/B test over two weeks. Each day, each sales agent is assigned a list of the contacts to call through the company's sales management system. During the test, the operations director randomly assigns each scheduled call to either "video" or "audio." (We will talk more about how to make random assignments in the next chapter; for now, you can imagine that the operations director tossed a coin over and over to determine which calls will be audio and which will

Table 2.1 Data from sales calls test.

Customer ID	Sales (within one week)	Call type
1513	118.44	Audio
1514	124.23	Video
1536	141.19	Audio
⋮	⋮	⋮

be video.) The assignment to audio or video was tracked in the sales management system, and the sales agents were instructed to use the call type that the customer was assigned.

For this business, most of the customers make a purchase during the call so that there are very few, if any zeros in the data set representing customers that didn't purchase. Occasionally, though, some orders don't come in until up to a week after the call. So, they ran the test over two weeks and then waited an additional week to see what orders came in. The operations director pulled data on the sales for each customer who was included in the test from their sales management system.

In the next few subsections, we will walk through the process of analyzing data from this test. The overall process is quite simple, yet there are a few subtle points that we will cover. To begin, the raw data looks like that in Table 2.1. Notice there is not a column for the audio data and another column for the video data. These data are "stacked." When there is a separate column of numbers for each variable, the data are called "unstacked."

Some commands want the data stacked. Other commands want the data unstacked. Some commands can handle data in both formats. It is important to be able to convert data from stacked to unstacked and vice versa; we will discuss this in a few pages.

2.1.1 Initial Analysis and Visualization

The first step in any analysis is to compute summary statistics (in this case, the mean of the response variable for each treatment group) and make some plots and/or graphs. When the operations director computed the average sales for the two groups, he found that the average sales were $120.73 for the video group, while the average sales for the audio-only group was $110.29. An average increase in sales of $10.44 per customer seemed like a huge gain in sales when multiplied over hundreds of customers each month.

A useful way for the operations director to present these data to the sales manager and his team is using a plot like the one in Figure 2.1, which is called a box

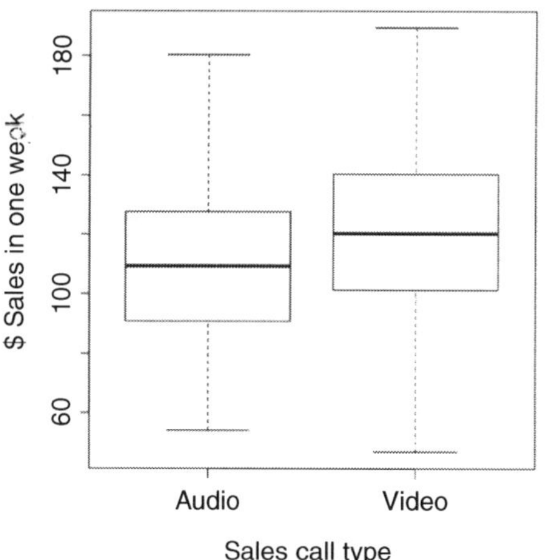

Figure 2.1 Box plot of one-week sales resulting from audio versus video sales calls.

plot. This plot shows the median sales for each group (the horizontal bars) along with some visual information about how much spread there is in the sales data. The boxes represent the interquartile range, i.e. the middle 50% of customers have sales in the range of the boxes. The whiskers show extremes in each group – for some packages the extreme values are the minimum and the maximum. For other packages they are the 1st and 99th percentiles, with outliers beyond the extreme values represented by dots.

Another common way to visualize the results of the test would be to use a bar plot comparing the average sales for the two groups, but we discourage this. While those who are new to box plots might find them a bit intimidating at first, they convey a lot more information about the data than a bar plot would. This idea is explored in Exercise 2.1.2.

Software Details

Create the box plot in Figure 2.1...

The R boxplot command can handle stacked or unstacked data.

```
df <- read.csv("AudioVideo.csv")
boxplot(sales_one_week~call_type,data=df)
# for stacked data

newdf <- unstack(df[,c(2:3)])  # unstack the data
head(newdf) #  see that the data are not stacked
boxplot(newdf) # for unstacked data
```

Table 2.2 Summary statistics for audio and video sales calls.

	Average sales ($, within one week)	StDev sales	Number of customers
Audio (1)	110.30 $(=\bar{x}_1)$	28.7 $(=s_1)$	100 $(=n_1)$
Video (2)	120.70 $(=\bar{x}_2)$	27.4 $(=s_2)$	100 $(=n_2)$

We can see from the box plot that there is a lot of "spread" in the data – sales amounts vary from $60 to $180 – and there is a fair amount of overlap between the two groups. This should make you wonder whether we really have enough data to say that there is a difference between the video and audio groups. Put another way, is the sales lift due to video really $10.44 or something else?

2.1.2 Confidence Interval for Difference Between Means

We use confidence intervals and statistical tests to answer the question, "Do the data say that the A and B treatments produced different outcomes?" The goal of a statistical test is to determine whether the difference we see between two summary statistics could have been produced by "noise" (i.e. "random variation") in the data or is due to a real difference in the average sales for the two groups. In this section we'll focus on the difference in average sales between the two groups using what is called a "confidence interval for the difference between two means" to compare the average sales between the groups.

The confidence interval compares the difference in average sales in each group to the *variation* in the data. Recall that the sample *variance* is a measure of the "spread" in a data set and the *standard deviation* is the square root of the variance. The average sales and the standard deviation in sales are summarized in Table 2.2.

To provide the formula for the confidence interval, it is helpful to introduce some statistical notation. We typically refer to the number of observations (in this case

Figure 2.2 Unstacked (left) and stacked (right) data.

```
> unstackeddf
  A B C
1 1 4 7
2 2 5 8
3 3 6 9
```

```
> stackeddf
  numbers subscripts
1       1          A
2       2          A
3       3          A
4       4          B
5       5          B
6       6          B
7       7          C
8       8          C
9       9          C
```

customers) in each group as n_1 and n_2. The averages of each group (in this case average sales) are denoted $\bar{x}_1$ and $\bar{x}_2$, and the standard deviation of each group is written as s_1 and s_2. We've used these same labels in Table 2.2, although we would remove them before presenting that table to a business audience. Using this notation, we can write the formula for the confidence interval for the difference between the two means as

$$\begin{aligned} \bar{x}_1 - \bar{x}_2 &\pm t_{(1-\alpha/2,n-2)} s_p \sqrt{\frac{1}{n_1} + \frac{1}{n_2}} \\ s_p^2 &= \frac{(n_1 - 1)s_1^2 + (n_2 - 1)s_2^2}{n_1 + n_2 - 2} \end{aligned} \tag{2.1}$$

Try it!

If it has been a while since you've computed means and standard deviation, you can practice using the file `AudioVideo.csv`. The relevant R commands are `mean()` and `sd()`. Reproduce the results in Table 2.2.

Software Details

Data are not always stored in the way the software wants...

Two common methods of storing data are "stacked" and "unstacked," as shown in Figure 2.2. Sometimes a program wants data stacked; other times a program wants data unstacked. You should be able to convert data from one format to the other easily.

To stack:`stack(df)` where "df" is a data frame corresponding to the left side of Figure 2.2.

To unstack:`unstack(df)` where "df" is a data frame corresponding to the right side of Figure 2.2. This data frame should have only two columns.

As a practical matter, sometimes data frames do not have only two columns, and we will have to choose the two columns that we want to unstack. As an example, consider the datafile `AudioVido.csv`, which has stacked data in three columns. The relevant data are in columns two and three, so if we wanted to unstack it, the command would be `unstack(df[,c(2,3)]`.

If you issue the command `unstack(df)`, all you get is an error message, because the `unstack` command wants a data frame with only two columns.

where $t_{(1-\alpha/2,n-2)}$ is the $1 - \alpha/2$ percentile of the t distribution with $n - 2$ degrees of freedom. Formally, this estimator is called the "two-sample test of means," but practicing statisticians often refer to it as the "difference in means" estimator or even the shorthand "diff in means" estimator. Some introductory textbooks

Try it!

Create data sets like those in Figure 2.2 and be sure that you can stack and unstack them.

suggest that if $n > 30$, then the z-distribution is a good approximation to the t-distribution, but this is not true as shown by Boos and Hughes-Oliver (2000) (see the Learning More section for further details). Since the t-distribution is more accurate and is easy to use with modern computers, we'll use the t-distribution regardless of the sample size.

Note the subscript "p" in the formula for the variance above. This stands for "pooled," and the formula gives what is called the "pooled estimator" of the variance. Pooling is done when the variances of the samples are roughly the same. In the present case, 28.7 and 28.4 are close enough that we'll consider them to be practically the same. For now, we just look at the box plots to determine whether the variances are the same. In a couple pages we'll use a formal statistical test to do this. Figure 2.1 supports this conclusion.

If the sample variances were different, we'd have to use the unpooled estimator (often called "Welch's two-sample t-test"), which is given below:

$$\bar{x}_1 - \bar{x}_2 \pm \frac{t_{1-\alpha/2,\nu}}{\sqrt{\frac{s_1^2}{n_1} + \frac{s_2^2}{n_2}}}, \text{ with degrees of freedom } \nu = \frac{\left(\frac{s_1^2}{n_1} + \frac{s_2^2}{n_2}\right)^2}{\frac{s_1^4}{n_1^2 \nu_1} + \frac{s_2^4}{n_2^2 \nu_2}} \tag{2.2}$$

where ν_1 and ν_2 are the degrees of freedom for the first and second samples, respectively. Interestingly, ν is not constrained to be an integer; it can have a decimal part, which means that the usual t-table in a book cannot be used; using a computer with a fractional degrees of freedom is mandatory! Of course, we won't calculate either Equation 2.1 or Equation 2.2: we'll use the computer to perform these calculations.

When we apply Equation 2.1 to the sales data, the 95% confidence interval for the difference in means is (−18.26, −2.62) where audio is the first sample and video is the second sample. This means that the plausible values of the true difference between the groups are between −18.26 and −2.62. The difference is negative, which simply means that sales for audio customers are less than sales for video customers, or, in statistical language, $\bar{x}_1 - \bar{x}_2 < 0$, which implies $\bar{x}_1 < \bar{x}_2$. If we switch which group is labeled 1 and which is 2, we would get the interval (2.62, 18.26). In the future, when dealing with data, we will not use "1" and "2" for subscripts, as it gets too confusing. Instead, we will use mnemonic subscripts such as $\bar{x}_A$ for the sample mean of audio and $\bar{x}_V$ for the sample mean of video. When presenting a theoretical result and no data are being used, we will still have to use "1" and "2" for subscripts.

The key question we are trying to answer with the confidence interval is whether the data say that the two groups have different average sales, i.e. whether the difference in average sales is something other than zero. Since the confidence interval for the difference between audio and video does not include zero, we conclude that the means are not the same and that video sales are greater than audio sales. The manager who suggested it will be happy to hear that, and, more importantly, the company can conclude that switching to video calls will likely produce more sales.

Software Details

To perform a two-sample test of means...

If the data are not stacked in the data frame "df," then use the command `t.test(df$audio,df$video,var.equal=TRUE)` or `t.test(df$audio,df$video,var.equal=FALSE)` where `audio` and `video` are vectors containing the sales data for customers in the audio and video groups. Actually, the default for this command is `var.equal = FALSE`, so executing `t.test(df$audio,df$video)` is the same thing as executing `t.test(df$audio,df$video,var.equal=FALSE)`.

A nice feature of using this command is that the results will be in terms of audio-video so that you can easily interpret the results.

If the data are in a stacked data frame, you can get these vectors by subsetting to pull out the elements of the `sales_one_week` column according to `call_type = "audio"` or `call_type = "video"`.

```
audio <- df$sales_one_week[df$call_type=="video"]
video <- df$sales_one_week[df$call_type=="audio"]
```

The command will accept stacked data, in which case a two-sample test can be conducted via `t.test(sales_one_week~call_type,data=df)` where "df" is the name of the data frame containing the variables, but look at the resulting CI: Is it for audio-video or for video-audio? It's hard to tell. So we prefer not to perform *t*-tests this way.

Overall, it's probably easier to unstack the data and perform the test:

```
df1 <- unstack(df[,c(2,3)])
# df[,c(2,3)] is the second and third columns of df
t.test(df1$video,df1$audio)
```

No less important than knowing whether the difference is significant, we have some idea of what the magnitude of the difference is: somewhere between \$2.50 and \$18 per call. Even if the truth is at the bottom of this range (and it's probably closer to \$10), multiplying \$2.50 per call by hundreds of calls makes a nice addition to the top line. Statistical tests tell you that an effect is "significant" – meaning that we have enough data to say that the difference is not zero – yet give little indication of the *size* of the effect, which is important to businesses. For this reason, we prefer confidence intervals to formal hypothesis tests for business analysis.

If Equations 2.1 and 2.2 are a bit daunting, most statistical packages include routines to compute the confidence interval for the difference in means. As we have discussed, there are two types of such tests: one test for when the two samples are from populations with equal variances (which uses the "pooled estimator" of the sample variance) and another test for when the two populations have unequal variances (which uses the "unpooled estimator" of the sample variance). The formula in (2.1) is the pooled estimator used when the variances are equal (which they are for the sales calls experiment as you can see in Figure 2.1 and Table 2.2).

To test whether the two variances are the same, we form a ratio and apply an F-test. Let the larger sample variance be sample 1 and the smaller sample variance be sample 2; it is customary to do this because the ratio is then greater than unity. $F = S_1^2/S_2^2$ has an F distribution with $df_1 = n_1 - 1$ and $df_2 = n_2 - 1$ degrees of freedom. The audio has the larger variance, so $S_1^2 = 821.61$ and $S_2^2 = 750.53$; therefore $F = 1.095$. Now $df_1 = df_2 = 99$ and the upper 5% tail of the $F(99, 99)$ distribution is 1.39. Since the calculated F-statistic does not fall in the tail, we do not reject the null of equal variances. Hence, when we do a two-sample test or calculate a confidence interval, we can use the pooled estimator of the variance. If we had rejected the null hypothesis, then we'd have to use the unpooled estimator.

Software Details

To test the equality of two variances...

```
var.test(df1$audio,df1$video)
```

gives the F-test in the above text. If the variables are entered in the order `var.test(df1$video,df1$audio)`, then the largest variance is in the denominator and the calculated F-statistic is less than unity. It is customary for this test statistic to be greater than unity, so put the variable with the larger variance first.

We have just given a standard textbook treatment of the two-sample test of means: test the variances, and, if the variances are found to be equal, then apply a pooled-variance (standard t-test) method. If the variances are found to be unequal, then an unpooled-variance method (e.g. Welch's test) is applied. It's really cheating to look at the data to decide whether to use a pooled or unpooled test. Good arguments have been made that one should just use the Welch method as a matter of course and forget about variance testing as a prelude to the two-sample test of means; see Moser and Stevens (1992) and Ruxton (2006).

The basic idea is that when the variances are equal, Welch gives roughly the same answer as the t-test and, when the variances are different, the Welch method is better.

In this chapter, for two-sample tests of means, we presented both the pooled and unpooled tests simply because they should be familiar. Henceforth, we will always use the "unequal variances" formula unless we know a priori that the variances are equal.

2.1.3 Reporting Results

In reporting test results to a broad audience, it is common to suppress details of the statistical analysis. Analysts often use keywords in their reporting that signal that they have done some statistical analysis and are confident that there is sufficient data to come to those conclusions. One of those keywords is the word *significant*, which is shorthand for "I did a statistical analysis and the data say that there is a difference." Another strategy that has worked well for us is to report the confidence interval in parentheses. Those who are familiar with statistical analysis will know what this means, while others can simply ignore it and trust that your analysis has led to the appropriate conclusion.

Following this advice, the test result could be summarized in a one-sentence email to the sales team:

> The test showed that video calling significantly increased sales per customer by about $10.44 per customer over audio sales call (95% CI = ($2.62, $18.26)).

The operations director might also want to provide a box plot like the one in Figure 2.1, which would allow the team to see the difference in the average sales and the range of sales amounts.

And that's really all it takes to report the results of an AB test. The results are straightforward and are *causal*: there are no lurking variables and no complicated models to explain, and it is a result that is easy to understand and hard to argue with. A result like this can galvanize the sales team to use video calling for all sales calls going forward, thus increasing sales by somewhere between $2.62 and $18.26 per call. By contrast, other analytics approaches like predictive modeling can be difficult to explain to those who are untrained in the methods and, in our experience, sometimes lead to inaction.

2.1.4 Hypothesis Test for Comparing Means

When we ask whether zero is contained in the confidence interval, we are actually performing a statistical hypothesis test. We can describe this test more formally by explicitly stating null and alternative hypotheses. For this test, the null hypothesis is that there *is no difference* in sales between the audio group and the video group, which represents a "conservative" view (but more on this later.) The alternative is that there *is a difference* in sales between the two groups. We can write this in statistical notation as

$$H_0 : \mu_1 - \mu_2 = 0$$
$$H_A : \mu_1 - \mu_2 \neq 0 \tag{2.3}$$

where μ_1 and μ_2 stand for the true means of the two groups. The goal of a statistical hypothesis test is to determine which of these two hypotheses is true. Assuming equal variances, we do that by computing the test statistic:

$$t = \frac{(\bar{x}_1 - \bar{x}_2)}{\sqrt{s_p \frac{1}{n_1} + \frac{1}{n_2}}} \tag{2.4}$$

Note that the test statistic formula includes all the same components as the confidence interval, just rearranged in a different way.

Once we compute the test statistic using Equation (2.4), the next step in the hypothesis test is to compare the test statistic, t, to a critical value. For a two-tailed test, there will be an upper critical value and a lower critical value, e.g. ± 2.10.

If t is above the upper critical value or below the lower critical value, then the data suggest that we should reject the null hypothesis in favor of the alternative. If t is between the two critical values, then we accept the null hypothesis as true.

You may remember that that critical value is taken from the t-distribution with degrees of freedom $= n_1 + n_2 - 2 = 100 + 100 - 2 = 198$ and $\alpha = 0.05$. Since the test is a two-tailed test (meaning that we aren't sure in advance whether sales will be higher for audio or video sales calls), we use $\alpha/2$ in each tail, which implies critical values of ± 1.972 (which we looked up in the t-distribution). We can compare this to the test statistic that we compute using Equation (2.4), which is $t = -2.336$. Since this value falls to the left of -1.972, it falls in the left tail, and we reject the null hypothesis.

Alternatively, we could compute the p-value of the t-statistic, which would be 0.00912 (to three significant digits, regardless of whether variances are assumed equal or not). Since the p-value is less than α, we reject the null hypothesis. Either way, it is clear from both the confidence interval and the hypothesis test that the result is significant. With 100 observations in each group, we have enough data to say that video calls produce higher sales (within one week) than audio calls.

Of course, we only have data from 100 customers in each group, so even though our statistical analysis leads us to conclude that there is a significant difference between the audio and video groups, we don't actually *know* the true long-run mean sales for each group. It could be the case, though it's not likely, that just by chance, some of the better customers may have been assigned to the video group, and so we may erroneously conclude that video calls result in higher sales.

This type of mistake – rejecting the null hypothesis when the null hypothesis is true – is called a type I error. A type I error – declaring that video calls are more effective than audio calls, when that, in fact, is not true – could lead the organization to invest a lot in training and telecommunications to switch the sales team over to video calls, so the statistical test is designed to avoid it.

When we choose α in the hypothesis test, we are setting the probability of a type I error. Computing a 95% confidence interval (as we did above) or setting $\alpha = 0.05$ in the hypothesis test means that we are willing to make the type I mistake (declaring there is a difference in means when there is no difference) 5% of the time. We say that the confidence of the test is $(1 - 0.05) \times 100\% = 95\%$.

What about the other type of mistake? A type II error occurs when we accept the null hypothesis, when, in fact, the alternative hypothesis is true. In this case, that would mean that we decide there is no difference between video and audio calls, which could lead us to stick with audio sales calls when we could make more money by using video sales calls (a lost opportunity). As we discuss in Section 2.1.5 and in Chapter 3, by making a few assumptions about the true population means and standard deviations, we can actually compute the probability that a type II error will occur and this is often referred to as β. The probability that we will not make a type II error is referred to as the *power*, which is $1 - \beta$. In the same way that, absent guidance to the contrary, we typically choose $\alpha = 0.05$, we also choose power $= 0.80$. So if you have to tackle a power problem and it doesn't tell you what value power should be, choose 0.80.

The definitions of type I and type II errors are summarized in Table 2.3. The type I error amounts to seeing something that really isn't there, while the type II error amounts to being blind – being unable to see something that is really there. Keep in mind that a type I error only happens when the null hypothesis is true and a

Table 2.3 Errors in hypothesis tests.

	H_0 is true	**H_A is true**
Accept H_0	No error "confidence" $= (1 - \alpha)$	**Type II error** probability $= \beta$ "false negative"
Reject H_0 in favor of H_A	**Type I error** probability $= \alpha$ "false positive"	No error "power" $= (1 - \beta)$

type II error can only happen when the null hypothesis is false. The main point you should recognize is that sometimes we can get unlucky and our data will lead us to the wrong conclusion.

It is also important to remember that the consequences of type I and type II errors can be very different. Consider a test for cancer where the null hypothesis is that the patient doesn't have cancer. A type I error means telling a person who doesn't have cancer that he does have cancer and then ordering a more sophisticated test, like a biopsy, which presumably will show that he doesn't have cancer. A type II error means telling a person who has cancer that he doesn't have cancer and sending him home to let his cancer grow. The consequence of a type I error is the patient's anxiety and fear until the biopsy can be performed; the consequence of a type II error is death. Since the cost of the type II error is very high, we would like to make the probability of a type II error low; when we make the probability of a type II error low, the probability of a type I error is going to be high, and we will get lots of false positives.

Unfortunately, there is no way to eliminate all errors; given some data, minimizing the chance of one type of error maximizes the chance of the other type. In making the trade-off between type I and type II errors or, equivalently, in balancing α and β, the default is usually to choose high confidence, which often results in lower power. In medical statistics (where many of these methods were developed), drugs or other new treatments must be proven to work before they are widely distributed. Since the type I error – declaring a drug to be effective when it is not – is considered the worse mistake, α is deliberately made small at the expense of tolerating many type II errors – which may lead companies to drop development of drugs that might, in fact, be beneficial. The preference is for high confidence and attendant low power, and so $\alpha = 0.05$ is the default in most statistical software.

In business, this default is not always desirable. Consider the case of choosing between two different versions of an ad. The cost of deploying one ad versus the other is trivial. If we make a type I error and conclude that the two ads are different when they really are not different, the business impact is negligible. It didn't cost us any money to make the switch, and using the new ad isn't costing us any revenue, so we're really no worse off. However, if we make a type II error and conclude that there is no difference between the ads when there really is a difference, then we are foregoing a lot of potential revenue by running a poorer ad. For this type of business problem, it makes sense to choose β to be quite small and not worry much about the type I error. (Choosing β to be small is the same thing as choosing power to be high.)

Of course, there are other business settings where the cost of a type I error is very expensive. Returning to the sales calls test comparing audio and video calls, a type I error that leads the company to conclude that video calls are better would result in a lot of effort and expense in switching over to video calling when there is no benefit – a failed investment. A type II error would result in the company foregoing that investment when it would increase sales – a lost opportunity. Depending on the costs involved and the potential opportunity, one type of error may be preferable over the other, and you should consider this when analyzing A/B tests.

It is also important to consider whether the initial test will be followed up with more tests. For example, suppose we are running a series of screening tests to find the most appealing flavors for a new product. Flavors that seem successful in the initial screening test will be investigated in more detail in subsequent tests. In this case, we don't want to miss any potentially good flavors, and so we might accept low confidence so that we can have high power; committing a type I error doesn't matter because we're going to follow up the first test with more tests that will reveal the error.

2.1.5 Power and Sample Size for Tests of Difference of Means

Hypothesis tests and confidence intervals are constructed assuming the null hypothesis is true, but what if the null hypothesis is false? It is important to know how good our tests are when the null hypothesis is false, and for this we need the concept of *power*. Recall that $\beta = P(\text{Type II})$. Power is simply $1 - \beta$. Since β is the probability of not rejecting the null when the null is false, power is the probability of correctly rejecting the null when the null should be rejected. Since power is a probability, it is always between zero and unity. Power close to zero is bad, and power close to unity is good. (Think about it.)

In order to calculate β or power, we need to know "the truth," i.e. what the true difference is. *Ceteris paribus*, a small difference is hard to detect and will result in low power, while a large difference is easy to detect and will result in high power. Power is something that should be calculated before any data are collected, so you'll have to have some idea of what the standard deviation is and what size of a difference you want to detect. Let's calculate power for the video sales example. Suppose that we have experience with this type of data, and we know the standard deviation is about 30, and we'll have 100 observations in each sample for our two-sample test of means. Suppose further that the value of the true difference is 5. What is the probability that we will correctly reject the null hypothesis that the two means are the same? The software will tell us that power = 0.216; we'll only correctly reject the null about one in five times.

Suppose, however, that we were interested in detecting a $15 difference. Then power would be 0.9404, and the probability of a Type II error would be 0.06, meaning that for this test, there was a 6% chance that we would have decided there was no difference when there is actually a difference of $15 in sales between the two groups. That means that we included enough subjects in the test. A probability of a type II error of 45% would suggest that we needed a larger sample size to decrease the probability of a type II error.

Power is the probability that we will detect a difference when there is actually a difference. Power is a function of the sample sizes in each group, the standard deviation (which measures the "noise" in the data), the confidence level α, and the actual difference between the two groups. Using statistical software, we computed that the power of the sales call test was 0.94; based on an assumption that the true difference in average sales is 15, we have a sample size of 100 for each sample and a standard deviation in the sales amount of 30 and a confidence level of $\alpha = 0.05$. (For our power calculation, we choose the difference and the standard deviation based on the numbers that we used to generate the data, which you generally won't have when you are analyzing a real test. More on this later.) This number means that given a true difference of 15, we will correctly reject the null hypothesis that the difference is zero 94% of the time.

What would have happened if the true difference had been only $5 per call? We can repeat the calculation to find that the power is 0.22, meaning that if the difference in sales between the audio and video groups was just $5, then four out of five times we would not reject the null hypothesis, even though we should reject it. Yikes! That's a big missed opportunity if we fail to find a significant difference between the two groups. While it is traditional to focus on confidence in statistical tests, it is important also to consider power, particularly in business where both types of mistakes can lead to lost revenue.

For most software packages, you won't have to specify the sample mean for each group, but only the difference between them. Therefore, you will have to ask yourself, "What size difference do I need to detect?" For a mean of $100, probably you don't want to detect a difference measured in pennies or even dimes, but a difference of just one dollar might be meaningful if it's one dollar per transaction and you're going to make thousands of transactions.

In the calculations we just performed, we happened to know the true standard deviations and the true means. In real life, we never have this luxury. We will have to make educated guesses about the standard deviations and the means, and this will require effort. Reading reports of similar studies can be useful, as can experience. Unless you have a great deal of experience with a particular type of data, you will not be very sure of your guesses, and you will rarely feel satisfied with the guesses you make. But make them, because power calculations are very important; when the available sample is really too small, power calculations based on rough guesses will help discover that.

Software Details

To compute power for a two-sample test of means...

```
power.t.test(n=100,delta=5,sd=30,sig.level=0.05,
power=NULL)
```

Pay attention, however, to whether the sample size you have to enter is the size of each sample or the total size of both samples! In R, it is always the size of each sample.

As already mentioned, there are five arguments for this command: difference, sample size, standard deviation, confidence level, and power. You enter four of these and the package will return the fifth. If you want to know what sample size you need to achieve a given power, simply enter values for the other four arguments. The above is for a two-sample test, which is the default. If you have a one-sample test, you must add the option `type = "one.sample"`.

To compute power for unequal sample sizes, it may be necessary to use a website that performs such calculations. Two such websites are DSS Research Power Calculators[a] and Power and Sample Size[b]; the former may be easier to use. Specialized software to install on your computer is also available; two free versions are G*Power from the University of Duesseldorf and PS from Vanderbilt University Department of Biostatistics.

In[1] general,[2] the same way we set $\alpha = 0.05$ by default, we also set power $= 0.80$ unless we have a reason to choose a different number.

Try it!

Consider testing "audio - video." Compute power for values −15, −10, −5, 5, 10, 15 and plot power on the y-axis against difference on the x-axis. Why are you not asked to compute power for a difference of zero? What number should you use for this value?

Once you have run the test, it may be tempting to use the observed difference in means when computing the power of the test. Don't! That is, if you observe a difference of 12.52 between the two groups, then you should not use this as the "difference to detect" in the power calculation. The reason is that if there really is a difference between the two groups of 20, but you observe a much smaller difference (because you got unlucky), then the "observed power" computed based on

1 https://www.dssresearch.com/resources/calculators/
2 http://powerandsamplesize.com/Calculators/

the observed difference will always be low. Instead, power calculations should be based on a difference to detect that would be meaningful to the business. This idea is discussed in much greater detail in Section 4.4.

The most common way to increase the power of the test is to increase the sample size, but we can also decrease the variation in the response measure, decrease our confidence level (and accept more type I errors), or test treatments that produce larger differences between the two groups. We'll discuss this more in Chapter 3 when we discuss the design of A/B tests.

We can also use the power command to compute sample sizes. The power command has five arguments: n, delta, sd, sig.level, and power. To compute power we filled in the other four. To compute a sample size for a specified level of power, simply leave sample size blank (or NULL) and fill in the other four arguments.

Try it!

How many observations do you need to achieve 80% power to detect a difference of 5 when the standard deviation is 30 and $\alpha = 0.05$?

```
power.t.test(delta=5,sd=30,sig.level=0.05,power=0.8)
```

All of the above methods for determining sample size require you to have some knowledge of the standard error. Sometimes this knowledge comes from experience or prior studies, but if you have no knowledge of the standard error, a pilot study is in order. A pilot study is used solely for the purpose of getting an estimate of the standard deviation to be plugged into a sample size calculation.

If the estimate is too small, the computed sample size will be too small, and the experiment will be underpowered. Even if the estimator of the standard deviation is unbiased, there's a 50% chance that the estimate will be too small (assuming a symmetric distribution for the sample estimate). To avoid this situation, good advice from Browne (1995) is to compute a one-sided upper-bound CI for the variance and use that value.

For completeness, we first give the usual $(1 - \alpha) * 100\%$ two-sided interval for the variance:

$$\left(\frac{(n-1)s^2}{\chi^2_{1-\alpha/2,n-1}}, \frac{(n-1)s^2}{\chi^2_{\alpha/2,n-1}} \right) \tag{2.5}$$

whence it is obvious that the one-sided upper-bound interval is

$$\left(0, \frac{(n-1)s^2}{\chi^2_{\alpha,n-1}} \right) \tag{2.6}$$

Software Details

To compute an upper-bound one-sided CI for the variance...

```
nn <- 10
s2 <- 100
upperbound <- (nn-1)*s2 / qchisq(0.05,df=9)
```

where, of course, the degrees of freedom equals $n - 1$.

Take the square root of the upper bound and use it for "sd=" in the `power.t.test` command to determine the sample size.

We have begged the question, "How many observations does the pilot study need?" When the sample size doubles, the precision does not double, but increases only by a factor of $\sqrt{2}$; quadrupling the sample size only halves the standard error. As the sample size is increased, the gains in precision decline. Based on such consideration, Julious (2008) and van Belle (2008) both recommend sample sizes of 12 (when observations are expensive), though if observations are cheap then the rule of thumb from quality control is that 30 is enough observations to get a good estimate of a variance (Hoerl and Snee, 2012, p. 155).

2.1.6 Considering Costs

Returning to the audio/video example, suppose the operations director had estimated that the costs of doing a video call were \$5.00 more than audio, due to the additional bandwidth required, more preparation time, and more time required during the sales call. Because video costs more, it is not enough to be sure that video produces higher sales than audio. We have to be sure that it produces at least \$5 more in sales than video to ensure that switching to audio has positive return on investment (ROI).

One way to determine whether video produces at least \$5 more in sales is to redefine the response variable for the test as sales less additional costs, i.e. $(x_A - 0)$ for audio and $(x_V - 5)$ for video. The null and alternative hypotheses are

$$H_0 : \mu_A - (\mu_V - 5) \geq 0$$

$$H_A : \mu_A - (\mu_V - 5) < 0 \tag{2.7}$$

The null hypothesis says that, net of costs, audio and video produce the same revenue, while the alternative says that, net of costs, audio provides less revenue than video. (A little bit of algebra shows that the alternative can be rewritten as $\mu_V - \mu_A > 5$.)

This is clearly a one-sided test, because video will *only* be adopted if it's worth at least $5 more than audio. Since this is a one-sided test, we'll need a one-sided confidence interval. We will address one-sided confidence intervals in great detail in Sections 4.1.1 and 4.10. For now, we will simply compute the one-sided confidence interval.

R provides an option for a one-sided confidence interval; you can use it to find that the 95% one-sided confidence interval for the difference in audio-(video-5), which is $(-\infty, 1.11)$.

Software Details

To get the one-sided CI...

```
newdf <- unstack(df[,c(2:3)]) # unstack the data
newdf$vminus5 <- newdf$video-5  # create new video-5
variable

t.test(newdf$audio,newdf$vminus5,conf.level = 0.95,
    alternative = "less")

  Welch Two Sample t-test

data:  newdf$audio and newdf$vminus5
t = -1.3725, df = 197.6, p-value = 0.08572
alternative hypothesis: true difference in means
is less than 0
95 percent confidence interval:
     -Inf 1.110423
sample estimates:
mean of x mean of y
 110.2920  115.7342
```

The output does state that the alternative is that the "true difference in means is less than 0" but it doesn't tell you whether the difference is audio-vminus5 or vminus5-audio, so you have to puzzle it out for yourself. The order in which the "data": are presented ("audio" then "vminus5") is helpful, and you can deduce that the difference is audio-vminus5.

From Equation (2.7) the null-hypothesized value is zero, and we observe that zero falls in the CI, so we do not reject the null. Thus, this one-sided interval suggests that the difference in sales does not exceed the $5 difference in costs. We note in passing that in some cases, when the data is just on the margin of significance, the one-sided test will produce a significant result when the two-sided test does not.

We can also use a two-sided confidence interval to produce the one-sided interval as follows. In the software, set α equal to twice the desired α, which is $\alpha = 0.10$ in this case. The software will report a 90% two-sided CI, which is: $(-11.99, 1.11)$. We then ignore the lower limit to obtain the interval of $(-\infty, 1.11)$. Given this analysis, our report on the test should say:

> The test showed that video calling does not increase sales per customer over audio calling by more than \$5 per call (95% CI = $(-\infty, \$1.11)$). We should not switch from audio to video calls.

In addition to suggesting that we continue the test, a good analyst might consider whether we could tolerate a higher risk of a type I error, which corresponds to concluding that video is better when it is not. In this case, a manager might be willing to tolerate a much higher chance of incorrectly switching to video, especially given that some of the team wanted to switch to video without any data supporting that decision.

Try it!

Reproduce the above results for one-sided and two-sided confidence intervals by creating a new variable for $(x_V - 5)$. Be sure that you are subtracting 5 from the video sales (not from the audio sales!) when you create the new variable. After you have done that, recompute the confidence interval at 80% confidence to see how it changes the result.

Exercises

2.1.1 Use data file `AudioVideo.csv` to reproduce the box plot in Figure 2.1, the summary statistics in Table 2.2, and the confidence interval for the difference in mean sales between the two groups, $(-18.26, -2.62)$.

2.1.2 Use data file `AudioVideo.csv` to make a bar plot. Compare this to the box plot from the previous exercise. What differences do you see?

2.1.3 Use software to compute the test statistic ($t = -2.36$), the critical value (1.972), and the p-value (0.000 912) in Section 2.1.4.

2.1.4 The management of a chain of convenience stores is considering increasing sales by reconfiguring the interior layout of all its stores. State the null and alternative hypotheses for the test. What is the type I error? What is the type II error? What are the consequences of each type of error? Should this be formulated as a one-tail or two-tail test?

2.1.5 For the audio/video example, suppose the standard deviation is 30 and you want to detect a difference of $5. How many observations in each sample do you need? Do this for both one-sided and two-sided tests. Hint: You need some value for power in order to solve this problem. What value of power should you choose? Another hint: To do this for one-sided or two-sided tests, use the option `alternative = "two.sided"` or `alternative = "one.sided"`.

2.1.6 Give an example of a business experiment where you would want high power and low confidence.

2.1.7 Give an example of a business experiment where you would want low power and high confidence.

2.1.8 Replicate the analysis in Section 2.1.6, but assuming that video costs $3 more per call than audio. Make sure to state a clear recommendation to the company based on your analysis.

2.1.9 For the audio/video example, suppose the standard deviation is 30 and you want to detect a difference of $5. What is power when the sample size is 100,200, 300, … , 1000? Plot this curve.

2.1.10 Suppose your website gets 2500 visitors a week, and you want to conduct an A/B test. Your clickthrough rate, a proportion between zero and one, is 0.68. The standard deviation is 0.4665. You want to detect a lift of 2%, i.e. you want to be able to tell if the clickthrough rate increases by $0.02 * 0.68 = 0.0136$ to $0.68 + 0.0136 = 0.6936$. How many weeks will it take to conduct the test? N.B. Technically this is a proportion problem and should be analyzed using the methods of the next section. However, we appeal to the normal approximation to the binomial and estimate this as a mean instead of a proportion to be consistent with Section 4.7; especially see the Learning More details for that section.

2.2 Case: Email Response Test (Two-Sample Test of Proportions)

A small clothing boutique that maintains an email list is planning to send an email to its customers, inviting them to call in for a special offer. The marketing specialist has come up with two versions of the email, but can't decide which one to send. To determine which one to send out to all the clients, a random sample of 400 client email addresses is drawn; 200 will be sent version A, and 200 will be sent version B. Version A asks the customer to call one telephone number, and version B asks the customer to call another telephone number; in this way it is easy to keep track of responses. A sample is shown in Table 2.4.

Summarizing these data we find that the sample size is 200 for both versions. The number of yes responses for A is 23 and the number of yes responses for B is 28.

We need to see if the percentages differ, so we start by computing the sample proportion for each group, which is just the fraction of people who responded to the offer. Using statistical notation, if x is the number of successes in n trials, then $p = x/n$. In Group A is $p_A = 23/200 = 0.115 = 11.5\%$ of customers responded, and in Group B is $p_B = 28/200 = 0.140 = 14.0\%$ of customers responded, suggesting that the B email is more effective at getting customers to call.

A *binary response* variable has only two possible outcomes: 0 or 1, yes or no, default or non-default, etc. The email response is a binary response, meaning that each subject either calls or does not call. The treatment is also binary; customers were randomly assigned to either email A or email B. We can visualize the relationship between two binary variables using a mosaic plot as in Figure 2.3. Observe that the two software packages produce different, but equivalent plots. The mosaic plot makes it easy to see the size of each group because the width of the bars is proportional to the size of the groups. This makes it easy to see that the sample size is the same for each group. It also makes it easy to compare the response rate for each

Table 2.4 Data from email response test.

Customer id	Email version	Response
750206	A	Yes
863086	A	No
076360	B	No
⋮	⋮	⋮

Figure 2.3 Mosaic plot of email response data.

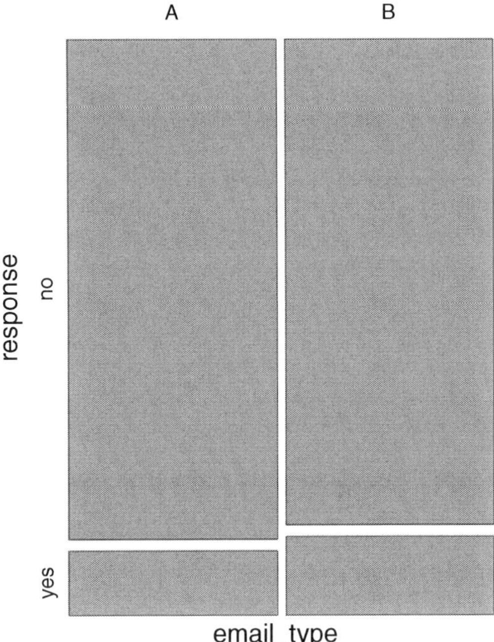

group, indicated by the height of the "yes" blocks, which we can see is similar for both groups. It also makes it immediately clear that a large majority of customers in both groups do not call in. Note that the mosaic plot can also be useful in situations where there are more than two treatments or when there is a categorical response with more than two levels.

Try it!

Summarize the data in `EmailResponse.csv` by computing the proportion who respond for each group.

Software Details

To make a mosaic plot...

```
table1 <- table(email_type,response)
mosaicplot(table1)
```

2.2.1 Confidence Interval and Hypothesis Test for Comparing Two Proportions

As in the sales calls test, we can compute a confidence interval, but in this case we need to use a different formula to compare proportions rather than means. The confidence interval for the difference between two proportions presented in introductory textbooks is given by

$$p_1 - p_2 \pm z_{\alpha/2}\sqrt{\frac{p_1(1-p_1)}{n_1} + \frac{p_2(1-p_2)}{n_2}} \tag{2.8}$$

where p_1 and p_2 are the sample proportions for each group and $z_{\alpha/2}$ is the cumulative standard normal distribution at $\alpha/2$.

Applying this to our own data with group A corresponding to sample 1 and group B corresponding to sample 2, we get

$$-0.025 \pm 0.0653 = (-0.0903, 0.0403) \tag{2.9}$$

This interval tells us the range of plausible values for the difference between the two population proportions. According to this interval, zero is a plausible value for the difference, so we do not reject the null hypothesis that the population proportions are the same. With just 200 respondents in each group, we do not have enough data to conclude that there is a difference in the response rate between groups.

2.2.2 Better Confidence Intervals for Comparing Two Proportions

The interval given in Equation (2.8) is called the Wald interval. We present it here because you may have seen it in an introductory stats books, but you should know that Wald intervals perform very poorly in many situations that come up frequently in practice. We recommend that you always use the "better" intervals that are available in software packages, which we discuss below.

Suppose we have an Internet display ad that has been served to 200 users. Those 200 impressions have resulted in 3 clicks on the ad (leading the user to our website). That means our ad has resulted in 3 "successes" and 197 "failures" in 200 tries, so the average clickthrough rate is $p = 0.015$. The problem is that successes are rare. Had there been one more success in the 200 tries, then the average clickthrough rate would have been 0.02, which is a third again as large. Clearly, if the event is rare, then the sample proportion p is not a very stable estimate of the true long-run clickthrough rate π; we would need a much larger sample size to get a stable estimate. In such an unstable situation, it is important that we provide managers with a confidence interval so that they understand the degree of uncertainty about what the long-run clickthrough rate actually is.

However, the Wald confidence interval for the clickthrough rate for a single ad isn't a very good representation of our uncertainty about the true clickthrough rate. So far, we have been focusing on the confidence intervals for differences between two proportions, but it is also useful to compute confidence intervals for an individual proportion. The Wald interval for a single proportion (not a comparison between two proportions) given in most introductory books is

$$p \pm z_{\alpha/2}\sqrt{\frac{p(1-p)}{n}} \tag{2.10}$$

When we observe 3 successes in 200 observations, the resulting confidence interval is (−0.00185, 0.031846). This confidence interval should raise your eyebrows. It extends into the negative range and it is impossible for a success rate to be negative. Truncating the interval on the left at zero is not a theoretically sound solution. Moreover, the Wald interval is very unreliable when p is small; see the Learning More section of this chapter for a discussion.

Several better alternatives to the Wald interval have been suggested, but they are not as easy to compute by hand. We strongly suggest that you use a software package that offers an alternative approach to compute confidence intervals for proportions. While there is general agreement that the Wald interval should not be used, there is no consensus on one interval that should be used in its place. A few of the better alternatives are Clopper–Pearson, Wilson score, Jeffreys, and Agresti–Coull. For practical purposes, it doesn't really matter which method your computer software uses, as long as it is not the Wald interval.

Returning to the email test, we can also use these improved intervals for comparing two proportions. Table 2.5 gives the 95% intervals for the difference between the email A and email B response rates produced by the Wald and R methods. As you have more observations, the Wald interval will get closer to the other methods, but can be quite bad as the sample size gets smaller.

Table 2.5 Confidence intervals for the difference in response rate between email A and email B.

Method	Lower limit	Upper limit
Wald	−0.0903	0.0403
R	−0.0953	0.0453

Software Details

To compute a 95% interval for a two-sample test of proportions...

```
prop.test(x=c(succ_1, succ_2),n=c(n_1, n_2))
```

where `succ_1` and `succ_2` are the number of successful outcomes in each group and `n_1` and `n_2` are the total number of observations in each group. The function documentation cites papers by Wilson and by Newcombe for appropriate adjustments to the Wald interval. By default, this command applies something called "Yates' correction." If you don't want it applied, then add the option `correct = FALSE`.

We failed to reject the null, and now we have no idea whether there is no effect, or there is an effect and we couldn't see it. Why are we in this predicament? Because we didn't do a power analysis before we conducted the study. We'd better learn how to do a power analysis for proportions.

2.2.3 Power and Sample Size for Tests of Difference of Two Proportions

The confidence interval for the difference in proportions tells us that there is no difference between the two emails' response rates (with 95% confidence). However, we actually know that the two emails do produce different response rates, because we generated the data ourselves. We used a random number generator to create the two sequences of 200 observations where the response rate for email A was 10% and for email B it was 15%. This difference of 5% would be very meaningful to the business. We are committing a type II error by concluding that there is not a difference. What went wrong? Why couldn't we detect such a large difference? Isn't $200 + 200 = 400$ a large sample? Quite possibly, the problem is that the test lacks power.

The power of the test comparing proportions can be computed using most statistical packages. The results of the power calculations for three different sample sizes are given in Table 2.6. There are minor algorithmic differences between software packages, but they all agree when results are rounded to two decimal places. At a sample size of $n = 200$ in each group, the power to detect a difference between

Table 2.6 Power for two-sample comparison of proportions, $\pi_1 = 0.10, \pi_2 = 0.15$.

$n = 200$	$n = 400$	$n = 800$
0.3266	0.5709	0.8570

0.10 and 0.15 is around 30%. That means that even with a 5% difference in response rate between email A and email B, with 400 customers, we would only be able to detect it about 1/3 of the time. In hindsight, this seems like unacceptably low power. If the sample size is doubled, power increases to slightly better than a coin flip. If it is doubled again, it increases to about 85%.

Software Details

To compute power for a difference of proportions...

```
power.prop.test(n=200,p1=0.10,p2=0.15,sig.level=0.05)
```

which returns a value for the power of 0.3266. Note that the command only has an argument for n, not for n_1 and n_2. As was the case with testing means, this command assumes the same number of observations in each sample. If you need a different number of observations in each sample, avail yourself of the previously mentioned suggestions.

The moral of this story is that if you fail to reject the null hypothesis, be sure that your test doesn't have low power! Low power plagues many studies, especially those in which the researcher negligently failed to do a sample size calculation before conducting the test. For example, when comparing two groups to test the safety of a new medication against an old medication, the conclusion frequently is that "there is no statistically significant difference between the groups," yet the authors never bothered to check whether the sample size was large enough to detect a difference! Consequently, physicians think the two medications are equally safe when one might actually be dangerous. This is a common problem, as noted by Reinhart Reinhart (2015, p. 16):

> In one sample of studies published between 1975 and 1990 in prestigious medical journals, 27% of randomized controlled trials gave negative results, but 64% of these didn't collect enough data to detect a 50% difference in primary outcome between treatment groups. Fifty percent! Even if one medication decreases symptoms by 50% more than the other medication, there's insufficient data to conclude it's more effective. And 84% of the negative trials didn't have the power to detect a 25% difference.

Remember that you must do the power calculation *before* you do the test; it can save you the trouble of conducting an underpowered test. How would you feel if you ran a test that found no difference and then discovered that your test had very low power? The resources devoted to an underpowered test that does not reject are wasted resources! As R. A. Fisher has been quoted, "To consult the statistician after an experiment is finished is often merely to ask him to conduct a post mortem examination. He can perhaps say what the experiment died of."

Try it!

For a two-sample, two-sided test of proportions, let $p_1 = 0.10$ and $p_2 = 0.12$, so we want to detect a difference of 0.02. How many observations do we need in each sample?

```
power.prop.test(n=NULL,p1=0.10,p2=0.12,sig.level=0.05)
```

As before, the default is for a two-sided test. For a one-sided test you must use the option `alternative = "one.sided"`.

The observant reader will have noticed that in all the examples above, both samples had the same sample size. This is because most software packages only calculate power for equal sample sizes in each group. This is not very realistic – many times we will want to have unequal sample sizes. For example, if we wish to test a completely new look for our emails, we might not want to do a 50–50 experiment. Instead, we might want to do a 5–95 test, where only 5% of visitors see the new email. That way, if the new version turns out to be a disaster, the damage is limited.

As was the case for means, we can also do sample size calculations using the power command for proportions. All that is necessary is to omit the sample size argument (or specify it to be NULL) and fill in a number for power.

We still have the same problem we had in determining sample sizes for means. There, we needed an estimate of the standard error to plug into the sample size calculation. To solve that problem, we conducted a pilot study. Here, we need some knowledge of the population proportion to plug into the sample size calculation, but we do not have to conduct a pilot study. If you really have no idea what the true population proportion might be, just use $\pi = 0.5$, as this will maximize the calculated sample size and thus give you a conservative estimate of the number of observations you need. This idea is explored in Exercise 2.7.

Exercises

2.2.1 Reproduce the mosaic plot in Figure 2.3.

2.2.2 Compute the confidence intervals in Table 2.5 by hand using the Wald interval (of course use software to compute means and standard deviations) and again using R. Is there a difference between the two intervals?

2.2.3 Compute a confidence interval for the difference between the treatment and control proportions who have contracted polio for Table 1.4 (i.e. ignore the "no consent" proportion).

2.2.4 Reproduce the power estimates in one column of Table 2.6.

2.2.5 Check the R documentation. What method does it use to compute intervals for proportions?

2.2.6 Table 2.6 shows a power calculation for $\pi_1 = 0.10, \pi_2 = 0.15$ at various samples sizes. Fix the sample size at 400 for each group, and $\pi_2 = 0.15$. Vary π_1 from 0.05 to 0.25 and plot power as a function of the difference $\pi_1 - \pi_2$. How does power change with the difference? Is the effect of the size of the difference symmetric?

2.2.7 For the scenario in Table 2.6, calculate power when $n_1 = n_2 = 50$; $n_1 = n_2 = 100$; and $n_1 = n_2 = 200$. Continue increasing the sample size, if necessary, to get a clear picture of what happens.

2.2.8 For the email test, suppose you wanted to detect a difference of 0.01, i.e. you wanted to tell the difference between a 10% response rate and an 11% response rate. How large a sample size for each version would you need (assume each version gets the same number)?

2.2.9 Suppose the proportion of unvaccinated children who get polio is 0.0005, and suppose the treatment and control groups are planned to have 200 000 observations each. If you want to be able to detect a reduction of 1/2 (e.g. from 0.000 5 to 0.000 25), how much power does the test have? This is why the Salk vaccine trials required such large sample sizes.

2.3 Case: Comparing Landing Pages (Two-Sample Test of Means, Again)

In this section, we apply the concepts from the previous sections to another example data set. An online retailer ran a test to determine which of two landing pages for their website produces more sales. The retailer's website testing tool randomly assigned users to either see version A or version B, and the analytics platform recorded which version each visitor saw and how much that visitor purchases. The data from the test consists of 11 563 rows where each row represents a site visit and indicates which version the visitor saw and how much was purchased.

The first step in analyzing any data set is graphing it. In this case, side-by-side box plots comparing sales for visitors who saw version A and visitors who saw version B are a natural choice and are shown in the left side of Figure 2.4.

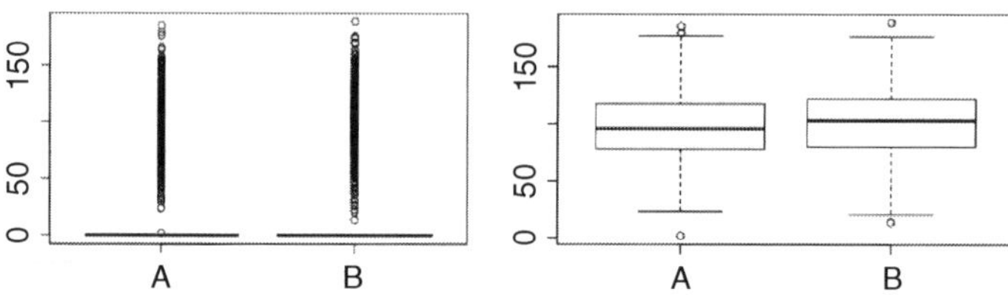

Figure 2.4 Box plots comparing sales for alternative landing pages. Left panel is all customers; right panel includes only customers who made a purchase.

Unfortunately, these box plots are not very informative, due to the large number of zeros in the data set (this is typical of customer purchase data – most customers don't buy anything on a given website visit). The horizontal line indicating the median of the data is at zero (because most customers don't buy), and most of the data points we can see are plotted as "outliers" outside the box.

A histogram (not shown) reveals the bimodal nature of the sales data. Most customers don't buy, but the average sale to customers who do buy is around $100. One way to improve the visualization is to restrict the box plots to nonzero values, as shown in the right side of Figure 2.4. Looking just at the customers who made a purchase, we can see that the median buy amount for version B is a bit more than that for version A and the dispersion of both is about the same. Judging from the box plots, there doesn't appear to be a substantial difference in the distribution of sales for the two groups.

Try it!

Practice by analyzing the data for this test in `LandingPages.csv`. Produce a histogram of the sales data. There is a spike at zero, representing customers who don't buy, and an almost imperceptible bump at 100, representing the customers who buy. Be sure that you can clearly see the second mode. You may have to make a second histogram and restrict the range of the data to get a good view of the second mode. Advanced users might try producing violin plots or bean plots comparing the two groups.

Having taken a look at the data with the box plots, the next step in the analysis is to compute the mean and variance of the sales as in Table 2.7. Eyeballing these summary statistics, the sales difference between version A and version B is around 0.40 when we look at all customers, and the standard deviation is over 20. Looking just at the customers who purchased, the difference in purchase amount is around 4, though the standard deviation exceeds 30. Either way, it doesn't look like there is much of a difference in the sales response between the versions.

Table 2.7 Summary statistics for sales.

	All customers			Customers who purchase		
	Customers	**Average sales**	**Std. Dev. sales**	**Customers**	**Average sales**	**Std. Dev. sales**
	n	$\bar{x}$	s	n	$\bar{x}$	s
Version A	5793	5.49	23.57	328	97.01	30.53
Version B	5770	5.91	24.92	336	101.41	31.30

Before we proceed with the analysis, we have to decide on which of two possible response variables we want to analyze: the average sales in dollars per visit (including the visitors who did not purchase) or the proportion of visits that result in a sale. Which response variable is most relevant depends on the intended goal of the landing page. Is the landing page intended to get more customers to buy or to get those who do buy to buy more? Most analysts would look carefully at both response variables. You could also analyze the average sales among just those customers who made a purchase; however, because customers were not randomly assigned to that that sub-condition, causal interpretations become more difficult. For now, let us focus on the average sales in $ per visit for all customers, leaving sales incidence to Section 4.5.

Using the formula in Equation (2.2), the 95% confidence interval for the difference in sales from R is $(-1.30, 0.47)$. It appears that the difference between the two versions is not significant, suggesting that the two landing pages perform similarly and the company can choose either.

To report this result to a broad audience, the website manager would describe the test, including images of both of the landing pages and then say:

> The test showed that Landing Page A and Landing Page B produced similar sales per website visit (95% CI = $(-1.30, 0.47)$).

And that is it! Analyzing and reporting the test results is extremely simple. For this test, the nonsignificant result should lead the team to look elsewhere for potential website improvements or brainstorm new ideas about how to improve the landing page.

But we made one mistake, and it's a big one. We failed to reject the null, and we have no idea whether it's because there is no effect or there is an effect and we couldn't see it. We really should have done a power analysis before conducting this study. We have to stop making this fundamental mistake.

Note that many website testing managers do not analyze their raw data directly and instead rely on website testing software such as Optimizely, Google Experiments, or Adobe Test&Target to determine which test results are significant. This is fine and can make the testing team more efficient, but it is good to walk through the process by hand at least once. Knowing how to analyze the test yourself will also allow you to design and analyze tests that are conducted in other domains where testing software is not available.

Exercises

2.3.1 Using `LandingPages.csv`, plot the histograms of sales for version A and for version B. Be sure to zoom in on the smaller mode by adjusting the scale so that you can get a good look at it.

2.3.2 Reproduce the box plots in Figure 2.4 and the summary statistics in Table 2.7.

2.3.3 Reproduce the confidence interval for the difference in sales among all customers.

2.3.4 The results of the test in Section 2.3 were for all visitors. Reanalyze the data for only those visitors who actually made a purchase, with $\alpha = 0.1$, and give a 90% CI. Report the results as you would to the website team. You cannot assert that one version of the ad produces more sales than the other. Why not? (Hint: Think of the Big Three of Causality.)

2.3.5 Identify three companies that offer A/B testing for websites, e.g. Optimizely. What features do these suites offer? What are the differences between them? Are they transparent about their methods? Do they use the Wald interval for a two-sample test of proportions?

2.3.6 Another way to approach the landing pages problem is via proportions. Suppose we were interested in whether A or B is more likely to lead to a purchase, ignoring the amount of the purchase. Create a dummy variable for whether or not the person made a purchase, and then do a two-sample test of proportions. Is there a difference in the proportions?

2.4 Case: Display Ad Clickthrough Rate

Display ads are the ads shown on websites such as CNN.com or Facebook. When a user visits a webpage that shows the display ad, it is called an *impression*, and when the user actually clicks on that ad to go to the advertiser's website, it is called a *clickthrough*. Advertisers want to know what the clickthrough rate is for different ads displayed at different websites, different times of day, etc. The typical display ad might have a clickthrough rate well under 1%. When dealing with extremely small proportions, it is sometimes useful to have access to specialized statistical models, e.g. the beta-binomial model.

2.4.1 Beta-Binomial Model

As we discussed in Section 2.2.2 the Wald confidence intervals for proportions works especially poorly in cases like this where π is near zero (or near one), even when n is relatively large. In this section we introduce a method for obtaining intervals based on a model called the *beta-binomial model*. It is better than the Wald interval and on par with the other intervals that are better than the Wald interval. We introduce it here because it is used in popular software such as Google Analytics Experiments. All the methods discussed thus far are exact when the data are normally distributed. *Exact* means that a 95% confidence interval will actually contain the true parameter 95% of the time. Of course, most data are not normally distributed, and when it is not, these methods are justified *asymptotically* – which means that the confidence intervals get close to being exact when the sample size is large. A nominal 95% interval that is not exact might contain the true parameter more or less than 95% of the time when the sample is not large enough. A common "rule of thumb" is that a sample size of 30 is large enough, but such advice can be misleading.

The *beta-binomial model* provides an alternative way to assess our uncertainty about the sample proportion, p, when the data are binary. The key idea behind this approach is that we can represent our beliefs about the true proportion π using a beta distribution. The beta distribution is defined on the interval $[0, 1]$ and is very flexible, so it is a good distribution to represent our beliefs about the true success rate. It has two parameters that control its shape, a and b. Typically this distribution would be denoted $\beta(a, b)$. The formula for it is given by

$$Prob(\pi) = \frac{\pi^{a-1}(1-\pi)^{b-1}}{B(a, b)}, 0 < \pi < 1$$

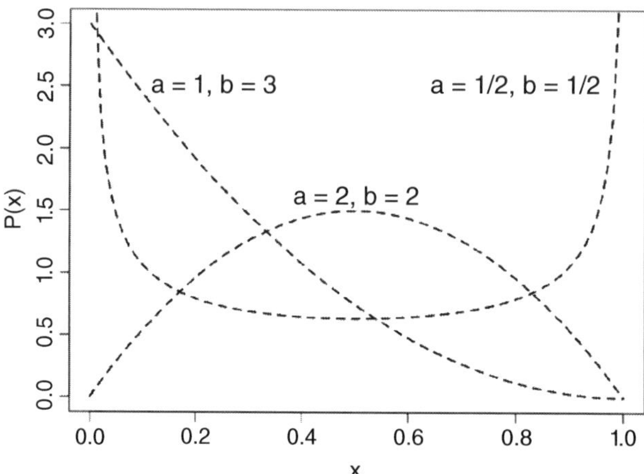

Figure 2.5 Beta distributions.

where

$$B(a,b) = \frac{\Gamma(a)\Gamma(b)}{\Gamma(a+b)} \text{ and } \Gamma(z) = \int_0^\infty x^{z-1}e^{-x}dx$$

If you don't know calculus, don't be intimidated by the above equations. We aren't actually going to use them. Some examples of the beta distribution for various values of a and b are shown in Figure 2.5.

Once we have this distribution that reflects our beliefs about the true clickthrough rate, we can do many useful things. For example, if our beliefs about π are represented by a beta distribution with $a = b = 2$, then we can compute the probability that the clickthrough rate is less than some number. The left 2.5 percentile of this distribution is 0.0943, meaning the chance that the true clickthrough rate is less than 9.43% is 2.5%. Similarly, the 97.5 percentile of this distribution is 0.9057, meaning that there is a 97.5% chance that the true clickthrough rate is less than 0.9057. Thus, the credible interval is $(0.0943, 0.9057)$.

Software Details

To peform calculations for the beta distribution...

The command to compute the percentile of the beta distribution is `qbeta(p, shape1, shape2)` where `p` is the probability of the desired tail area (or a vector of desired tail areas), e.g. 0.025 or 0.95; `shape1` is the a parameter; and `shape2` is the b parameter.

In a spreadsheet, the percentiles can be computed using a formula like `BETA.INV(p,a,b)`.

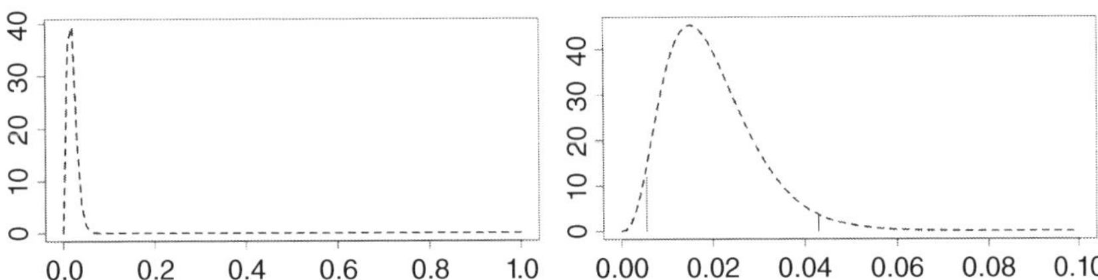

Figure 2.6 Beta-binomial distribution (left) and blowup of region $X \in (0, 0.1)$ (right).

Here is the key part. If we have data with 3 success and 197 failures, then a reasonable representation of our beliefs about the true success rate is a beta distribution with a = number of successes + 1 = 3 + 1 = 4 and b = number of failures + 1 = 197 + 1 = 198, which is shown in the left panel in Figure 2.6. Notice how the distribution spikes on the left; to get a better view of the relevant region, we restrict the range of the x-axis, resulting in the right panel of Figure 2.6.

The distribution in Figure 2.6 represents our beliefs about what values the true π might take. Using this distribution, we can calculate lower and upper limits by computing percentiles from this distribution. For this distribution, the 2.5 percentile is 0.005 45 and the 97.5 percentile is 0.043 00, resulting in a *credible interval* of (0.005 45, 0.043 00). Note that this interval is *asymmetric* around p; the success rate might be a lot higher than the clickthrough rate we have observed ($p = 0.015$), but it can't be much lower.

Consider an ad that has 37 551 views and only 66 clickthroughs. The proportion of views that get clickthroughs is $p = 66/37\ 551 = 0.001\ 76$. What is the credible interval for the true population proportion? The beta-binomial model yields a credible interval of (0.001 38, 0.002 23). Since the sample size is large, the Wald interval from formula (2.8) is quite similar: (0.001 33, 0.002 18).

The analysis we have described here is *Bayesian*, which is an alternative form of statistical analysis that is growing in popularity. There are many technical (and some deeply philosophical) differences between Bayesian methods and the methods usually taught in introductory stats classes. This is why we refer to the interval from the beta-binomial model as a *credible interval* rather than a confidence interval. This signals that this analysis is Bayesian. In practice, the results from both types of analysis are often very similar, but Bayesian analysis has a few other advantages besides it being exact for small samples: it allows for repeated analysis of data that streams in over time, and the distributions like those in Figure 2.6 can be used to compute other important quantities, like the expected ROI of an ad accounting for costs.

Those familiar with Bayesian inference may have noticed that we used a uniform *prior* by adding 1 to both the successes and failures. For large sample sizes, the prior won't make a large difference in the results, but for smaller sample sizes, you may want to use a more informative prior based on, say, past data on display ad clickthrough rates.

2.4.2 Comparing Two Proportions Using the Beta-Binomial Model

So far in this section, we have discussed Bayesian one-sample credible intervals, but in an A/B test we want to compare two different ads. The Bayesian approach to comparing two treatments is a bit different than the two-sample test of proportions that we have discussed previously.

Let's reconsider the email response test from the previous section, where version A got 23/200 and version B got 28/200. As we saw earlier, the usual two-sample test of proportions produced the following 95% CI: $(-0.0953, 0.0453)$. An analyst who simply "disregards" the negative part of the interval and reports $(0, 0.0453)$ will be throwing away most of the interval. This is not a legitimate statistical practice. This is definitely a situation for the beta-binomial.

Try it!

Plot the beta distribution that characterizes our beliefs about the clickthrough rates for version A and version B. How are the two distributions different?

```
# here is the plot for A; you make the plot for B
  x <- seq(0,1,0.01)
  y <- dbeta(x,23+1,177+1)
  plot(x,y,type="l")
```

We could compute the beta distribution for each version where a = number of successes +1 and b = number of failures +1, but our ultimate objective is to figure out which version has a better response rate. To do this, we will compute the probability that A has a better response rate than B, i.e. $P(\pi_A > \pi_B)$, based on those beta distributions for π_A and π_B.

The simplest way to do this (avoiding integrals!) is to use a random number generator to generate a random draw from the beta-binomial model for A and a random draw from the beta-binomial model for B. For each pair of draws, we will check whether the draw from the distribution of π_A is greater than the draw from the distribution of π_B. We will repeat this procedure a few thousand times and then count up how many times our draw of π_A is greater than the draw of π_B. The number of times that the draw from A exceeds the draw from B, divided by the number of repetitions, is a good approximation of $P(\pi_A > \pi_B)$. This can be done in a spreadsheet, in a statistical programming language like SAS or R, or in a general programming language like Python. When we computed this, we found that

$P(\pi_A > \pi_B) \approx 0.23$, meaning that there is a 23% chance that the true success rate for A is greater than the true success rate for B. Despite the fact that B is doing better based on 400 emails, there is still nearly a one in four chance that A performs better than B in the long run. The email manager might want to send out another 100 emails of each type to gain greater confidence before she sends the email B to the rest of the mailing list.

Software Details

To compute probabilities of the beta-binomial...

To compute the probability that the success rate for treatment A is greater than the success rate for treatment B, we set up a spreadsheet where each row in the sheet contains the draws for π_A and π_B. The first row should contain:

In cell A1 enter `=BETA.INV(RAND(),23+1, 177+1)`, in cell B1 enter `=BETA.INV(RAND(),28+1, 172+1)`, and in cell C1 enter `=IF(A1>B1,1,0)`.

Your sheet will look something like this:

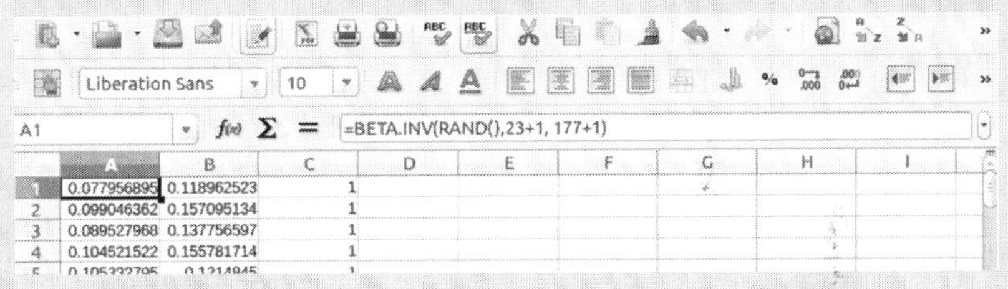

Note that the numbers that appear in each cell in your spreadsheet will not match ours, since we are generating random draws from the distributions for π_A and π_B.

Select these three cells and drag them down to row 999 so that you have 999 rows filled in. To compute the probability that A has a better success rate than B, we average the cells in column C by entering `=SUM(C1:C999)/1000`. The result will be something around 0.23.

If you recalculate (by hitting `F9`), this number may change slightly, because we are averaging over random draws. (If it changes a lot, then your number of draws is too small.) The accuracy of this estimate will vary depending on how many draws you take. In this example, we found 999 draws to be too few, producing estimates that ranged from 0.214 to 0.244. We leave it as an exercise to determine how many draws is needed to get estimates that are accurate to the third decimal place.

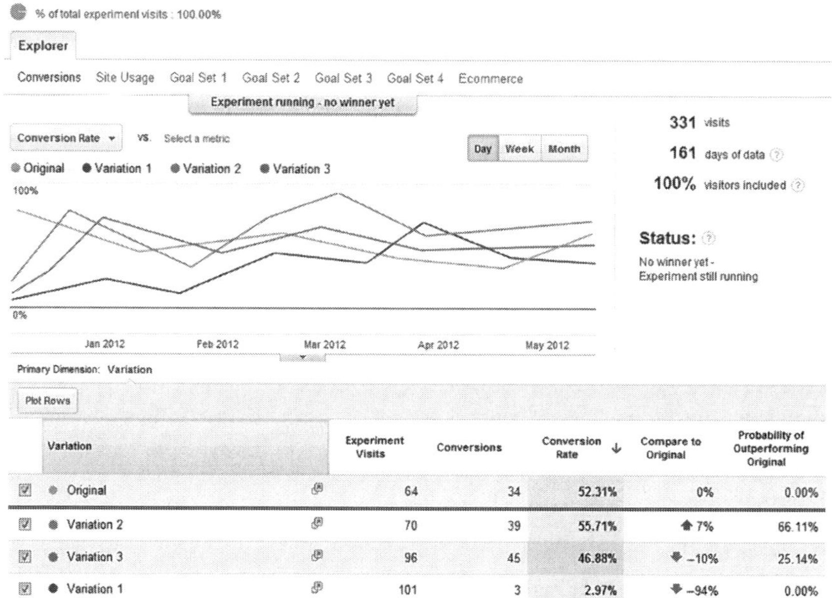

Figure 2.7 Example report from Google Analytics Experiments shows analysis based on beta-binomial model. Source: Reproduced with permission of Google Analytics.

Unfortunately, this analysis is not found in most statistical packages, but it is implemented in some website testing tools including the Experiments tool in Google Analytics. Figure 2.7 shows an example of a report from this tool. As you can see, the report gives the number of experimental visits and the number of *conversions* for each of four versions of a webpage. "Conversion" is simply another word for "success," and it could be a click or a purchase or a download – whatever the main goal is for the webpage. The report shows that Variation 2 has a 66.11% chance of outperforming the original version and this computation was based on the analysis we described in this section.

Exercises

2.4.1 Consider a beta distribution with $a = 4$ and $b = 6$. What does this distribution look like? What is the upper 5% tail? What is the lower 2.5% tail? This is *not* a confidence interval or a credible interval; we are just picking percentiles on a distribution.

2.4.2 For the example in which we asserted that $P(\pi_A > \pi_B) \approx 0.23$, how many replications do you need to be sure that this probability can be estimated accurately to two decimal places? First, compute this quantity several times for 999 rows to get an idea of the variability of the answer. Then increase the number of rows to 1999, 2999, etc., until several repetitions gives the same answer to two digits. How many replications are necessary to get three decimals?

2.4.3 Comparing two versions of a webpage, $P(A) = 132/1431$ and $P(B) = 212/2019$. Write a program in a statistical programming language like R or Python to compute $P(\pi_A > \pi_B)$ using the beta-binomial. If you don't know any such language, then use a spreadsheet. Remember, you want your estimate to be stable to at least a couple decimals.

2.4.4 An ad has had six clickthroughs in 114 impressions. Compute the beta-binomial credible interval for the clickthrough rate. Compute a traditional Wald confidence interval. Compare the two intervals.

2.4.5 An A/B test results in 18 successes in 223 trials for version A and 26 successes in 287 trials for version B. What is $P(\pi_A > \pi_B)$?

2.5 Case: Hotel Ad Test

A hotel was interested in testing two types of display ads, one that focuses on its benefits and the other that focuses on its loyalty program. Representative ads are displayed in Figure 2.8. Unlike the previous case in this chapter, this is a real case with real data from a major hotel chain, but they want to remain anonymous, so we can't show you the actual ads.

The brand manager for the hotel chain who, of course, likes more bookings, thinks that version A of the webpage will produce an increase in the number of customers who make a reservation. The manager of the loyalty program, which serves multiple hotel chains within the company, thinks that version B will produce an increase in the number of customers who register for the loyalty program. The two business units' separate quest for profits creates two separate response measures of interest: (i) bookings and (ii) sign-ups for the loyalty program. In a situation such as this, before the experiment begins, it is important to choose a primary response, the one that will govern actions in case the responses yield disparate results. If you wait until after the experiment is concluded and it turns out that both managers were right – the ad on the left increases reservations and the ad on the right increases loyalty program sign-ups – then the hotel manager and the manager of the loyalty program will argue forever about what to do. By setting the response variable in advance, you help the organization to establish its goals.

ACME
HOTEL
Book Now and Save!
• complimentary continental breakfast
• free highspeed internet
• free chocolate on your pillow
Book Now!
ACME
HOTEL
Stay Longer, Earn More Points
double, triple or quadruple your points!
sign up here

Figure 2.8 Display ads for Acme hotel, versions A (left) and B (right).

Table 2.8 Results of hotel display ad test.

	Loyalty ad	**Benefits ad**
Impressions	10 312 452	17 323 981
Bookings	1 × normal	2 × normal
Registrations	1 × normal	3 × normal
Ad cost ($)/booking	53.44	25.94
Ad cost ($)/registration	142.19	64.15

Results of the experiment are presented in Table 2.8. In a great surprise to all, the benefits ad produced twice as many bookings as the loyalty ad and three times as many registrations in the loyalty program ("1 × normal" means "1 times the normal rate"). Moreover, when the total ad cost is divided by the number of responses, the cost per booking was $53.44 for the loyalty ad, while it was only $25.94 for the benefits ad. In this case, both responses pointed in the same direction: run the benefits ad. And that is what the company did, using their advertising budget much more efficiently.

We end the chapter with this example, because it illustrates a few points. First, it shows that experimental results are often *surprising*; in this case no one expected that the ad featuring hotel benefits would produce more loyalty program sign-ups. Second, it shows a situation where you would want to collect and analyze two different response variables. We'll talk more about choosing response variables in the next chapter. Finally, it is a perfect illustration of the galvanizing effect that a clean experiment can have on an organization. What was once a contention issue – should we spend our advertising budget on ads for the hotel or ads for the loyalty program – became a decision that the whole organization could agree on. Clean data from clean experiments is the high-octane fuel of today's "data-driven" organizations.

2.5.1 Tips on Presenting Experimental Findings

For an experiment to have business impact, the results have to be communicated to those in the organization who are making the decisions. If decision makers don't understand the results of the experiment, they can't act on it. While many business analytics students focus on how to analyze data, the business analysts who often have the greatest impact in organizations develop strong skills at *communicating* their findings. Throughout this chapter, we have worked hard to demonstrate good practices in communicating results. While becoming a great data communicator takes practice and experience, these tips might be helpful to new analysts:

1) **Know your audience**: Consider who is (or should be) interested in the result of the experiment? Are they marketing managers? Engineers? Human resources executives. Information technologists? Finance staff? Each of these audiences will have a different level of domain knowledge, comfort with data, understanding of statistics, and patience for technical details. Your presentation should be tailored to the skills and interests of your audience.
2) **Summarize the experiment**: You should always provide a short summary describing how the experiment was conducted so that your audience knows exactly where this data came from.
3) **Provide context**: Your description can also include background about why getting this information was important to the organization.
4) **Lead with your recommendation**: Decision makers want to know what to do – they have hired you to do the analysis for them and they usually trust you. So, give 'em what they want by leading with a specific recommendation for what they should do.
5) **Support with analysis**: While you don't need to drag the decision maker through every detail of your analysis, try to drop enough clues that another analyst could replicate your findings. Practice summarizing the analysis as succinctly as you can while still providing relevant details.
6) **Use a graph**: Data visualizations are often highly effective at communicating results, often giving the viewer a more nuanced and complete sense of the data than simple averages reported in a table.
7) **Check your labels**: It is easy to fall into statistical jargon when reporting the results of a test. Your statistical software will use generic statistical terms like "average response," "success rate," or "factor." These terms will become very natural to you but can be very confusing to readers who are less familiar with statistics. Look for these generic terms – in your writing and especially in your graph and table labels – and replace them with terms that are specific to your experiment and meaningful to decision makers. It may seem like tedious work to do this translation, but it will help decision makers quickly process and act on your analysis – which is your ultimate goal.

Applying these principles to a specific test, we would draft a written report for human resource managers as follows:

Summarize the experiment	To determine which incentives work best for increasing technician retention, we ran a test where technicians were assigned to receive either a cash retention bonus ($2000 at 12 months) or increased training opportunities starting at 6 months (also at an additional cost of $2000).
Provide context	Our company currently spends $20K in the first year to recruit and train a new technician, so it is important to find ways to increase retention.
Lead with the commendation	The test findings strongly suggest that we should provide increased training opportunities, since they result in greater retention at the same cost.
Support with analysis	Only 12.3% of technicians who were randomly assigned to increased training opportunities left the company before they reached 18 months tenure versus 18.3% of those who received the cash bonus (95% CI for difference = (3.2%, 8.8%)).

Of course, the reporting would vary based on how it is communicated (e.g. an email, a presentation, a webpage) and who the intended audience is. We leave it as an exercise to create an appropriate visualization to support the conclusions above.

Exercises

2.5.1 In this case, we couldn't show you the actual conversion rates, because the hotel chain did not want to reveal the conversion rates for their ads. Suppose the number of bookings was 10 348 for the loyalty ad and 34 767 for the benefits ad. Present your analysis as you would the hotel manager and the loyalty program manager following the guidelines in this section. For statistical comparisons, you may use either the confidence interval for comparing proportions (but not the Wald interval) or the beta-binomial model.

2.6 Chapter Exercises

2.1 A firm wishes to compare two banners ads. Banner A has 7642 impressions and 85 clickthroughs. Banner B has 11 212 impressions and 122 clickthroughs. What are the clickthrough rates for each ad? Is this difference statistically significant? What is the power? (This is a trick question; power has to be calculated *before* the experiment is conducted.)

2.2 How can one assess the quality of the results from power calculation websites or free software? One way is to compare their results to those of reliable software packages. Use the equal sample size results from a reliable software package and compare them to the results from the website calculator. If they agree, then faith in the unequal sample size results from the website calculator may be justified.

2.3 The file `polio.csv` has the number of polio cases and "under 18" US population by year for several years. For convenience assume that all the polio cases are children in the "under 17" population. Compute the proportion of children who contract polio each year for the years 1948–1955. Imagine that all children were vaccinated after the polio season was over in 1949. Do a two-sample test of proportions to compare the incidence of polio in 1949 with the incidence of polio in 1950. Repeat this exercise for the years 1951–1952. Remember that this is a one-sided test, but the p-values are so small that you can use two-sided intervals. Suppose that the vaccine was really a placebo! Which year was a type I error? Which year was a type II error? How could these errors have been avoided? (More precisely, how could the probability of their occurrence been dramatically reduced?)

2.4 Do some sample size calculations on polio. Recall that the incidence rate is 0.0005. Suppose there are 20 000 observations in the control group and 20 000 in the experimental group. The expected number of cases in the control group is 10 (do you see why?); suppose there were 10 cases in the control group. Would 5 cases in the vaccinated group be compelling evidence? Suppose that each group had 40 000 cases, the number of cases in the control group was the expected number, and the number of cases in the vaccinated group was again one half the number in the control group. Would this be compelling evidence? How about if each group had 100 000 cases?

2.5 Use the beta-binomial model to compute the probability that the treatment is more effective than the control for the data in Table 1.4. Compare this result to a two-sample test of proportions.

2.6 Suppose we wish to detect a difference of \$0.094 (just under a dime) between two different online ads. Suppose the standard deviation of the response (sales) is \$103.77 (the standard deviation will be large because most clicks don't produce sales, so there are lots of zeros in the data set). For an A/B test, how many observations do we need in each sample?

2.7 We suggested that setting $\pi = 0.5$ provides a conservative estimate of the sample size for a test of proportions (i.e. the largest sample size). For a two-sample test, where p_1 increases from 0.1 to 0.9 and p_2 equals $p_1 + 0.01$ (so you're trying to detect a difference of 0.01). Of course, set power = 0.80.

2.8 As part of a class assignment in an experimental design course, a student wants to test two different pots to determine which one boils water faster, steel bottom or copper bottom. In order to do a sample size calculation, he needs an estimate of the standard deviation. To do so, he took one of the pots and boiled water in it six times. The times to a roiling boil were 133, 148, 137, 142, 145, 150. What sample size should he use to detect a difference of 15 seconds? How about a difference of 5 seconds?

2.7 Learning More

A good quote about statistics and causality as you learn about experimental design:

> You sometimes hear it said that "You cannot prove causality with statistics." One of my professors, Fred Mosteller, would often say: "You can only prove causality with statistics." When he said this, he was referring to randomization as the reasoned basis for causal inference in experiments. (Rosenbaum, 2010, p. 35)

• Exercise 2.6 is taken from section 5.2 of the paper by Feit and Berman (2019).

• Google Experiments, Optimizely, and other large platforms for A/B website testing do not use the tests described in this chapter. They use much more sophisticated methods, but the basic ideas are the same. For example, to perform a valid statistical test, one must specify the sample size in advance and then wait until that sample size has been achieved and only then look at the results. In practice, A/B testing on the web is often corrupted by users looking at the results before the proper sample size has been reached and sometimes taking decisions before the proper sample size has been reached. To deal with this reality, Optimizely developed a method that allows users to evaluate tests before the proper sample size has been reached. The math is too technical for this book, but the interested reader can consult the paper by Johari et al. (2017).

Section 2.1 "Case: Improving Response to Sales Calls"

• The argument that the t-test is better than the z-test even when $n > 30$ is technical and can be found in the article by Boos and Hughes-Oliver (2000). The fact is that the difference between the z and the t can be substantial, even for "large" sample sizes, especially when the data are skewed. The $n > 30$ rule is a holdover from the precomputer days when the only recourse most persons had was to use tables in the back of a book. Now we have computers to calculate percentiles of distributions, and the tables are no longer necessary.

• We covered the basics of analyzing an A/B test with a continuous response variable. We began with a graphical comparison, which we believe is a vital first step for any business analyst. The graphical analysis of data can be divided into two categories: exploration (finding out new results) and presentation (communicating results to others who might not be statistically sophisticated). Stephen Few has written excellent books on these topics that are at a similar technical level as this book. Anybody who has to present his/her results graphically or with tables to a nontechnical audience should read *Show Me the Numbers* (Few, 2012). Those who are interested in the visual exploration of data to obtain new results should read *Now You See It* (Few, 2009).

• As to CIs and p-values, the inadequacy of the p-value has been recognized. The flagship organization of statisticians, the American Statistical Association, and several journals have recommended against it (Wasserstein and Lazar, 2016). One of the primary reasons we eschew the use of the p-value is that when the null hypothesis is not rejected, the p-value is basically uninterpretable, whereas the CI still conveys useful information.

• A sophisticated reader may notice that our approach to statistical analysis sometimes leans Bayesian and this was intentional. For example, some of the interpretations of the confidence interval we provide in Section 2.1.2 are more consistent with a Bayesian credible interval than a classical confidence interval. We generally find the Bayesian approach more accessible to a business audience, and there is a growing use of Bayesian inference in business practice (for example, in Figure 2.7), yet we don't want to bog the novice business analyst down with an extended discussion of the philosophy of inference.

• The ability of a test to uncover an effect requires a balancing act between α and β, along with a priori knowledge of the standard deviation and likely sample sizes. Conducting tests without considering power can be a big mistake and is illustrated by this quote from Andrew Gelman's blog (Bohn, 2018):

> What I found was that small process improvements were almost impossible to detect, using the then standard experimental methods. For example, if an experiment has a genuine yield impact of 0.2 percent, that can be worth a few million dollars. (A semiconductor fabrication facility produced at that time roughly \$1 to \$5 billion of output per year.) But a change of that size was lost in the noise. Only when the true effect rose into the 1% or higher range was there much hope of detecting it. (And a 1% yield change, from a single experiment, would be spectacular.)

• Determining the proper sample size can be confusing at first. Lenth (2001) offers some guidelines:

1) Specify a hypothesis test on a parameter θ.
2) Specify the significance level α of the test.
3) Specify an *effect size* – how big of an effect do you want to be able to detect?
4) Obtain historical values or estimates or educated guesses of other parameters needed to computer power, e.g. the standard deviation.
5) Specify a target value for power; how much power do you want to have?

Section 2.2 "Case: Email Response Test"

• The defects of the Wald interval for the one-sample test first gained widespread notice in Brown et al. (2005), and its defects for the two-sample test were first presented in Brown and Li (2005); both articles offer better methods than the Wald. The formulae for Agresti–Coull and Clopper–Pearson can be found in the Brown articles. Textbook presentations often say that the Wald interval is safe to use if p is not near zero or unity and n is large; this is wrong. For example, if $p = 0.106$ and $n = 100$ and you compute a 95% interval, you get a 95.2% interval. But if $p = 0.107$ and $n = 100$, you get a 91.1% interval. Figure 2.9 depicts this situation for a nominal 95% CI with a sample size of $n = 50$ and various values of p. As you can see, for values of p between 0.1 and 0.2, actual coverage ranges between 0.85 and 0.95.

This brings up an important point about statistical software packages. Different software packages can give different answers to the same problem, and these differences can be small or large. Sometimes there are different methods for conducting the same test. Sometimes the same method can yield slightly different answers in

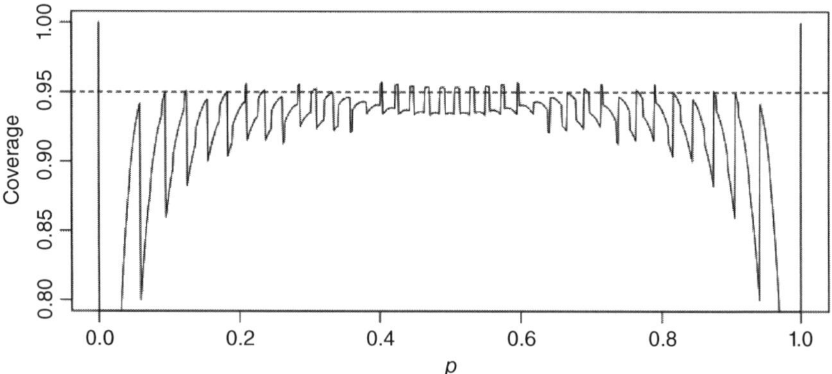

Figure 2.9 Wald interval coverage for a nominal 95% CI.

different packages because of algorithmic choices made by the software developer. For example, one package might use Agresti–Coull intervals, while another uses Clopper–Pearson intervals. Sometimes, one method might be better than another. All of these reasons come into play for the two-sample test of proportions.

• For further details on why post hoc power calculation is not a good idea, see the article by Hoenig and Heisey (2001).

Section 2.3 "Case: Comparing Landing Pages"

• Sometimes it is useful to zoom in on some part of a histogram. In R, it is very easy to do this. Consider a histogram of 100 000 random normals. Suppose we want to take a look at the lower tail of this distribution, e.g. values of x less than −2.

```
x <- rnorm(100000)
hist(x)
hist(x[x < -2])
```

• Bean plots and violin plots are modern alternatives to box plots that are more informative. Not all software packages produce these types of graphs, but R does. To make a violin plot, install the "vioplot" package, load it, and the execute `vioplot(audio)`. To make a bean plot, install the "beanplot" package and execute `beanplot(audio)`.

Section 2.3 "Case: Comparing Landing Pages"

• As another example of how to zoom in on a histogram, consider the histogram of sales from the landing pages data set. To zoom in on the bump around 100, use the command `hist(df$sales[df$sales>20])`.

Section 2.4 "Case: Display Ad Clickthrough Rate"

• A modest exposition of the beta-binomial model is given by Boldstad (2007). A primer on testing the difference between two proportions is http://www.ncbi.nlm.nih.gov/pmc/articles/PMC3307549.

When calculating percentiles from the empirical distribution of the beta distribution, we used 999 simulations and divided by 1000 to compute the probability. Why not 1000 simulations and divide by 1000? We are calculating, say, the number of times that $P_A > P_B$. What are the possible values? Zero is a possibility. So is one. So is two.… So is 999 for a total of 1000 possibilities. Therefore, when we run 999 simulations, there are 1000 possible outcomes, which is the denominator for the probability calculation. If we ran 1000 simulations, we would have to divide by 1001. Suppose we found that $P_A > P_B$ a total of 34 times. Which would you rather report, $34/1000 = 0.034$ or $34/1001 = 0.033\ 966\ 033\ 97$?

Section 2.5 "Case: Hotel Ad Test"

- Presenting data is an art, and in Section 2.5 we provided several basic tips for good reporting. Analysts who want to further develop these skills should take a look at *The Wall Street Journal Guide to Smart Information Graphics* (Wong (2013)), which provides excellent tips for making compelling data visualizations, and *Storytelling with Data* (Knaflic (2015)), which focuses on the art of leading decision makers to conclusions.

3

Designing A/B Tests with Large Samples

We know now how to analyze an A/B test. How do we set up an A/B test so that we can credibly believe its results? It is much more difficult than one might imagine, because there are many pitfalls that await the experimenter. We have seen that the first trial of the Salk polio vaccine – involving hundreds of thousands of children and costing an enormous sum – was really just an exercise in collecting observational data, because the experiment was not designed well.

> Controlled experimental studies are typically regarded as the gold standard for which all investigators should strive, and observational studies as their polar opposites, pejoratively described as "some data we found lying on the street." In practice, they are often closer to each other than we would like to admit. The distinguished statistician Paul Holland, expanding on Robert Burns, observed that "All experimental studies are observational studies waiting to happen." (Wainer, 2016, p. 30)

What that last sentence means is that if there a single mistake in setting up the experiment, then the experimental results will be no better than observational data, and that's precisely what happened to the first Salk trial. Hence, experiments must be set up with great care and attention to detail.

In this chapter, we learn the basics of setting up an experiment so that its results do not become just observational data. In essence, what we cover in this chapter is a series of laundry lists of "things to watch out for" or "things to consider" when setting up your experiment.

After reading this chapter, students should:

- Be able to define internal and external validity and recognize features of a design that increase internal and external validity.
- Recognize the key decisions that need to be made to design an experiment.

Business Experiments with R, First Edition. B. D. McCullough.

Companion Website: www.wiley.com/go/mccullough/businessexperimentswithr

- Know the two forms of randomness that are critical to ensuring valid statistical results from an experiment.
- Understand the importance of communicating their planned experimental designs to analysts, those executing the test, and decision makers who will act on the results of the test.

3.1 The Average Treatment Effect

It is important to be able to recognize the ways that various disciplines talk about experimental design. One such way comes from the economics field experiment literature, for example. Let T be the treatment group, and let C be the control group. When we compute $\overline{x}_T - \overline{x}_C$, this is an estimate of $\mu_T - \mu_C$, which is formally called the average treatment effect (ATE). This is a technical term and we don't use it in this book, but many other literatures (e.g. health care, medicine, field experiments in economics) use this terminology, so you should be able to recognize it when you hear it and know what it means. It is the difference in the mean outcomes between units assigned to a treatment and units assigned to control. In the context of A/B testing, the ATE is difference between the means of the A and the B groups. We want to make sure that we can get unbiased estimates of the ATE for an experiment, and doing so requires three assumptions:

1) *Random assignment*. All units have the same probability of being assigned to either group. When this is not true, then there is the potential for selection bias. As we know from Chapter 1, selection bias means we cannot trust the results.
2) *Excludability*. The result of a treatment depends only on whether that unit received the treatment, and not on anything else. As an example, consider an A/B test to determine the effect of a discount on purchasing a specific item. If persons who receive the discount also get an email reminding them of the discount, then the difference between the treatment and control groups cannot be ascribed to the discount – it could be due to the email. Excludability is a short way to say, "The experiment should be designed so that the treatment is not confounded with anything else."
3) *Noninterference*. This means that the treatment affects only the treated unit, and not any other units. In the literature this is sometimes called the "Stable Unit Treatment Value Assumption" (SUTVA), and it requires that the treatment has no spillover effects that affect other units. Suppose a "get out the vote" campaign targets registered voters by sending them literature on the importance of voting. A control group of registered voters does not get the literature. After the election, the researchers can check the voting rolls to see who voted and who did not and determine whether the literature had an effect. If you are in the control group and your roommate is in the treatment group, it might happen that you see the literature and wind up voting. If that happens, the SUTVA

is violated. Of course, if this happens just once, it won't really affect the results. But if it happens in many households, it could vitiate the results.

The above perspective is commonly applied in economics, especially in field experiments. A nontechnical article that elaborates on this methodology and provides many examples is Harrison and List (2004). Another perspective on ensuring the usefulness of experimental results comes from the psychology literature, which focuses on two types of validity, internal and external. There is obvious overlap between the two approaches.

Exercises

3.1.1 Take any of the experiments presented in Chapters 1 or 2, and assess the extent to which it could violate random assignment, excludability, and noninterference.

3.2 Internal and External Validity

The validity of an experiment assesses the extent to which the results of the experiment are useful. First, in order for an experiment to be useful, it must be conducted correctly. Second, if the study is conducted correctly, the results of the study must be applicable in the real world. Internal validity concerns itself with the former, and external validity with the latter.

1) *Internal validity.* Can we attribute changes in the dependent variable to changes in the treatment? (Do we have causality?) Laboratory experiments are very conducive to internal validity, because it is much easier to control the environment.
2) *External validity.* Can we generalize the results of the experiment to other people or settings? Laboratory experiments are less conducive to external validity, because the laboratory usually is not at all like the real world, where we need the results to matter.

Counterpoint to these concepts of validity are possible reasons that validity may fail to obtain, of which there are many. They are collectively referred to as "Threats to the experiment." In Chapter 2 we preferred to give three requirements for causality, while other researchers give as many as eight. Again, different researchers have different lists of internal and external threats. We give some common ones and eschew the specialized threats that are only relevant to some disciplines.

3.2.1 Threats to Internal Validity

1) *History.* Any event that occurs while the experiment is being conducted can adversely affect the results. Example: When conducting the effect of a store promotion at two different stores on the same day, if it rains at one store but

not at the other, this could affect the number of customers who come to the store and thus bias the results.

2) *Maturation*. Subjects change during the course of an experiment. Example: When analyzing the effect of a stimulus on memory, an improvement in memory may be due to practice during the experiment rather than the stimulus.
3) *Testing*. The subject's response may be affected by the testing procedure in addition to the treatment. Example: A new method of teaching mathematics is assessed by an exam each week. It is possible that the student improves just because he gets better at taking tests due to having so much practice taking the tests. In this case, the researcher cannot be sure that any improvement is due to the teaching method and not the repeated test-taking.
4) *Instrumentation*. If different measuring devices are used before and after, or if the persons using the instrument to take measurements are poorly trained or become bored, measurements can be affected. Example: Test subjects in different locations or at different times are read the instructions. One of the test examiners speaks loudly in a clear voice. Another examiner mumbles in a low voice. Poor performance in the latter group may be due to the examiner and not the treatment.
5) *Regression to the mean*. If only subjects who have high or low scores on some measure are selected for an experiment, subsequent performance may be affected by regression to the mean. If you're not familiar with the concept of "regression to the mean," visit the Learning More section at the end of the chapter. If the example doesn't make sense, do likewise. Example: Students who perform poorly on a math placement test are given special math instruction. Improved scores after the instruction may not be due to the instruction, but due to regression to the mean.
6) *Selection*. If experimental units are not randomly selected from the population, then bias can occur. In particular, some aspect of how units were selected affects the results. If selection is truly random, done with a random number generator, then you don't have to worry about this. Example: A business-productivity software developer gets volunteers to test a new software product. The volunteers are randomly assigned to treatment and control. The results show difference between the two groups. It's quite possible that only highly motivated employees seeking to improve their productivity would volunteer for this study. Their productivity is already near maximum, so there was nothing for the software to improve.
7) Mortality. Some experimental units may drop out of the experiment. Example: A chain store is testing a new incentive scheme for its store managers. The experimental group has 18 managers, and the control group has 18 managers. Over the six months of the study period, four managers leave the experimental group, and five leave the control group. The problem here is not that the sample

sizes have decreased, nor is the problem that the sample sizes are now unbalanced – statisticians know how to deal with these problems. The problem is that the managers did not drop out randomly. If the weaker or less capable managers dropped out of the study, then the sample does not look like the population and the results of the experiment cannot be generalized to the population. An exit interview with each dropout might allay these fears.

8) *Diffusion of treatments*. Sometimes the control group gains access to the treatment. Then the treatment affects both groups, and the effect of the treatment cannot be cleanly estimated. Example: A fast-food chain wants to test a training program to help its cashiers handle customers more quickly. Suppose some cashiers in the control group learn of the methods and copy the methods so they can do their jobs more efficiently. The results of the experiment might well show no effect when, in fact, there is a substantial effect.
9) *Failure of randomization of treatments*. If treatment assignment is not truly random, then unexpected or unknown confounds can occur. If you use a random number generator, it's hard to make this mistake. Other seemingly random mechanisms can actually induce severe bias. Example: A group of individuals are assigned to control or treatment based on whether the first letter of their last name is in the first or second half of the alphabet. Because of ethnic differences in last names, the makeup of neither group will look like the population.

3.2.2 Threats to External Validity

The concept of external validity is largely focused on the question: To whom do the results of this experiment apply? A pair of examples might help firm up this idea.

A large chain convenience store has good reason to believe that a remodel of its stores (which are standardized) will increase sales and profits. An experiment consisting of control and treatment stores indicated that the redesign increases sales and profits by much more than the cost of the redesign. Should the redesign be rolled out to all stores in the chain? This is the question of external validity. It worked in the sample; will it work in the population?

An international corporation instituted a new ad campaign in one country, to great success. Will it work in another country? This is also a question of external validity. Do the results of the study generalize to other times, places, and populations?

There are threats to the generalization of experimental results, i.e. threats to external validity:

1) *Population*. Random selection of units ensures generalization to the population. Sometimes you cannot sample from the target population.... "Selecting a sample that is not representative of the target population." Small samples

have a high probability of not reflecting the target population due to what is called a "lack of balance." The solution to this is matching. These concepts are discussed in Chapter 4.

2) *Hawthorne effect*. When persons are aware they are begin studied, they often change their behaviors so that their responses in the experiment cannot be generalized to the real world. The effect is named after the Hawthorne Works, a Western Electric factory outside of Chicago. A study was performed on the workers to determine if their productivity was affected by higher or lower levels of lighting. Productivity increased during the experimental period, but, after the experiment ended, productivity declined to its prior level. A later reanalysis of the data concluded that the workers were responding not to the treatments, but to the experiment itself. In particular, the workers were responding to the extra attention paid to them during the experiment.
3) *Ecological* (also called "environmental"). Do the results generalize to settings beyond those used in the original experiment? To the extent that the experimental setting is not like the "real world," the effect that is observed in the experimental setting may be attenuated in the real world.
4) *Temporal*. Do the results generalize to other times, and not just the times/seasons during which the experiment was conducted? An experiment concerning ice cream consumption that is conducted in the summer might be applicable in the winter.

Related to the ecological threat is a special case called the *context effect*. When some idiosyncrasy of the test environment produces the effect, but that idiosyncrasy does not exist in the target environment, when the test is rolled out, the desired effect does not appear. As an example, showing ads to test subjects in a lab setting may indicate that ad A is much better than ad B. Yet, when ad A is rolled out as a web ad on various websites, it doesn't work. The reason may be that persons in a lab pay much more attention to ads than persons surfing the web.

Exercises

3.2.1 For any experiment in Chapters 1 or 2, analyze it for internal and/or external validity, as appropriate (some experiments only have enough information to assess one or the other).

3.3 Designing Conclusive Experiments

Each experiment is different and has its own pitfalls, so there is no way to warn against all of them. However, there are certain areas where problems tend to crop up, and we can identify these areas.

The most important thing is to approach experiments with a design mentality. Don't just conduct an experiment, but *design* an experiment and *then* conduct it. Ask yourself, "What could go wrong?," and then guard against those things. Show the design to other persons and ask them to poke holes in it.

Think of this as a preflight checklist: it doesn't guarantee you'll have a safe flight (successful experiment) or even that your plane will get off the ground (the experiment will be completed), but it's the best way to begin.

Defining Treatments

The immediate business question, properly formulated, will go a long way toward defining possible treatments. By properly formulated, we mean a causal question. We do not mean a general question, like "How do we increase sales?" or "What will bring more visitors to our website?" We mean a specific question, like "Will a 10% off sale increase revenues by more than 10%?" or "Is spending money on Google Ads cost effective?"

For yes/no questions, the treatment/control approach is called for, for example, "Should I run this ad?" or "Should I deploy this training protocol?" Either/or questions lead to A/B tests, for example, "Which ad should I run?" or "Is this training protocol more effective than the other one?"

These questions don't just jump out at you, but you have to find them. One website manager likes to go to company happy hours and ask coworkers, "How can we make our website better?" Another way is simply to ask the boss, "What keeps you up at night?"

Choosing a Test Setting

There are two prototypical test settings, the lab and the field. In the lab there is more control over the environment, and usually there is better measurement, which is good for internal validity. This level of control might be bad for external validity, where extraneous influences might be important. Consider exposing website viewers to display ads. In a lab, subjects might be more attentive and recall might be high. In the field, subjects might not even see the ad, let alone be able to recall it.

There is a continuum of settings between the field and the lab. A chain of convenience stores might conduct an experiment in a single store, over which it is much easier to exert control, e.g. training of the staff, monitoring the experiment, etc. In this sense, the single store (which exists in the field) is really a type of lab.

Even an entire city might constitute a lab of sorts. A retailer might have extensive experience with, say, the city of Peoria, and we might use that city as an experimental site. Whether the results of such an experiment generalizes to other cities is an external validity question.

Choosing Response Measures

Often there is a range of key performance indicators (KPIs) from which to choose, and there is a tension between the extremes. One KPI might be close to the treatment and easy to measure, and another might be close to the desired outcome (farther from the treatment) and difficult to measure. A good example is the use of statins. A close KPI might be blood cholesterol level, but is this really the appropriate measure? Don't people take statins to live longer? Then the KPI should be death, but what kind? All deaths, or only heart attacks? And what should be the time horizon for counting deaths? Five years, 10, or 20 years? When deploying web ads, it's easy to measure clicks, but is that why web ads are purchased? Aren't they purchased to increase sales? Sales of what? Over what time period? Choosing a response measure is not necessarily easy.

A useful example comes from Section 3.1 of Kohavi et al. (2009a). An online seller changed the design of the "click to purchase" button from version A to version B, thinking it would increase sales. Tracking the actual purchase was complicated, so instead of counting sales, the KPI was "clicks on revenue generating links." The new method had 64% fewer clicks! Was the experiment a success? Realize that "clicks on revenue generating links" is a reasonable approximation to sales only if the conversion rate for both A and B versions is the same. However, version B also mentioned the price of the product – fewer people clicked, but those who clicked were prequalified not to be turned off by the price. Version B had a significantly higher conversion rate and ultimately more sales.

There should be only one KPI, and there are two reasons. The first is statistical; if there are several KPIs, then you can go fishing and declare success any time one of the KPIs is significant, but how will you know it's not a type I error? The second is organizational. Different KPIs will have different importance to different bureaucracies. Remember the Hotel Ad Case from Section 2.5? The brand manager and the loyalty program manager had different aims, but, fortunately for the analyst in that case, one ad worked better for both purposes. What if one had been good for the loyalty program but bad for bookings, while the other ad had been bad for the loyalty program but good for bookings? How would that have been resolved? And pity the poor analyst, stuck in the middle. You can have secondary KPIs, but these should not be used to determine the success or failure of the experiment.

Not only should there be one KPI, but it should be chosen before the data are even collected. To stop infighting is a sufficient reason, but it also adds clarity to the experiment. If you wait until the data are in, you really didn't design the experiment well.

Selecting the Unit of Analysis

The unit of analysis is whatever the treatment is applied to. Usually this is obvious, but sometimes not. A campus promoter wanted to know what style of poster would generate more interest in an upcoming event. Four different types of posters were made and placed at various points around the campus. Each poster had a QR code linked to a different website, so the analytics team knew which poster had been scanned how many times. After the data were collected, the team had an argument over the unit of analysis: Is it the user who scans the poster, or the poster? It's the poster, since the treatment was applied to the posters.

Sometimes the unit of analysis can be a store (Slushie example from Section 5.3.5) or even an entire city (DigiPuppets from Section 5.2 and Harry's Razors from Section 5.3.3).

Selecting Subjects

A basic principle of statistical analysis is that the sample has to be drawn from the target population. If the sample isn't random, then the laws of probability cannot be used to extrapolate from the sample to the population. Therefore, knowing your target population is of critical importance. If you don't know the target population, you don't know where to sample. A Cadillac dealer might wish to conduct a survey to help him figure out how to sell more cars. He has at his disposal his customer list, and if he only wants to sell more cars to his existing customers, then taking a random sample from the customer list is a good idea. On the other hand, if he wants to sell more cars to all consumers of luxury cars, then he'll have to find some way to purchase a list of local Mercedes and BMW customers and sample from the combined three lists. As another example, if a publishing company wants to evaluate whether one promotion strategy works better than another and decides to run a field experiment, they have to consider the target population from which to sample. If their goal is to learn how their current customers respond, they might focus on customers from their current mailing list. However, if they hope to learn about how potential customers respond to the promotions, they might choose to sample customers from a larger list of avid readers.

The sample has to be randomly drawn from the population. It's easy enough to generate a random sample if you have a list of customers, just assign each customer a number from one to n, and then use a random number generator to produce the sample. At the other extreme, consider a web experiment. If the target population is actual visits (not visitors) to your website, then sampling long enough to avoid hour of day and/or day of week effects might be sufficient. If the target is visitors, then sampling from the visits is actually a convenience sample! In online experiments you should keep in mind that mostly heavy users of the website or app are more likely to be included, compared with light users. Since most online tests include in the sample all the visitors in a fixed period, this group will naturally include more of the frequent users and fewer of the infrequent users. To overcome this issue, you should consider test designs that assign treatments to users (rather than to visits), track users across visits, and cap the number of times each user is exposed to the treatment. There are technically sophisticated ways to do that that include cookies and canvas fingerprinting, for example.

Random sampling from target population is the best was to achieve strong external validity, and be careful about selecting the target population.

Assigning Treatments to Units

There are two necessary forms of randomization in an experiment. The first, just discussed, is the selection of the sample from the target population. The second concerns the assignation of treatments to the experimental units. Randomization of treatment assignment ensures that external influences are randomly distributed between the treatment and control groups and a particular external influence does not crop up mostly in either the treatment or the control group. There is an important caveat to this: if the sample size is small, then there can be a high probability that the external influence is not balanced between the treatment and control groups (see Section 5.3 for an extended discussion of this phenomenon). When the sample size is large enough, randomization of treatment assignment balances the treatment groups, achieving strong internal validity.

Choosing Factor Levels

If the factor is categorical and binary, choosing factor levels is trivially easy. On the other hand, if the factor is numeric, you have to make a choice for the hi and lo levels. Absent experience with this particular problem, it may be necessary to run a few experiments in order to puzzle this out. Credit card companies, for example, have been doing this for years, so they already have a good idea whether the teaser rates for an introductory credit offer should be 5 and 10% or 13 and 18%. It is better to be aggressive than timid. If you make the difference between hi and lo too small, consumers might not care about the difference, and that would be like not having a factor at all. Even if consumers would respond to a small difference, it is still advisable – on statistical grounds – to choose a larger difference. The reason we want a larger difference is because our goal is to get a precise estimate of the effect – e.g. the slope in a regression of Y on X. Look at Figure 3.1. Suppose we can only take nine observations. If we collect them over a small range of X, say, from X_0 to X_1, we could get many possible regression lines, ranging from a slightly negative slope (line B) to a steeply positive slope (line A). With a small difference, we are asking for an imprecise estimate of the effect. On the other hand, for a large difference, say, from X_0 to X_1, all the estimated slopes are constrained to be about the same. It should now be intuitively obvious that a larger range for X implies a more precise estimate of the slope (effect).

If runs are not particularly expensive, a nice idea is to run a preliminary experiment with only two runs, one with all variables set to minimize the response and another with all variables set to maximize the response. If the results are

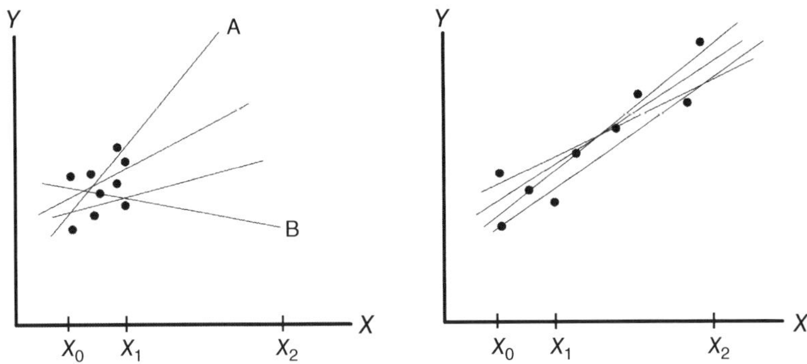

Figure 3.1 Range of X and precision of slope estimates.

approximately the same, it suggests that the high and low levels were not set aggressively enough.

Making Sure You Will Have Enough Data

To go to the time, trouble, and expense of conducting an experiment that is underpowered is an exercise in futility. Know what effect size you want to detect, and know how large a sample you need to detect an effect of this size. This has already been covered for A/B tests of means and proportions in Sections 2.1.5 and 2.2.3, respectively. Necessary sample sizes can be reduced if the observations can be paired (Section 4.2) or a test is a one-sided test (Section 4.1.1).

Avoiding Other Problems

There actually is a third use of randomization in experimental design, and that's randomizing the order of the experiments: do use execute the experiments in the standard order!

For web experiments, there is no need to worry about blinding, but for other types of experiments, single blinding or even double blinding may be necessary.

Consider generating synthetic data prior to running the test and analyzing it. Never collect data without knowing how you're going to analyze it, and if the experiment is novel, analyzing synthetic data may lead to improvements in the design as you stumble across something you hadn't thought of before.

Similarly, consider running an A/A test, especially with web experiments. It can be used to collect data so that the entire system can be assessed for any flaws. If you run it only once, the results should fail to reject the null, and if the null is rejected, you may have a problem somewhere! You can also run the experiment several times to ensure that the null is rejected 5% of the time (if $\alpha = 0.05$).

Kohavi et al. (2009a) discuss five pitfalls in conducting randomized experiments on the web.

Communicating the Design to Others

You should prepare summary forms to communicate the design to all the parties involved – each party will need a different summary:

- Other analysts who may be asked to comment on the design and offer constructive criticism or who may be involved in the analysis of the data.
- The persons who will be conducting the experiment.
- Managers, who may not know much statistics, who have to approve the design or use the results of the experiment.

A perhaps apocryphal tale illustrates not only the need for well-thought-out protocols but also the need for making sure the protocols are followed. A new breast cancer treatment for moderate to severe cases had already gone through two rounds of trials with impressive results. During the third and presumably final trial, the randomization process was to treat every other person who signed in to the clinic where the trial was being conducted. Well, the attendants doing the sign-in figured this out and, out of compassion, started leaving blank lines on the sign-in sheet so that they could make sure that the severe cases could get this promising new treatment. This induced a strong bias in the treatment assignment. The formal result of this trial was that the promising new treatment actually reduced survival time. This greatly surprised the researchers (since this was third trial for the treatment and the first two had turned out very well), who launched a top-to-bottom investigation. Finally, they discovered the compassionate acts that sabotaged the experiment. The point is that you need well-defined protocols *and* you need to make sure the protocols are followed.

Exercises

3.3.1 The "treatment" in an experiment is: (a) dependent variable, (b) independent variable, (c) lurking variable, and (d) confounding variable.

3.3.2 Random assignment is crucial for: (a) internal validity, (b) external validity, and (c) measurement validity.

3.3.3 Random sampling is crucial for: (a) internal validity, (b) external validity, and (c) measurement validity.

3.3.4 Based on the information in this section, create a form that you can fill out for each experiment. This form should describe the research question, the response variable, treatments, selection of subjects, etc. This can be your preflight checklist.

3.4 The Lady Tasting Tea

In the late 1920s, in Cambridge, England, at a gathering for afternoon tea, a woman claimed that the taste of the tea depended on whether the milk was added to the tea or the milk was first poured and then the tea was added. R. A. Fisher, the godfather of experimental statistics, was at that tea and took up the challenge. He engaged the group in a discussion of how they might test this hypothesis. If she is given just a single cup of tea, she has a 50–50 chance of being correct. Of course, she might have such an ability, but she is a human who might make a mistake, especially if testing many cups. So if she is presented with 10 cups, should one mistake disqualify her? If not one, should two? Within a short time, the group had taken up the challenge and its members were making tea, infusing it, and presenting the various cups to the woman so that she could make her determination.

This tea-tasting scenario is described in detail in Chapters 2 of Fisher's magnum opus, *The Design of Experiments* Fisher (1971). He figures out how many cups she should taste, in what order they should be presented, and works out the probabilities of different outcomes. Indeed, Fisher considered many questions about setting up this experiment:

- How many cups should be used? (to ensure statistical validity)
- Once the number of cups has been determined, how many of each type? An equal number of each type? An unequal number?
- How many cups can she test before her taste buds fail her?
- Should the cups be paired? (present two at a time, one with milk added to tea and the other with tea added to milk)
- How should chance variation in temperature or strength of the tea be handled?
- How much should the lady be told about the order of presentation? If it is paired, should she be told that they are paired?
- In what order should the types of tea be presented? First all of one type and then all of the other? Alternated, first a cup of one type, then a cup of the other type, and then back to the first type? Or should some other arrangement be used?
- What type of tea should be used? Should the same type of tea be used for every cup? Maybe different teas should be tested.

Each one of the above points has the potential to alter the design or analysis of the experiment.

Exercises

3.4.1 Try to come up with your own answers to the questions Fisher posed about the tea-tasting experiment.

3.4.2 Can you think of anything Fisher missed?

3.5 Testing a New Checkout Button

This problem illustrates the type of fun you can have when you really think about *each* aspect of the design of an experiment.

You manage a website where 10% of the visitors get to the checkout page with items in the cart, but only half of these actually complete the purchase. You want to test a new button that, you hope, will increase the percentage of actual purchases. The new button is "A" and the current button is "B." To keep the math simple, you'd like to see 10% lift at the usual power (0.80) and significance (0.05) levels.

You want half the visitors who get to the checkout page assigned to group "A" and the other half assigned to "B." Here's the question: Do you assign visitors to "A" or "B" when they first get to your website or when they actually get to the checkout page?

Think about it. You'll have to crunch some numbers to shed some light on this question. Refer to the exercises.

Exercises

3.5.1 If you assign treatment when visitors first get to the website, then the difference you want to detect is a change from 5.0 to 5.5%. How many visitors need to come to website so that this test can be conducted?

3.5.2 If you assign treatment when visitors finally get to the checkout page, the difference you want to detect is a change from 50.0 to 55.0%. How many visitors need to come to the website so that this test can be conducted?

3.6 Chapter Exercises

3.6.1 In Exercise 8 a student boiled water to conduct a pilot study to compute a variance in order to perform a sample size calculation. That was an experiment. Criticize the design of that experiment.

3.7 Learning More

Section 3.1 "The Average Treatment Effect"

Many disciplines estimate the ATE via the *Rubin causal model*. Let Y_{i1} denote the potential outcome for unit i if treatment is applied, and let Y_{i0} denote the potential outcome for unit i if it is in the control group. Let T_i be a treatment indicator. Then

$$\text{ATE} = E(Y_{i1}|T_i = 1) - E(Y_{i0}|T_i) = 0$$

While it is indispensable at higher levels of experimental design, we eschew the Rubin causal model because, in the words of the eminent statistician Howard Wainer, "Although the fundamental ideas of Rubin's Model are easy to state, the deep contemplation of counterfactual conditionals can give you a headache" (Wainer, 2016, p. 11). We think it much easier to stick with $\mu_T - \mu_C$ to estimate the average treatment effect.

Section 3.2 "Internal and External Validity"

Everybody who works with statistics should know about regression to the mean. Our brief mention couldn't begin to do justice to this fascinating topic. It's all over the place, and most people don't recognize it. A classic example is given concerning Israeli flight instructors. In training, pilots fly many, many maneuvers. Each pilot has a mean level of performance; some days they fly a bit better, other days a bit worse. The flight instructor had started out praising pilots when they flew better than average and criticizing pilots when they flew worse than average. He concluded that praise didn't work and criticism did, because pilots who got praise did worse after receiving praise and pilots who were criticized did better after the criticism. Thereafter, he ceased praising good performance and only criticized poor performance. In fact, praise and criticism had nothing to do with pilot performance; the pilots' performances were simply regressing to the mean. Some layman's (nontechnical) articles are by Senn (2011), which uses graphs to get across the idea, and Smith (2016), which has many colorful examples of the phenomenon.

Section 3.3 "Designing Conclusive Experiments"

For planning experiments we recommend the following: Vining (2013), Freeman et al. (2013), and Simpson et al. (2013).

Section 3.4 "The Lady Tasting Tea"

Senn (2012) has written a nice explanatory article on this famous experiment. It includes the following: "There are two important features of Fisher's prescription that are sometimes overlooked. The first is that the sequence of the cups should be chosen at random, and the second that the lady should be informed that this will be so. Let us consider these in turn." You can probably intuit the assertion that the sequence should be chosen at random, but if you can't figure out why the lady should be informed, then you need to read this brief article. It will help improve the design of your experiments.

Section 3.5 "Testing a New Checkout Button"

The section "Testing a New Checkout Button" is adapted from https://www.evanmiller.org/lazy-assignment-and-ab-testing.html

4

Analyzing A/B Tests: Advanced Techniques

This chapter focuses on analyzing A/B tests using techniques that go beyond the basics we discussed in Chapter 2. We begin with consideration of one-sided confidence intervals (CIs). Two-sided tests and intervals are standard in introductory textbooks, and one-sided tests are sometimes covered, yet one-sided intervals are all but ignored. However, when presenting results to persons not fluent in statistics, it is much better to present a CI than a test. Therefore, it is important to know about one-sided CIs for those occasions when it is necessary to conduct a one-sided test and then present the results to others. We introduce the one-sided CI and test in Section 4.1 where we revisit the audio/video case from Section 2.1.

Matching observations in the A sample with observations in the B sample can reduce variance and enable the analyst to detect effects that would otherwise be missed due to a type II error. Advanced matching methods are discussed in Section 5.3.1, but to comprehend these methods, one first must be acquainted with the elementary matching method, which we cover in Section 4.2. We know how to compare two alternatives, A and B, but what will we do when we have three alternatives, A, B, and C, or four alternatives, A, B, C, and D? These are collectively referred to as "A/B/n" tests, which is the subject of Section 4.3.

We know that when designing a test, well before the data are collected, we should compute power. Not everyone is aware of this, and it may come to pass that you have to analyze an experiment conducted by someone who did not do a power analysis as part of designing his experiment. In case the experiment fails to reject the null, you may be asked, "Did it reject because there is no effect or because the sample size was too small?" There is no good answer to this question, but there are bad ways and worse ways to wrestle with this question. The bad way is the subject of Section 4.4; the worse way is to do a power analysis *after* the experiment has been conducted using the sample mean and sample standard deviation in the power command. This is sometimes called "ex post power" or "observed power."

Business Experiments with R, First Edition. B. D. McCullough.

Companion Wcbsitc: www.wilcy.com/go/mccullough/businessexperimentswithr

Especially when we have more data on our test subjects – perhaps we mail catalogs and found no difference in the sales generated by two different versions of the catalog – we can ask questions about specific types of subjects. For example, we may say, "All right, overall there was no difference, but maybe that was a difference for persons who live in the suburbs of Eastern cities in neighborhoods with an average income exceeding $100 000." In this case, we have singled out a specific subgroup of the test subjects, and we can retrospectively apply the experiment to them. This is called "subgroup analysis" and is discussed in Section 4.5. A not uncommon phenomenon in subgroup analysis is that there may be an overall effect but, when the subgroups are examined, the effect is not found in any of the subgroups! How can it be that the effect does not exist in any of the subgroups, yet exists in the overall data? This phenomenon is known as "Simpson's paradox" and is the subject of Section 4.6.

We have seen that the data demands of an A/B test can be quite severe in the sense that a very large number of observations may be required, especially when testing for very small effects. In such a situation, it may be advisable to use the "test and roll" method that doesn't require so many observations, which is presented in Section 4.7.

By the end of this chapter, readers should know how to:

- Conduct a one-sided test and calculate the correct one-sided interval.
- Analyze a test where there are more than two levels of a treatment, for example, an A/B/C test.
- Increase power by matching pairs of observations in an A/B test.
- Investigate whether the data collected in an A/B test were capable of finding an effect, if there really was an effect, using the concept of the *minimum detectable effect*.
- Analyze an A/B test within subgroups to determine whether there are any significant effects within subgroups.
- Understand Simpson's paradox and its implications for analyzing subgroups.
- Use the test and roll strategy when necessary.

4.1 Case: Audio/Video Test Reprise (One-Sided Tests)

One-sided tests are sometimes discussed in introductory texts, but when they are discussed, it is only in the context of conducting a test, not for computing one-sided CIs. Yet, one-sided tests are important and so are one-sided intervals. In this section we take another look at the audiovisual test so that we can more properly consider it as a one-sided problem, perform the appropriate test, and compute the requisite CI.

4.1.1 One-Sided Confidence Intervals

Many textbooks suggest that if you think an effect might be in a particular direction, then you should use a one-sided test or CI. We suggest a more stringent criterion. In order to use a one-tail test or compute a one-sided CI, it must be the case that the other tail is functionally equivalent to no effect, i.e. a large difference in the other direction must lead to the same action as no difference at all. Suppose $H_0 : \mu_1 - \mu_2 \leq 0$ vs. $H_A : \mu_1 - \mu_2 > 0$. It must be the case that both $\mu_1 - \mu_2 < 0$ and $\mu_1 - \mu_2 = 0$ lead to the same course of action. To make these ideas clear, let us consider two cases.

The Acme company has created an additive that, it believes, will increase the shelf life of a chemical.

Example 4.1 The company wants to test whether or not the additive increases the shelf life. The company is not indifferent to finding that the additive really decreases the shelf life. If there is no effect, then they just have to add some more of the special ingredient. If the effect is negative, then they need to completely rework the chemistry or perhaps abandon the idea. The company will conduct a two-tailed test.

Example 4.2 The company has offered the additive to a prospective purchaser. The prospective purchaser will only purchase the additive if it really increases the shelf life. Whether the additive has no effect or has a negative effect makes no difference to the prospective purchaser, who will conduct a one-tailed test.

Let us return to the audio/video case from Section 2.1. Recall that $\bar{x}_V = 120.73$ and $\bar{x}_A = 110.29$. All the sales agents in the company currently use audio. The operations director is only going to switch to video if video is demonstrably better than audio. Hence, the two-tailed CI we computed previously is really not appropriate, and we will compute a one-tailed CI. Even though we will focus on the CI and not on the hypothesis test, it is nonetheless useful to write out the null and alternative hypotheses so that, when the CI is used as a test, it is clear what is being rejected or not rejected.

If you are unfamiliar with one-sided confidence tests, we suggest that you peruse the appendix to this chapter before proceeding.

We have the choice of writing $\mu_V - \mu_A$ or $\mu_A - \mu_V$. Let us use the former, since it is more intuitive: we are looking for evidence that video generates more revenue than audio, so we want $\mu_V - \mu_A$ to be positive (what we're looking for tells us how to write the alternative – once we write the alternative, the null is easy to write). The one-sided null and alternative hypotheses are

$$\begin{aligned} H_0 &: \mu_V - \mu_A \leq 0 \\ H_A &: \mu_V - \mu_A > 0 \end{aligned} \tag{4.1}$$

The usual graphical representation of the null and alternative hypotheses in the form of rejection and non-rejection regions is given in Figure 4.1. All the α is placed on the right-hand side. A large value of $\bar{x}_V - \bar{x}_A$ (video is much better) will fall in the right tail, constituting strong evidence in favor of the alternative and making us disbelieve the null.

Tests and CIs, while representing the same information, express the information differently. Consequently, we need to think about CIs differently than we think about tests. Now, any two-sided CI (L, U) can be viewed as the intersection of two one-sided intervals: (L, ∞) and $(-\infty, U)$. The only question is, "Which of these two intervals do we want?" To answer this question, simply state the type of confidence statement you wish to make (assume $\alpha = 0.05$ so that the confidence level is 95%):

- To be 95% confident that that population parameter is *at least* some number implies that the interval should be (L, ∞) (an upper interval).
- To be 95% confident that the population parameter is *at most* some number implies that the interval should be $(-\infty, U)$ (a lower interval).

As should be clear, our "population parameter" is the mean difference between the two methods, in this case $\mu_V - \mu_A$. What will convince us that video generates more revenue than audio is $\bar{x}_V - \bar{x}_A$ being some positive number that is far above zero; this will make us confident that the population difference is at least zero; then we want an interval of the form (L, ∞). To get a visual intuition, imagine the one-sided test, as in Figure 4.1. We want a one-sided interval that will cover the tail. The interval $(-\infty, U)$ does not cover the tail; so we want (L, ∞). Just as in the two-sided case, if the null value is in the interval, then the null is accepted, and if the null value does not fall in the interval, then the null is rejected.

Getting this interval is another matter. Depending on your familiarity with one-sided intervals, it may be easier to derive one-sided intervals from two-sided

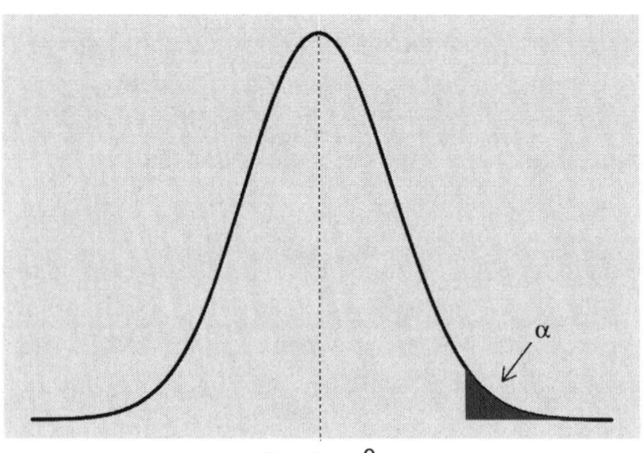

Figure 4.1 One-sided, two-sample test.

intervals; this method is easier to understand, and it is harder to make a mistake. To get a 95% one-sided interval, α must be set at 0.10 so that each tail of the two-sided interval has 5%.

In the present case, let's use a two-sided interval to get a one-sided interval. To get a 95% one-sided interval, we'll specify $\alpha = 0.10$ and get a two-sided interval of $(3.89, 16.99)$.

The pair of one-sided intervals that make up this two-sided interval are $(3.89, \infty)$ and $(-\infty, 16.99)$. As we saw above, we want a one-sided interval of the form (L, ∞). Hence, the one-sided 95% CI for $(\mu_V - \mu_A)$ is $(3.89, \infty)$. Since the null difference (zero) does not fall in the interval, we reject the null and conclude that video does produce more revenue than audio.

When presenting a one-sided CI to persons who don't use statistics regularly, do *not* say something like "We think likely values of the difference are between 3.89 and infinity." Using the word "infinity" in such a situation is just asking for trouble. Instead, say something like "We are confident the difference is at least 3.89."

Try it!

Reproduce the above result for a one-sided interval of $(3.89, \infty)$. Are you assuming variances are equal or unequal? Does it matter? Then get the 90% one-sided CI.

Software Details

Let's use the software to get the one-sided intervals directly...

It's probably easier if the data are unstacked, because then you can choose which variable is first and which is second, and it's easier to interpret the output. So unstack the data.

```
t.test(df1$video,df1$audio,alternative = "greater")
```

gives the one-sided 95% CI (3.89, Inf).

If you instead use the formula version to perform this test on the stacked data: `t.test(sales_one_week call_type,alternative="greater")` you really aren't sure whether "greater" refers to "audio-video" or "video-audio" and the output doesn't tell you. It may be easier to use the unstacked formulation.

In Section 2.1.6 video was more expensive than audio, and we quickly gave a one-sided interval. Let us reconsider this problem in more detail. When using CIs for hypothesis testing, it is often advantageous to write out the null and alternative to make clear what the null-hypothesized value is; it is this value that is going to fall in the interval or not, leading to rejection or not rejection. This time we write the alternative to > rather than <. Compare Equation (4.2) to Equation 2.7:

$$\begin{aligned} H_0 &: \mu_V - \mu_A \leq 5 \\ H_A &: \mu_V - \mu_A > 5 \end{aligned} \tag{4.2}$$

We want to be 95% confident that the difference is at least \$5, which implies an interval of the form (L, ∞). If the null difference (= \$5) falls in this interval, then we accept the null and believe the difference is less than or equal to \$5. If the value \$5 does not fall in this interval, then we reject the null hypothesis and conclude that the population difference is more than \$5.

The 90% CI for the difference $(\mu_V - \mu_A)$ is $(3.89, 16.99)$. So the 95% one-sided upper interval is $(3.89, \infty)$. The null difference \$5 falls in this interval, so we do not reject the null. There is no evidence that video is five dollars better than audio.

When doing this in terms of $V - 5$, the null and alternative for this formulation are

$$\begin{aligned} H_0 &: (\mu_V - 5) - \mu_A \leq 0 \\ H_A &: (\mu_V - 5) - \mu_A > 0 \end{aligned} \tag{4.3}$$

which makes clear that the null-hypothesized value is 0, not 5.

After forming the variable $x_V - 5$, we compute the 90% CI for the difference $(-1.110, 11.995)$, so the one-sided upper 95% interval is $(-1.11, \infty)$. The null difference, zero, falls in this interval, so we accept the null. We do not have evidence that the video generates \$5 more per call.

4.1.2 One-Sided Power

One-sided power is just like two-sided power, except only one tail is marked off, instead of two. Consider the audio/video case, where the null and alternative are given in Equation 4.1. Based on experience, suppose the standard deviation of each group is 27.5, and we will have a sample size of 100 for A and 100 for B. With $\alpha = 0.05$, the decision rule then is: reject H_0 when $\bar{x}_V - \bar{x}_A > 6.4 (= 1.645 * 3.89)$ (computing the common standard deviation as $\sqrt{s_1^2/n_1 + s_2^2/n_2} = 3.89$ for simplicity). Suppose we want to detect a difference of four dollars. One-sided power will be the answer to the question: "If $\mu_V - \mu_A = 4$, what is the probability that the null is rejected? i.e. what is the probability that $\bar{x}_V - \bar{x}_A > 6.4$?" Visually, see Figure 4.2, where power will be the shaded area, which (assuming normality) is easily seen to be 0.2676. Observe that 6.4 is $(6.4 - 4)/3.89 = 0.62$ standard deviations above the mean. Using a z-table, looking up $z = 0.62$ yields an upper-tail area of 0.2676.

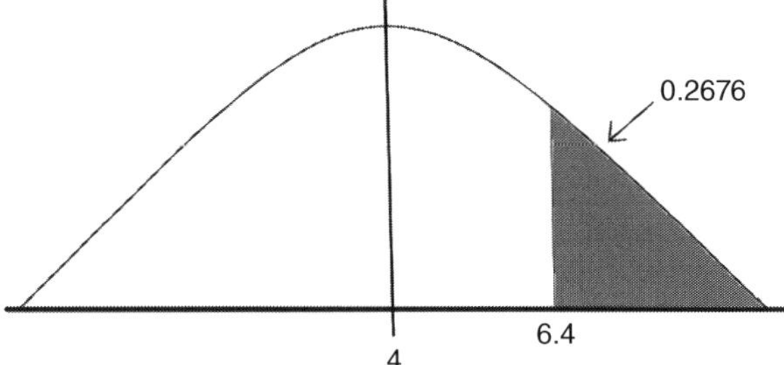

Figure 4.2 One-sided power calculation.

Try it!

Use the below R command to obtain the same answer that hand calculation yielded for the one-sided power shown in Figure 4.2.

```
power.t.test(n=100,delta=4,sd=27.5,sig.level=0.05,
             type="two.sample", alternative="one.sided")
```

What is power for detecting a difference of \$6?

Exercises

4.1.1 Suppose that video calling would be profitable if it generated \$2 more, rather than \$5 more. Should video calling be adopted? (That is, do we reject $H_0 : \mu_V - \mu_A \leq \$2$? This hypothesis can be tested or a CI can be computed in order to answer this question.)

4.1.2 Review the one-sided test for video calling presented in Section 2.1.6. Is it *really* a one-sided test? Is it the case that $\mu_A - (\mu_V - 5) > 0$ is functionally equivalent to $\mu_A - (\mu_V - 5) = 0$?

4.1.3 Consider a one-sided test:

$$H_0 : \mu_X - \mu_Y \leq 10$$
$$H_A : \mu_X - \mu_Y > 10$$

We want to detect an effect size of 10. The standard deviations of X and Y are both about 50. How many observations should be in each sample? If we can only have 20 observations in each sample, what is the power? What is power if we can have 100 observations in each sample?

4.1.4 For the one-sided test in the previous problem, suppose we wanted to detect a difference of 2. How many observations would we need?

4.1.5 An analyst determined that she needed 476 observations in each group for a two-sided test with power = 0.8, $\alpha = 0.05$, and sd = 22 to detect a difference of 4. Then she realized it's a one-sided test. Observations cost $250 a piece. How much money did she save?

4.2 Case: Typing Test (Paired *t*-Test)

The Acme company has created a typing program that it believes will increase typing speed. Acme certainly would want to know if its program actually made typists slower, so this will be a two-tailed test. In order to generate evidence as to the efficacy of the program, 20 employees who type regularly as part of their jobs are to take a training program that will help them type faster. To determine whether the program really works, the employees are first given a standard typing test to measure their typing speed, the "before" test. Then they go through the training course, followed by another test of typing speed, the "after" test. As we discussed in Chapter 3, this is a *within-subjects* design, where we measure the effect of both conditions ("before" and "after" training) for each subject.

The file `typing.csv` contains two columns of typing speeds, before and after, and if we compute the average response rate for both groups, we find that the average test score was 48.05 ($\overline{x}_b$) before the training and 53.05 ($\overline{x}_a$) after the training. The box plot and dot plot shown in Figure 4.3 together display the full distribution of the data, and we can see that there is quite a bit of variation in the scores across the 20 employees in each group.

Software Details

To make a dot plot...

We know how to make a box plot, but dot plots are sometimes useful. To make a dot plot the "lattice" package is necessary, so install it (using the `install.packages` command) and load it (using the `library` command). The data have to be stacked, and the data frame has three columns: "employee," "before," and "after." We only want to stack the "before" and "after" columns.

```
df <- read.csv("typing.csv",header=TRUE)
library(lattice)
df1 <- stack(df[,c(2,3)])
dotplot(ind~values,data = df1,horizontal=TRUE)
```

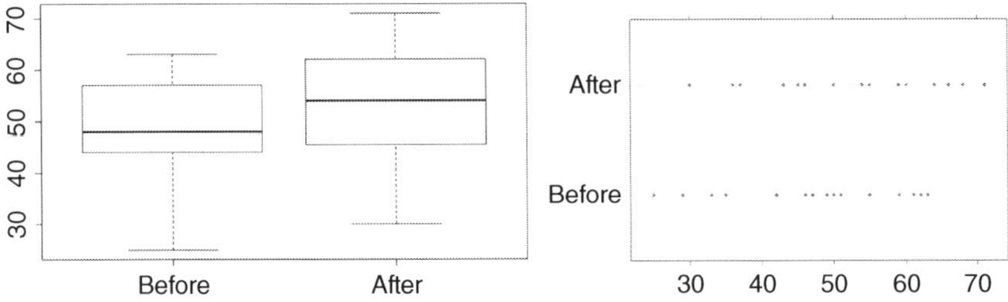

Figure 4.3 Box plot and dot plot of typing test performance.

We can apply the CI for comparing means from Section 2.1.2. The 95% CI for the difference in means is $(-12.07, 2.07)$ (this interval does not assume equal variances). Apparently the training has no effect on typing speed.

> **Try it!**
>
> Look at Figure 4.3. The variances of before and after appear to be about the same. Recompute the above CI, this time assuming that variances are equal. What do you find? (This is why we say it doesn't hurt to use the "variances unequal" method when variances are equal.)

4.2.1 Matched Pairs

An important assumption of the two-sample CI from Section 2.1.2 is that each observation is independent of every other observation in the sample. If we reflect on this for a moment, it is clear that the "before" and "after" observations are not independent. Each "before" observation is linked to a specific "after" observation because they are the test scores for the same employee. An employee who has a higher than average "before" score is probably going to have a higher than average "after" score. In statistical parlance, there is a positive correlation between the before and after variables, and this violates the assumption of independence of the two samples. This is made clear in Figure 4.4, which shows clearly that persons who score low on the before test also score low on the after test and similarly for persons who score high. Thus, the samples are not independent and are, in fact, positively correlated. As we will see shortly, this positive correlation will enable us to decrease variance and achieve better, more accurate results.

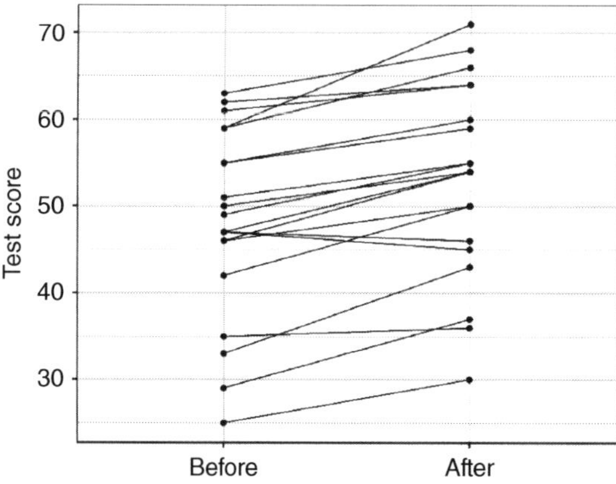

Figure 4.4 Before and after typing scores.

Software Details

To make the matched-pair plot shown in Figure 4.4...

The data will have to be stacked and will need a column for Label (whether the material is A or B), Data (all the numerical values), and ID. You should stack A and B and then add the third column for ID (which will be two copies of Employee, i.e. that has the numbers 1 through 20 twice).

Install the package "ggplot2" and stack the variables. One way to stack the variables is

```
df1 <- stack(df[,2:3])
df1$ID <- c(df$employee,df$employee)
```

and then build the plot with the following commands:

```
ggplot(df1, aes(x=ind, y=values, group=ID)) +
  geom_line(alpha=0.8) +
  geom_point(alpha=0.7,size=2.5) +
  theme_bw(base_size = 20) +
  xlab("") +
  ylab("test score")
```

To deal with this situation, statisticians have developed the "matched-pair" methodology, also known as the "two-sample test of means for correlated samples" and the "paired comparison design." Essentially, this method redefines the response variable as the *difference* between the "before" and the "after" scores for each employee, i.e. d = before - after. We then compute a CI for d to see whether the difference is significantly different from zero. Once we compute the difference, we no longer have two different sets of measurements for before and after, so we use a one-sample CI. First, we compute the mean ($\bar{d}$) and standard deviation (s_d) of the 20 differences, and then the CI for the difference is

$$d \pm t_{(1-\alpha/2,n-1)} \frac{s_d}{\sqrt{n}} \tag{4.4}$$

If the CI for the difference contains zero, then we do not have sufficient evidence to conclude that the training improves the test scores.

Software Details

To obtain a confidence interval for matched pairs, use unstacked data...

```
t.test(df$before,df$after,paired=TRUE)
```

The 95% CI for the difference is $(-6.6, -3.4)$. Since this interval does not contain zero, we can conclude that the training does increase typing test scores. We might report this to the team manager as follows:

> For the 20 employees who took the typing training, test scores improved by 5 points before versus after the training (95% CI = $(3.4, 6.6)$).

This finding that the training is beneficial might be surprising given that we had a nonsignificant result when we did the CI for the difference in means. When we take the differences, we end up with 20 data points, where before we had 40 data points, so you might think that with fewer observations we would be less likely to find a significant difference. However, if we compute the standard deviation for these 20 points, it is much smaller, since we have, essentially, "netted out" the differences in baseline performance between employees when we take "before minus after" differences (or maybe "after minus before" differences, depending on your preference). This lower variance in the data results in a narrower CI and a more powerful hypothesis test. This allows us to observe an effect that is really there and that we otherwise would have missed.

The standard errors for the unpaired and paired tests are 3.49 and 0.77, respectively (assuming unequal variances). R doesn't display these quantities, and we don't want to write code to get them. To get approximate standard errors, we can use the CIs from R. The 95% CI for the unpaired test is $(-12.08, 2.08)$, and we'll divide the width of this interval by 4 to get an approximate standard error of 3.5. The paired interval is $(-6.6, -3.4)$, and dividing the width by 4 yields 0.8. This is a huge reduction in variance! Note that the observed mean difference between before and after is always 5, regardless of whether the data are paired or not, so the signal is always the same. What changes is the standard deviation. The smaller standard deviation represents less noise. Using the paired test increases the signal to noise ratio, allowing us to see an effect that could not be seen with the noisier unpaired test. This illustrates the importance of variance reduction as an experimental goal.

Another way to look at the better performance of the paired test is to recognize that the independent samples method assumes there is no relationship between the before and after test scores and so overestimates the variance. The matched-pair method takes account of the dependence and correctly estimates the variance. You may recall from your first statistics course the rules for the variances of sums and products of random variables:

$$var(X + Y) = var(X) + var(Y) + 2 \cdot cov(X, Y) \tag{4.5}$$

$$var(X - Y) = var(X) + var(Y) - 2 \cdot cov(X, Y) \tag{4.6}$$

where $cov(X, Y) = E[(X - \mu_X)(Y - \mu_Y))$. A positive covariance implies that if X is above its mean, then Y is usually above its mean too and if X is below its mean, then Y is usually below its mean too. We have already noted that the "before" and "after" scores have a positive covariance. Under the assumption of independence, $var(X - Y) = var(X) + var(Y)$, which is necessarily larger than the correct variance of the difference when the variables have positive covariance. In practice, a positive covariance is more common than negative covariance, so we can usually expect matching to reduce variance and improve inference.

Of course, as we discussed in Chapter 3, this test design has two potential problems. First, because the employees take a test before and after the training, they can probably guess the objectives of the test and are not blinded. If they dislike the training and are thinking ahead, they might intentionally perform poorly on the second test to sabotage the experiment. A second potential problem with a within-subjects design is that there might be something about when the tests were administered that is causing the observed effect, instead of the training. For example, it might have been too hot in the room during the "before" test, and so the employees performed poorly. As a reminder, what we have described above is sometimes called a *within-subjects design*.

If instead we chose a *between-subjects design*, then we might have 20 employees acting as the control group who are asked to go about their workday normally, while 20 different employees are part of the treatment group and take the training. We would administer the tests to both groups at the same time to eliminate the time confound. With this design, instead of controlling for typing ability, we are randomizing over typing ability. We also avoid potential time confounds or demand effects. However, the data we collect would be more noisy, and it would take more subjects to determine conclusively whether the training works.

Software Details

To compute power for matched pairs...

We know that the standard deviation of each sample is about 11, so instead of computing the pooled standard deviation, we'll just use the number 11. Suppose we want to detect a difference of 5. For an unpaired test, we'd issue the command

```
power.t.test(delta=5,sd=11,sig.level = 0.05,power=0.8)
```

But for a paired test, we'll issue the command

```
power.t.test(delta=5,sd=11,power=0.8,sig.level=0.05,
          type="paired")
```

To detect a difference of size 5 with power = 0.80, the unpaired test requires 77 observations, whereas the paired test requires only 40. If observations are expensive, this can represent a substantial savings.

Exercises

4.2.1 You work for a small start-up brewery. The master brewer has come up with two new beers, but the brewery only has space and equipment to produce one of them. To decide between the two, he tells you that he'll have all 22 employees stay after work. Eleven will drink recipe A, and the other 11 will drink recipe B. Each person will rate the beer on a scale from 1 to 10. The brand that has the highest average rating will be the winner and go into production. Comment on this design. If it is satisfactory, say why. If it is unsatisfactory, say why and propose a better design.

4.2.2 For the typing test, the Acme Typing Company employs many typists and is considering the purchase of the training program, but will only do so if it actually increases the typists' speeds. The vendor agrees to train a small group of Acme's typists to prove to Acme that the training program really works. The null and alternative are $H_0 : d \geq 0, H_A : d < 0$ where d = before - after. (Convince yourself that this is the correct formulation of the null and the alternative, given the way d is defined.) Reanalyze the data.

4.2.3 This was discussed in the text, but code wasn't given for doing this. For the typing data, compute the standard error of the difference assuming the samples are independent – use the variances equal formula or simply perform a t-test and get approximate standard errors from the CI. Now compute the standard error of the difference for the paired data. How much of a reduction occurs?

4.2.4 Visualize the variance reduction that occurs when moving from the independent t-test to the matched pair. On the same graph make a box plot of the variables before, after, and before-after. After making the box plots, ask yourself: What might you do to make it easier to compare them visually?

4.2.5 A well-known experiment in the experimental design literature was reported by Box et al. (Box et al., 2005, p. 81). Two types of material for boys' shoes, A and B, are to be tested to determine how long they wear. The data are in `BoysShoes.csv`. Ten boys are randomly chosen for an experiment, and each gets a pair of shoes, one of which is soled with A and other with B. After a suitable period of time, each sole is measured, and the percentage of the sole that has been worn off is recorded. Each boy will have some amount of wear for type A and some amount of wear for type B. Which foot gets type A is randomly assigned to each boy (why is this important?) and is recorded in the variable Afoot (right now we have no way to determine the effect of this variable, but we will learn how to do so in Section 6.5 when we cover ANCOVA). Is there a difference in the wear between materials A and B? How much variance reduction does the matched-pair method effect? Graphically represent the data to show that one material wears better than the other.

4.2.6 For the boys' shoes experiment, suppose that a shoe is considered worn out when the wear amounts to 50%. Suppose further that material B is cheaper than material A. The manufacturer currently uses material A and would like to switch to the cheaper material, but not if the customers (many of whom are repeat customers) would notice that the shoe wears out sooner. Assume that shoes last for about a year. Should the manufacturer switch materials?

4.2.7 In the boys' shoes example, which is the experimental unit, the boy or the shoe? What are the consequences of each choice?

4.2.8 For the typing example, compute the variance of before. Compute the variance of after. Compute the variance of diff = before - after. Do you see how matching reduces variance? To gain further insight, compute the covariance of before and after, and apply Equation 4.6.

4.2.9 Take the lower bound of the 95% CI for the typing test as a conservative estimate of the increase in speed that results from the training. Assume that typists are paid \$20/hour with a mean typing speed of 50 words per minute. Assume that training costs \$500 per person, training takes four hours, and turnover is negligible. Should the company pay for training for its typists? What is the increase/decrease to the bottom line after the first year? In the second year?

4.3 A/B/n Tests

What happens when the response has more than two levels? Suppose you want to test three landing pages, or four, or n landing pages. Just as an A/B test is a two-sample test of means testing the null hypothesis that both means are equal, an A/B/n test is just a one-way ANOVA testing the null hypothesis that all n means are equal. It may be useful to review one-way ANOVA before proceeding; if the p-value for the F-statistic in the ANOVA table is small, then the null hypothesis is rejected. However, that is just the beginning, for we know only that the means are not equal; we do not know which mean is higher than another. Suppose we are conducting a test with four levels, A, B, C, and D, and the null has been rejected. There are many possible results – $A > B > C > D$, $A = B > C = D$, $A = B = C > D$, etc. – and we will have to determine which of the many possibilities is the case.

For this example we will use observational data, simply as a reminder that statistical methods can be applied to observational data to obtain useful information when conducting an experiment is too difficult. A car rental agency wonders whether different types of cars are rented for varying lengths of time. Data are collected for compact, sedan, luxury, and minivan rentals. The data are in the file `CarRental.csv`. These data are in columns, and some packages will want the data stacked ("stacked" means two columns, one for the response (numerical data) and another for the labels (sedan, luxury, etc.), so be prepared to do some data munging if necessary).

The ANOVA table is given below:

```
Analysis of Variance

Source         DF   Sum Sq   Mean Sq   F-Value    Pr(>F)
ind             3     2295       765     54.48    <2e-16 ***
Residuals     108     1516        14
—
Signif. codes: 0 '***' 0.001 '**' 0.01 '*' 0.05 '.' 0.1 ' ' 1
```

The p-value is essentially zero, so the null hypothesis that all the means are equal is rejected.

Software Details

To perform an ANOVA...

```
df <- read.csv("CarRental.csv",header=TRUE)
head(df) # check to make sure the data are read correctly
newdf <- stack(df)
head(newdf) # doublecheck that data are stacked properly
newdf <- na.omit(newdf) # drop NA values from the dataframe
summary(aov(values~ind,data=newdf))
```

Formally, the ANOVA model is

$$y_{ij} = \mu + \tau_j + \epsilon_{ij}, i = 1, 2, \ldots, n\, j = 1, 2, \ldots, k$$

where the i denotes the observations (rows) and j denotes the columns (treatments). The null hypothesis is $H_0 : \tau_1 = \tau_2 = \ldots = \tau_k = 0$ against the alternative that the treatment means are not all zero. Before we can rely on the ANOVA results, we have to make sure that the assumptions of the model are satisfied.

There are three primary assumptions of ANOVA, which we present in the order that they should be verified. The act of verifying these assumptions is called "model checking":

1. Independence of the errors. In a randomized experiment, this is usually automatic. For observational data, this cannot be verified, but we can find evidence that it is not true by examining plots of the residuals and looking for patterns. Plot the residuals against time, or the levels, or the fitted values, or any other relevant variables. However, ANOVA is robust to deviations from independence, so of the three assumptions, this is not the most important.
2. Constant variance, also called homoscedastic errors. This can be analyzed graphically by making a box plot of residuals for each level and seeing if the box plots are roughly the same. Any insight gained from the graphs should be validated by a formal statistical test. There are also many tests for homoscedastic errors that we will discuss anon.

Software Details

To plot make a box plot of residuals by level...

```
df <- read.csv("CarRental.csv",header=TRUE)
newdf <- stack(df)
newdf <- na.omit(newdf)
myaov <- aov(values~ind,data=newdf)
boxplot(myaov$residuals~newdf$ind)
```

3. Normality of the errors is not the most important because ANOVA is robust to departures from normality, as long as the departure isn't too severe. This can be checked by a normal probability plot (also called a quantile plot or a q-q plot) or a formal statistical test of the null hypothesis that the errors follow a normal distribution. If the number of observations for each level is large, then one plot can be made for each level; otherwise all the residuals can be pooled and just one plot need be made, provided that the variances are homoscedastic. At this stage we should also check for outliers among the residuals – if the residuals have outliers, then the residuals cannot be normally distributed. Occasionally, residuals are found to be non-normal and contain outliers, but, after outliers are investigated and remediated (typo corrected, observation deleted, etc.), the residuals are normal.

Software Details

To make a normal probability plot of the residuals...
First decide whether to make plots for each level, a plot for all the observations (or both). Adjust the below instructions accordingly.

```
qqnorm(myaov$residuals)
qqline(myaov$residuals)
```

For plots of the levels, the above may be modified to subset the data to draw a plot for each level.

There are many tests for constant variance, with names such as Levene's test, Bartlett's test, Forsythe's test, and many others. For the present problem, it may not be safe to assume that the variances are all equal, especially because we have not examined the normality assumption. With normality, we would use Bartlett's test to have more statistical power. With non-normal data, we would use the Brown–Forsythe test (also called the "modified Levene test"). If one variance is larger and the errors are non-normal and the sample sizes are different, then Cochran's test is a good choice. We will not further discuss verifying the assumptions, also called model checking. You may need to refer to your first statistics book for a refresher.

Software Details

To test the equality of variances for an ANOVA...

```
bartlett.test(values~ind,data=newdf)
```

And remember, the null hypothesis is that the variances are equal.

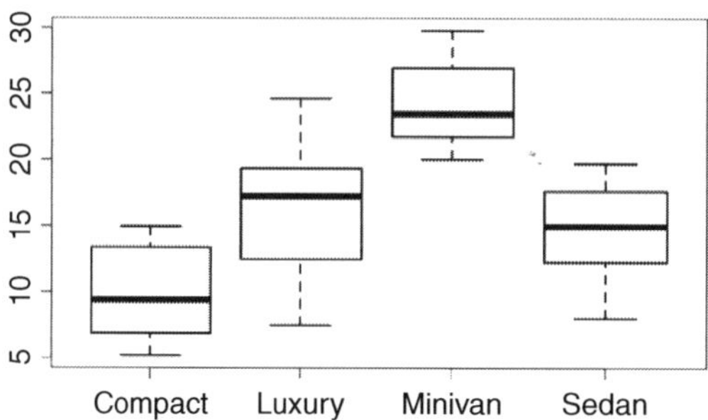

Figure 4.5 Box plot for the car rental data.

For sake of exposition, pretend that all the model assumptions have been verified; we will proceed with trying to figure out precisely why the null of "equal means" was rejected, bearing in mind that the variance for luxury cars is somewhat larger than for the other types of vehicles.

That the null of equal means was rejected is nothing we could not have predicted from seeing the box plot in Figure 4.5. As we know, however, visualization and formal statistics are complementary, and any conclusion reached with one should be verified with the other.

What we can deduce from the box plot is that compact has the lowest number of days in the rental period and minivan is the highest. What is unclear is whether luxury is greater than sedan or whether they are tied. There are many approaches to resolving this question, and they all come down to comparing all the pairs of means – taking account of the multiple hypotheses problem, to be sure! – and then figuring out the relationship between all the means based on the pairs. A few of these approaches are t-tests with an adjustment, often a Bonferroni adjustment; Tukey's honest significant difference (HSD) test; and Newman–Keuls test. Such methods usually give a CI for each of the differences, as shown in Figure 4.6 in the

```
  Tukey multiple comparisons of means
    95% family-wise confidence level

Fit: aov(formula = values ~ ind, data = newdf)

$ind
                     diff        lwr        upr     p adj
luxury-compact   6.284900   3.791872  8.7779278 0.0000000
minivan-compact 14.519357  11.482857 17.5558573 0.0000000
sedan-compact    4.683470   2.383909  6.9830305 0.0000034
minivan-luxury   8.234457   4.970318 11.4985967 0.0000000
sedan-luxury    -1.601430  -4.194161  0.9913007 0.3763601
sedan-minivan   -9.835887 -12.954765 -6.7170095 0.0000000
```

Figure 4.6 Tukey HSD intervals for the car rental data.

"lower CL" and "upper CL" columns. When interpreting these intervals, it can be helpful to look at the box plot of means, e.g. Figure 4.5.

Focus attention on the lower ("lwr") and upper ("upr") bounds for the CIs on the differences. The interval for luxury–compact is entirely positive, suggesting that the difference is positive so that $\mu_{\text{luxury}} > \mu_{\text{compact}}$. Similarly we see that $\mu_{\text{minivan}} > \mu_{\text{compact}}$ and $\mu_{\text{sedan}} > \mu_{\text{compact}}$. It is safe to say that compact has the shortest rental time.

Next we see that $\mu_{\text{minivan}} > \mu_{\text{luxury}}$, but the interval for sedan – luxury includes the origin, so we conclude $\mu_{\text{sedan}} = \mu_{\text{luxury}}$. Finally, that the interval for sedan – minivan is entirely negative implies that $\mu_{\text{minivan}} > \mu_{\text{sedan}}$. In sum, compact is the smallest, minivan is the biggest, and sedan and luxury are tied in the middle.

Software Details

To perform multiple comparisons for an ANOVA...

```
TukeyHSD(aov(values~ind,data=newdf),conf.level=0.95)
```

This type of analysis does not always result in such nice, neat conclusions. Sometimes the data are so noisy that it is difficult to reach any conclusion. The file `ABCD.csv` contains weights (in ounces) of a machine part from four different sources. Testing the null hypothesis that all four sources have the same mean weight results in an F-statistic of 2.756 with a p-value < 0.05, so the null hypothesis that all the means are equal is rejected. The next step is to figure out the reason for the discrepancy: Is just one of the means different, or just two, or all they all different? Tukey's HSD will show that all the differences have CIs that include zero. Which is it: Are all the means equal or are they not? The data are too noisy for us to tell! It appears the means are not all equal, but we can't tell which ones are different.

Another difficulty with all pairwise differences is that the results are not transitive. Normally, we expect that if $X = Y$ and $X = Z$, then $Y = Z$, and this is always true in a deterministic world where there are no random errors. In a probabilistic world filled with random noise, when analyzing pairwise differences, we can deduce that $\mu_X = \mu_Y$, $\mu_X = \mu_Z$, but $\mu_Y > \mu_Z$, a most logically unsatisfying result. This idea is explored in Exercise 4.3.3. But if that's all the data can say, then that's what the data say. In such a situation, what the data are saying is that you need more data.

The plain vanilla one-way ANOVA employed in this section for the A/B/n test is also called the "completely randomized design." The car rental example would not qualify as a completely randomized design, even though it was analyzed by one-way ANOVA, because the data were observational.

Exercises

4.3.1 Perform model checking on the CarRental data.

4.3.2 How would you set up an experiment to test whether different cars are rented for different lengths? Why wouldn't you conduct such an experiment?

4.3.3 Verify the claim about `ABCD.csv`.

4.3.4 Monthly sales for four stores are in `StoreSales.csv`. Compute Tukey HSD intervals to try and determine the sales ranks of the stores (which store sells the most, which store is second, etc.). Do you notice anything that seems strange?

4.4 Minimum Detectable Effect

When a test returns an insignificant result and tells you that "x does not affect y," some analysts mistakenly try to calculate "observed power" in an attempt to answer the question: "Was there really no effect or did my study simply lack sufficient power?" They take the mean and standard deviation from the results and plug these into the power calculation. Such an effort is also called a *post hoc power calculation*, and it's a bad idea. Recall the definition of power: the probability of correctly rejecting a null hypothesis that is false. Another way to think about power is this: power is the probability of detecting an effect, given that the effect is really there. Now, if the test result is insignificant, we don't know whether the null hypothesis is true or false, and calculating power in this situation is to assume that the null hypothesis *is* false without *knowing* that it is false. The practical implication of this follows: if a test fails to reject, a post hoc power calculation will *always* show low power because there is a one-to-one correspondence between post hoc power and the p-value. If the p-value is small, post hoc power is high; if the p-value is large, post hoc power is low. Post hoc power doesn't tell you anything that the p-value hasn't already told you.

On the other hand, there are times when someone else (not you, because you know better) has done a test without having done a power calculation in advance, the test has not rejected the null, and *you* get stuck with analyzing the results. In this bad situation, you'd like to have some sort of (imperfect) answer to the question, "Did it not reject because there's no effect, or because the test had low power?" An attempt to answer this question lies in two related concepts called the *minimum detectable effect* (MDE), which is measured in original units, and the *minimum detectable effect size* (MDES), which is measured in standard deviations; we will distinguish further between these concepts below.

Suppose we have a really large sample so we can use the normal distribution instead of the t-distribution (this just makes the math easier). For a two-tailed test, the lower and upper rejection regions for a test at the 5% level are defined by ± 1.96. Let us adopt the convention that $z_{0.025}$ is the (positive) value of the z-distribution that puts 2.5% in the upper tail. Suppose there really is an effect so that the null hypothesis is false, and let's suppose the effect is positive so that we want the test statistic to fall in the upper tail. Let the *effect size* be d. Suppose we want power to be 90%. What this means is that we want 90% of the distribution for $\mu = d$ to be above $+1.96$. The 10% of the distribution for $\mu = d$ that falls below $+1.96$ is β, the probability of a type II error when $\mu = d$. This scenario is depicted in Figure 4.7.

The MDES is given by

$$d = z_{\alpha/2} + z_{\beta} \tag{4.7}$$

Question: What is the numerical value of d? We know that it is some number greater than 1.96. How much greater? Recall that the distribution for $\mu = d$ is normal with a standard deviation of unity. To get 10% in the lower tail, that number will be 1.28 above 1.96. Therefore, $d = 1.96 + 1.28 = z_{\alpha/2} + z_{\beta} = 3.24$. This value is the MDES, and it is measured in terms of standard errors. This means that for a two-tailed test with $\alpha = 0.05$ and power $= 0.90$, *if* the null hypothesis is false, then the smallest effect we can reliably detect (with power $= 0.90$) is 3.24 standard errors above the null value of the mean (which is zero in this case). To obtain the minimum detectable effect MDE, simply multiply the MDES by the standard error; the MDE is measured in the original or natural units.

$$\text{MDE} = \text{MDES} * \text{standard error} \tag{4.8}$$

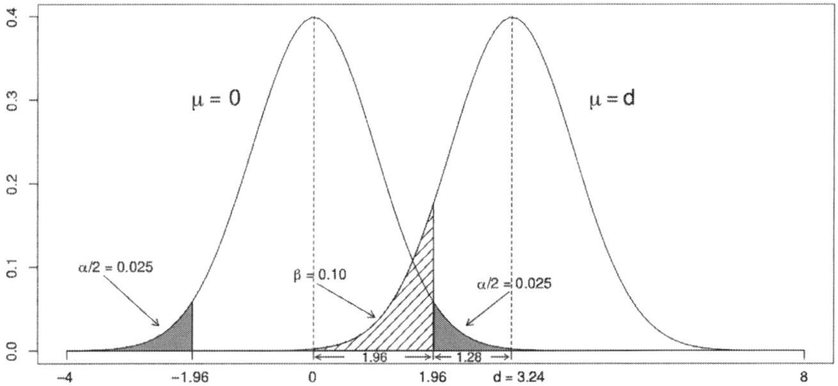

Figure 4.7 MDES when $\alpha = 0.05$ and $\beta = 0.10$: $1.96 + 1.28 = 3.24$.

As an example, let us consider a new version of the audio/video example from Section 2.1. We generated these data, so we know what the true means and variances are ($\mu_v = 115, \mu_A = 105, \sigma_A = \sigma_V = 30$), but let us pretend that we are ignorant of these quantities. The data can be found in `AudioVideo2.csv`, which is not formatted the same way as `AudioVideo.csv`. With these data, $\overline{x}_V = 110.89$ and $\overline{x}_A = 104.49$. As before, we desire the interval "video-audio" and not the other way around. Assuming unequal variances, the 95% CI for the difference between the means is $(-2.12, 14.92)$, which includes the origin, and does not reject the null hypothesis that the true mean is zero. We didn't do a power calculation before conducting this test, which was a mistake, so let us attempt a back-of-the-envelope MDE calculation. We need the standard error of the difference, and R does not provide it. So let's take the width of the interval and approximately write $17.04 = 2 * 1.96 * se$, which implies that the standard error of the difference is about $se = 4.34$. (The correct standard error equals 4.32; so our crude estimate actually is quite close to the correct number.) With $\alpha = 0.05$ and 90% power, the MDES = 3.24, so the smallest effect we could reliably detect is MDE $= 3.24 * 4.34 = 14.06$. We might then tell ourselves that these data are so noisy that we probably couldn't even detect a difference of 10 or 12 dollars with this sample size.

Everything worked out nicely in this example, but it is important to remember that the MDE is just a very rough approximation that is at best suggestive of the truth. Nonetheless, calculation of the MDE can sometimes shed some light on the situation when someone made a mistake and forgot to do a power calculation before conducting a test.

To review this concept, let us return to the email response case in Section 2.2.1, where we wrote: "We used a computer program to generate the two sequences of 200, for version A the population proportion was 0.10, and for version B the population proportion was 0.15 – this is a very large difference." Yet the CI $(-0.0903, 0.0403)$ included zero, so this large difference was not detected. Calculate the MDE to see if it sheds any light. We can deduce that the standard error is $0.0653/(2 * 1.96) = 0.0333$. If $\alpha = 0.05$ and power $= 0.90$, then MDES $= 3.24$ so MDE $= 3.24 * 0.0333 = 0.108$. We know the absolute value of the true difference (0.05) is much less than the MDE, so for this sample size we might not detect this difference.

Exercises

4.4.1 Show graphically that the MDES for a one-tailed test is $z_\alpha + z_\beta$.

4.4.2 What is the MDES multiplier for a two-tailed test when $\alpha = 0.05$ and power $= 0.8$? For a one-tailed test?

4.4.3 A two-sample test of proportions produced a 95% CI of $(-0.087, 0.086)$, but the incompetent analyst neglected to compute power before conducting the test. Let $\alpha = 0.05$ and power $= 0.90$. What is the minimum detectable effect?

4.4.4 A two-sample test of means produced a 95% CI of $(-4.5, 3.2)$. Let $\alpha = 0.05$ and power $= 0.80$. The rookie analyst forgot to computer power before conducting the test. What is the minimum detectable effect?

4.4.5 A two-sample test of proportions produced a 95% CI $(0.12, 0.22)$. Of course, a 95% CI implies $\alpha = 0.05$. Suppose that we now want $\alpha = 0.10$ and power $= 0.95$. What is the minimum detectable effect?

4.4.6 For the email response example from Chapters 2 where we know the true difference, compute power knowing the true proportions and then compute "power" using the estimated proportions. Note the difference.

4.5 Subgroup Analysis

The *Harvard Business Review* article by Anderson and Simester (2011) titled "A Step-by-Step Guide to Smart Business Experiments" lists seven steps, the fourth step of which is: "When the results come in, slice the data." Sometimes this activity is also called "slice and dice" and refers to selecting subsets of the data for further analysis. It is also called "subgroup analysis." There are two distinct purposes for subgroup analysis, hypothesis testing and hypothesis generation, and it is important not to confuse the two. The former tests an hypothesis on a subgroup, and the latter examines many subgroups in hope of finding an interesting result that merits further examination, perhaps via a follow-up experiment. As an example, let's return to the landing pages case from Section 2.3, in which an online retailer was testing which of two landing pages produces more sales. The 95% CI for the difference was $(-1.30, 0.47)$. It appeared that there was no difference between the two versions. Even when excluding zeros from the data and focusing only on those visits that resulted in sales, the CI still included zero: $(-9.1, 0.3)$. But suppose we have additional data!

When you visit a webpage, that webpage knows how you came to that page: perhaps you typed the URL into the browser, or you clicked a link on a page from some other website, or you clicked on a link in an email, or some other way. In the present example, it so happens that the company had sent out an email to its clients, and so we know whether the customer came to the landing page by clicking on a link in the email or from some other webpage. There is a third column in the data set, source, indicating whether the customer arrived from the email or from some other place. The variable source is a *covariate*, which is a measurement on an experimental unit that is not controlled by the experimenter. Note: Covariates are never randomized (because they are not part of the experiment). Remember, we have excluded zeros from the data and are focusing only on visits that resulted in sales.

It may be the case that customers who come from the Internet are different from those who come from email. To investigate this question, we will perform a "subgroup analysis." What this means is we will see whether the A & B versions make a difference to customers who came from the Internet and again for customers who came from the email. As a first step we will, as usual, produce graphs and statistics. Specifically, we will display box plots and compute summary statistics. For this example, let us exclude the zeros from the data set. The resulting summary statistics are presented in Table 4.1, and the resulting box plots are presented in Figure 4.8. Note the box plots correspond to the lower half of Table 4.1, not to the upper half.

The box plots are suggestive of a difference between the response rates, but hardly conclusive: for email and for Internet sources, version B appears to be slightly better than version A. Given the previously reported CI that showed no difference between A and B, we might just be seeing random variation. Looking at plots isn't going to answer the question, as the graphical results are not sufficiently striking. The summary statistics are a bit more informative. Looking just at version, we see no big difference between A and B. Looking just at source

Table 4.1 Summary statistics on sales for subgroup data, zeros excluded..

		n	Mean	Std dev
Version	A	328	97.02	30.53
	B	336	101.41	31.30
Source	Email	101	106.61	28.94
	Internet	563	97.92	31.17
Overall		664	99.24	30.98
Version A & email		50	98.87	28.28
Version B & email		51	114.20	27.80
Version A & Internet		278	96.69	30.96
Version B & Internet		285	99.13	31.39

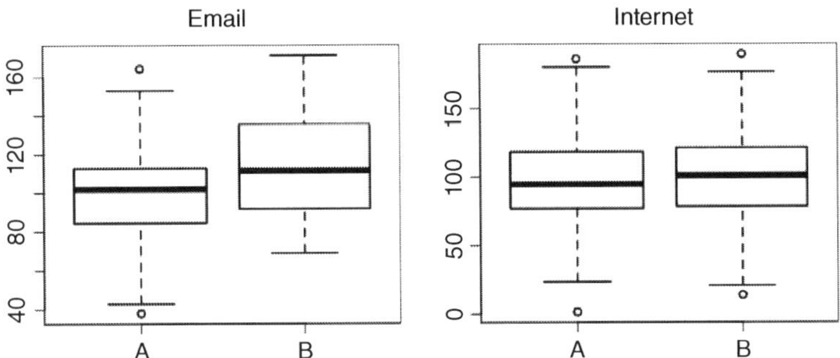

Figure 4.8 Subgroup box plots for landing pages (zeros excluded from data).

(email or Internet), we see a difference of almost 10 dollars between email and Internet, but that amount is still just one third of the standard deviation.

Remember, this difference between 106 and 97 would not be causal, even if statistically significant, because "source" was not a randomized treatment. The next two differences will be causal (if they are statistically significant) because "version" (A or B) was a randomized treatment.

Drilling down, looking only at the email source and then comparing versions A and B, we see a difference of over 15 dollars, which is half a standard deviation. Looking only at the Internet source and comparing versions, we see a difference of less than three dollars. We have finished poring over the summary statistics.

Now let's compute CIs for these differences to see which ones are significant and which ones aren't. The 95% CIs for $(\mu_A - \mu_B)$ are $(-7.60, 2.72)$ and $(-26.41, -4.26)$ for Internet and email, respectively. The former interval shows no difference between the two versions as far as Internet visitors are concerned. The latter interval indicates that, for email customers, version B is better than version A. While the two versions did not produce different sales overall, we may have found a subgroup where it does matter which version the user sees; this is what subgroup analysis is all about: finding the subgroup where the treatment *really* matters.

We have several important points to make about this result:

- First, the CI for $(\mu_A - \mu_B)$ for email customers when zeros are not excluded is $(-3.36, 1.30)$, which shows that adding all the zeros to the data set hides the effect; it was very important for us to focus on the data excluding the zeros.
- Second, because the treatment (version A or version B) is randomized within the group of persons who clicked on the email link, we can give a causal interpretation to the result: version B causes persons who click on the email link to spend more, conditional on the person spending in the first place. See Learning More for more details on this.

- Third, causality does not apply when the treatment is not randomized. Suppose we tested whether Internet or email had a higher mean for nonzero buys. The 95% CI for the difference ($\mu_{internet} - \mu_{email}$) is (−14.9, −2.4), indicating that email has a higher mean. However, this result is purely correlative, not causal, because we did not randomize on source.
- Fourth, subgroup analysis can suffer from the statistical problems of low power and multiple hypotheses, as discussed in Section 4.5.1; therefore, run another experiment before actually betting money on a subgroup result.
- Finally, only slice on variables collected before the experiment! Failure to follow this stricture comes back to bite in Example IV of Section 4.6.3.

4.5.1 Deficiencies of Subgroup Analysis

Subgroup analysis comes into play when you have covariates. Recall that in the above example whether a customer came from the Internet or from an email link was a covariate. If you conduct an experiment on customers that are in an existing CRM database, you likely have some additional information about the customers in that database suitable for conducting a subgroup analysis, i.e. you are likely to have many covariates.

However, as the number of subgroup analyses performed increases, the probability of a type I error increases, too. If you have a large database and perform, say, 20 subgroup analyses, even if there is no effect, you are quite likely to find a "significant" result due to a type I error. To see this,

$$\begin{aligned} P(\text{at least one Type I error}) &= 1 - P(\text{no Type I Errors}) \\ &= 1 - 0.95 \times 0.95 \times 0.95 \times \ldots \times 0.95 \\ &= 1 - 0.95^{20} \\ &= 1 - 0.3585 \\ &= 0.6415 \end{aligned}$$

This is the problem of testing multiple hypotheses on the same data set: if we conduct 20 tests and find one that is statistically significant, we don't know whether the rejection is probably due to the existence of a real effect or because we stumbled over a type I error. Therefore, if more than a very few subgroups are tested, all positive subgroup results have to be taken with caution, and these results may be better used as a means for generating new hypotheses to be tested/validated in subsequent experiments. As a general rule, if you are testing many hypotheses, you are engaging in subgroup analysis for hypothesis generation. If you are going to test only a very few hypotheses, be sure to specify the hypotheses before the experiment is conducted and be sure there is some justification for the hypothesis; then you can engage in subgroup analysis for hypothesis testing. This idea is explored in Exercise 4.5.7.

Power calculations and figuring out what sample size you need very often are predicated only on testing the main hypothesis. If you have a sample that is just large enough to test your main hypothesis, then your subgroups will not have sample sizes large enough to detect effects if they exist; i.e. they will be underpowered.

As an example, suppose the response is a binary variable and the experiment has 400 units, half getting the treatment and half getting the control. For convenience assume the proportion of responses in each group is 0.5. The standard error of the treatment effect is $\sqrt{0.5^2/200 + 0.5^2/200} = 0.05$. Suppose an interaction effect for the treatment group is considered. Assume 100 in each subgroup. The standard error for the interaction effect is $\sqrt{5^2/100 + 0.5^2/100} = 0.10$. The standard error for the subgroup is twice the standard error of the main effect. If the experiment is powered to just barely detect the main effect, then it will never see the subgroup effect unless the subgroup effect is much larger than the main effect.

Many analysts design tests, especially sample size calculations, thinking only of main effects. Then they analyze subgroups, not realizing the underpowered nature of these searches, and miss effects that are really there. Therefore, know in advance whether you wish to perform subgroup analysis and choose your sample size accordingly.

4.5.2 Subgroup Analysis of Bank Data

We conclude this section with a subgroup analysis of (almost) real data. The "Bank Marketing Dataset" contains 45 211 observations on nine variables for customers who were sent a loan offer: age (in years), job, marital status, education, whether or not the customer had defaulted previously, the amount of money (in dollars) in the customer's bank account, whether or not the customer has a housing loan outstanding, whether or not the customer has a personal loan outstanding, and finally, the variable y indicating whether he responded positively to the loan offer. These data are in the file `BankData.csv`. We will use the covariates to define subgroups and perform subgroup analysis.

Let us first determine whether the version affects the acceptance of the loan offer. The contingency table for this A/B test is given in Table 4.2.

Table 4.2 Contingency table for aggregate A/B test.

	No	Yes	Total
A	200 70	2 612	22 682
B	198 52	2 677	22 529
Total	392 922	5 289	

The 95% CI for the two-sample test of proportions $(\pi_A - \pi_B)$ is $(-0.0096, 0.0023)$ with a p-value of 0.2308, showing no difference between the two versions. The current version of R applies a "Yates correction" by default. If this is not desired, then the option "correct=FALSE" should be added to the command.

Next let us consider whether the version of the loan offer might have an effect on any of the subgroups. Subject to the problem of multiple hypotheses discussed above (if you test many hypotheses you're bound to get a type I error), these will be causal results. First let us consider whether having a personal loan matters. There are 7244 observations for which the variable "loan" takes on the value "yes." The contingency table for these observations is given in Table 4.3.

The 95% CI for $(\pi_A - \pi_B)$ in the subgroup "loan = yes" using R is $(0.0008, 0.0243)$ with a p-value of 0.036. It appears that version does make a difference for customers who have a personal loan and version A is better than version B.

Let us next consider whether having an amount in excess of 1500 makes a difference. There are 10 895 observations with an amount exceeding 1500. The contingency table is given in Table 4.4.

Using R, the 95% CI for $(\pi_A - \pi_B)$ in the subgroup "amount > 1500" using R is $(-0.031, -0.002)$ with a p-value of 0.0212. It appears that version does make a difference for customers who have an amount greater than 1500 and that version B is better than version A.

Caveat: Above we checked the subgroup for personal loans and found an effect. Suppose that you checked the subgroup for housing loans and found no effect.

Table 4.3 Contingency table for subgroup "loan = yes" A/B test.

	No	**Yes**	**Total**
A	3360	265	3625
B	3400	219	3619
Total	6760	484	7244

Table 4.4 Contingency table for subgroup "amount > 1500" A/B test.

	No	**Yes**	**Total**
A	4557	835	5392
B	4560	943	5503
Total	9117	1778	

Could you then assert that there is a difference between personal loans and housing loans? No! These are two individual tests, and they cannot be combined after the fact to make a comparison between the two subgroups. This is such a common mistake that Gelman co-authored an article about it (Gelman and Stern, 2006), to which we refer the reader. To make a comparison between subgroups, it is necessary to consider both groups at the same time in a formal model.

Exercises

4.5.1 Reproduce the box plots in Figure 4.8.

4.5.2 Reproduce the summary statistics in Table 4.1.

4.5.3 Whoever was in charge of the subgroup analysis erred by not doing a power calculation in advance. What are the minimum detectable effects for the two pairs of subgroups (i.e. Internet and email) corresponding to the CIs $(-7.60, 2.72)$ and $(-26.41, -4.26)$? (Choose α and β.)

4.5.4 Suppose that before we conducted the study, we knew we'd do a subgroup analysis, with one subgroup for email and another for Internet. We expected to get 50 observations of each version for email and 275 of each version for Internet. Based on experience, we knew the standard deviation of each subgroup to be 30. What is power if we wish to detect a difference of \$5? \$10? \$15?

4.5.5 Reproduce the confidence intervals for the Internet/email subgroups example, $(-7.60, 2.72)$ and $(-26.41, -4.26)$.

4.5.6 Suppose you want to have an overall α of about 15%. How many subgroup tests can you perform? (Choose $\alpha = 5\%$ for an individual test.)

4.5.7 The file `ttvariables.csv` contains 1000 observations on each of 23 variables. Y is a response variable corresponding to an experiment on X where 0 denotes control and 1 denotes treatment.

(a) The treatment had no effect; perform a test to verify this.

(b) The variables Z1 through Z10 are binary covariates. See if any of these subgroups has a significant treatment effect. For example, for Z1, do a two-sample t-test on the subgroup for $Z1 = 0$ and another test on the subgroup for $Z1 = 1$.

(c) The variables Z11 through Z20 are continuous variables. See if any of these subgroups has a significant treatment effect. Change the continuous variable into a categorical variable with categories "high" and "low" or "high," "medium," and "low." (Answers will vary depending on where the cutoffs between high, medium, and low are made. This may be difficult and time consuming. The work should be divided up among a few persons. If you're doing this by yourself, don't feel the need to do them all.) Here is some code to get you started:

```
attach(df) # to access variables by name with-
out "df$"
hist(Z11,prob=TRUE) # make histogram
# put a density plot on top to aid interpreta-
tion
lines(density(Z11))
# cuts will be labelled A B C, etc.
# the cuts are arbitrary but informed by the his-
togram
A <- (Z11 < 0.2)
B <- (Z11 > 0.2)&(Z11 < 0.8)
C <- (Z11 > 0.8)
t.test(Y[A]~X[A]) # no effect
t.test(Y[B]~X[B]) # no effect
t.test(Y[C]~X[C]) # no effect
```

(d) How many hypotheses have you tested?
(e) How many significant treatment effects have you found in the subgroups?
(f) If there are no significant treatment effects in the population, what is the expected number of "significant" results in this sample?
(g) How many of your significant results are real treatment effects as opposed to false positives? Which ones are they?

4.6 Simpson's Paradox

Simpson's paradox refers to a phenomenon whereby the results of an analysis of aggregated data are the reverse of the subgroup results. We have seen that there can be no effect in the aggregate, but that there still can be an effect in a subgroup. The reverse can happen, when there appears to be an effect in the aggregate but there is no effect in the subgroups. Even more striking, the subgroup effect need not be null (i.e. no effect), but can be the opposite of the aggregate effect. A practical example of Simpson's paradox and proportions follows. In the sequel, to minimize distractions, we will not bother with formal statistical tests – suffice it to say that the effects of Simpson's paradox can be statistically significant.

4.6.1 Sex Discrimination at UC Berkeley

In the fall of 1973, an observational study on possible gender bias was conducted at the University of California, Berkeley. In that year, there were 12 763 applicants for graduate admission; Table 4.5 is a two-way table that gives the data according to the variables outcome (admitted or denied) and sex (male or female).

Most persons unfamiliar with statistics and Simpson's paradox might conclude that this constitutes an "obvious" bias against women with respect to admission into graduate school at Berkeley. The school conducted an investigation into this supposed bias, and the results for 6 of the 101 departments are presented in Table 4.6.

As can be seen, women have a higher admission rate in four of the six departments. (It is mathematically trivial to alter the data slightly so that women have a higher rate in each department while having a lower "overall" rate.) The resolution of the paradox is that majors A and B are also easy to get in – about two-thirds of the applicants (men or women) get accepted. So although men and women have the same acceptance rate, 10 times as many men are accepted into programs A and B because 10 times as many men applied to easy majors. Majors C–F are more popular with the women – 1346 men applied vs. 1702 women. Majors C–F are hard to get in – about one-fourth of the applicants (men or women) get accepted. What

Table 4.5 Admission to Berkeley graduate school by sex.

Sex	Admitted	Denied	Adm+Den	Adm/(Adm+Den)
Male	3738	4704	8442	3738/8442 = 0.443
Female	1494	2827	4321	1494/4321 = 0.346

Table 4.6 Credit default rates for men and women.

	Male			Female		
Dept	**Appl.**	**Accep.**	**Perc.**	**Appl.**	**Accep.**	**Perc.**
A	825	512	0.62	108	89	0.82
B	560	353	0.63	25	17	0.68
C	325	120	0.37	593	202	0.34
D	417	138	0.33	375	202	0.54
E	191	53	0.28	393	94	0.24
F	373	22	0.06	341	24	0.07
Sum	2691	1198	0.45	1835	628	0.34

the data show is that many men and a few women applied to departments that are easy to get into, while few men and many women applied to departments that are hard to get into. The "overall" acceptance rate is misleading.

To see this more clearly, look at how the overall averages are calculated for men. In particular, see what weights have to be applied to each percentage to obtain the overall percentage:

$$\frac{825}{2691}0.62 + \frac{560}{2691}0.63 + \frac{325}{2691}0.37 + \frac{417}{2691}0.33 + \frac{191}{2691}0.28 + \frac{373}{2691}0.06 = 0.45$$

$$0.30 \cdot 0.62 + 0.21 \cdot 0.63 + 0.12 \cdot 0.37 + 0.15 \cdot 0.33 + 0.07 \cdot 0.28 + 0.14 \cdot 0.06 = 0.45$$

It is easy to see that the two departments that are easy to get into (the first two departments with acceptance rates near two-thirds) account for over 50% of the weight. Meanwhile, applying the same maths for women,

$$\frac{108}{1835}0.82 + \frac{25}{1835}0.68 + \frac{593}{1835}0.34 + \frac{375}{1835}0.54 + \frac{393}{1835}0.24 + \frac{341}{1835}0.07 = 0.34$$

$$(0.06)0.82 + (0.01)0.68 + (0.32)0.34 + (0.20)0.54 + (0.21)0.24 + (0.19)0.07 = 0.34$$

it is easy to see that the two departments that are easy to get into account for only 7% of the weight. This shows that Simpson's paradox is a weighting problem in the calculation of an aggregate.

Obviously, a necessary condition for Simpson's paradox is that the subgroups have differing sample sizes. Another necessary condition is the existence of a lurking variable that is related to both the treatment (sex) and response (admissions). In this case, the lurking variable is department. The salient point is that if we had the same sample size in each subgroup, the paradox would not occur.

When subgroups are created by a characteristic, there is the possibility that the characteristic is influenced by the treatment. If assignment to a subgroup is influenced by the treatment, it may be impossible to disentangle the effects of treatments and characteristics. Therefore, it is important to determine whether or not the treatment could influence the characteristic; usually this requires deep subject matter expertise. An example follows.

4.6.2 Do You Want Kidney Stone Treatment A or Treatment B?

Table 4.7 shows the results of two treatments for kidney stones and their success rates. These data are not experimental, but they are observational, collected by examining old records of kidney stone treatments. Looking at the marginal results in the far right column, we see that Treatment A was tried 350 times and achieved success 273 times for a rate of 0.78. It appears that Treatment B is the better treatment, since its success rate is higher, 0.83. However, after treatment, the stones were measured and categorized as small or large, thus

Table 4.7 Kidney stone success rates.

	Small stones	Large stones	Combined
Treatment A	81/87 = 0.93	192/263 = 0.73	273/350 = 0.78
Treatment B	234/270 = 0.87	55/80 = 0.69	289/350 = 0.83

allowing a subgroup analysis. The subgroup analysis shows that Treatment A is better than Treatment B for small stones and that Treatment A is also better than treatment B for large stones. Yet, as mentioned already, Treatment B is better than Treatment A overall.

You are the patient. An ultrasound indicates the size of the stones that plague you – they are small. Which treatment should you pick? Is this a simple case of Simpson's paradox, and should we trust the subgroup analysis rather than the overall result?

Is the category (small stones or large stones) influenced by the choice of treatment? Large stones are more painful than small stones, so it is fair to say that large stones are more severe than small stones. Suppose Treatment A really is better, but it's also more expensive, so it's typically given to more severe cases. Treatment B is less expensive, so it's typically given to less severe cases. What we are witnessing in the aggregate is *not* the superiority of Treatment B but, rather, Simpson's paradox. The aggregate result for Treatment A is dominated by the large stones cell, pulling its aggregate down toward 73%, while the Treatment B aggregate is dominated by the small stones cell, lifting its aggregate up toward 87%. This last sentence is important. When explaining your results to a manager who might know nothing of statistics, it is not sufficient to say, "Oh, this is just Simpson's paradox." You must be able to *explain* the result. By contrast, if the treatments had been randomized across the severity of the cases and if the sample sizes had been balanced, then the overall result would agree with the subgroup results. When considering whether to believe the aggregate results or the subgroup results, a thought experiment can help clarify the issue: if you have small stones or if you have large stones, would you want Treatment A or Treatment B?

Simpson's paradox can occur any time a treatment variable can be broken into subgroups by some third variable, so it is important to be able to recognize it. Another example may help solidify the concept. A medical study examined the relationship between 20-year survival rates and smoking. For 1314 women aged 45–74, 32% of the nonsmokers died, whereas only 23% of the smokers died. Does smoking really improve survival? When the survival rates are calculated for age ranges (e.g. 18–24, 25–34, etc.), the nonsmokers have a better survival rate in each category. The aggregate result simply reflects the fact that the smokers tended to

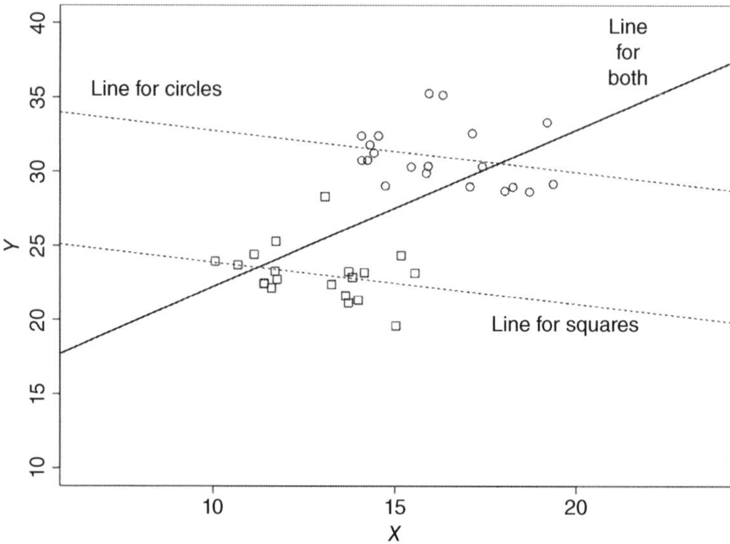

Figure 4.9 An illustration of Simpson's paradox in regression.

be younger and the nonsmokers tended to be older; the aggregate result reflects the distribution of the ages, not the effects of smoking.

Simpson's paradox can afflict more complicated statistical analyses such as regression. A graphical regression example is depicted in Figure 4.9. The regression line through the sample of squares has a negative slope, the regression line through the sample of circles has a negative slope, but the regression line through all the points has a positive slope.

4.6.3 When the Subgroup Is Misleading

Some authors assert incorrectly that the subgroup result is always correct and the aggregate result is always wrong. Simpson's original paper gives a counterexample (Simpson, 1951, p. 241) where, after a comparison of dirty and clean cards, he writes: "The investigator deduced a positive association between redness and plainness both among the dirty cards and the among the clean, yet it is the combined [aggregated] table that provides what we would call the sensible answer, namely, that there is no such association." Based on our reading of numerous examples of Simpson's paradox, we believe it is usually but not always the case that the subgroups are more informative than the aggregates; four counterexamples follow. Subject matter expertise and careful thought are necessary to decide whether the subgroups or the aggregates provide a more useful characterization of the data.

Example 4.3 In baseball, when comparing two players, it is possible for one player to have a higher batting average in each of two successive years, but for the other player to have a higher average over the two years. This is most likely to occur when one player has few at-bats during one of the years, as shown in Table 4.8.

When one asks, who is the better hitter over the two-year period, the answer is not David Justice, even though he had a higher batting average each year.

Example 4.4 Consider a legislative bill that passed both houses of Congress. The voting results are presented in Table 4.9.

Looking at the North and South groups, one might conclude that a higher percentage of Democrats voted in favor of the act, but this conclusion would be wrong, as the aggregate is clearly the correct result.

Example 4.5 Suppose that two editorial trainees are both on a two-week probationary period in contemplation of being hired permanently. Bart and Lisa edit articles for two weeks, and count is kept of how many articles each successfully edits, as shown in Table 4.10. The first week Lisa successfully edits zero of two articles.

While Bart had a higher proportion each week, who is the better at editing articles? The following thought experiment can clarify the issue: If Lisa had edited all 20 articles or Bart had edited all 20 articles, how many would each have successfully edited?

Table 4.8 Simpson's paradox in batting averages.

	Derek Jeter	David Justice
1995	$12/48 = 0.250$	$104/411 = 0.253$
1996	$183/582 = 0.314$	$45/140 = 0.321$
Overall	$195/630 = 0.310$	$149/551 = 0.270$

Table 4.9 Simpson's paradox in voting.

	Democrats voting yes	Republicans voting yes
Aggregate	$152/248 = 61\%$	$138/172 = 80\%$
Northern states	$145/154 = 94\%$	$138/162 = 85\%$
Southern states	$7/94 = 7\%$	$0/10 = 0\%$

Table 4.10 Simpson's paradox in editing.

	Week 1	Week 2	Overall
Lisa	0/2	6/8	6/10
Bart	2/8	2/2	4/10

Example 4.6 A new treatment (A) for septic shock is hypothesized to be better than the existing treatment (B) for emergency room visits. Two thousand patients are randomly assigned one of the two treatments. As the data in Table 4.11 show, 86% survive with treatment A, but only 70% survive with treatment B. However, after the experiment was over, a subgroup analysis was performed, separating the patients according to whether their diastolic blood pressure (DPB), taken 24 hours after admission, was over or under 50mmHg; treatments A and B are tied for efficacy in each of the subgroups. Which is it? Is treatment A superior or are the two treatments the same?

Observe that only one-tenth of the treatment A patients had their blood pressure drop precipitously, while half of the treatment B patients had such a drop. DPB was not measured before the treatment, but after, which strongly suggests that treatment B is the cause of the drop in blood pressure. If a subgroup is defined on the basis of a variable that could be correlated with the response (because it is measured *after* the experiment), the subgroup results must be interpreted cautiously. Subject matter expertise suggests that the aggregate result is the sensible interpretation of the data.

Table 4.11 Simpson's paradox: septic shock example.

	Alive	Dead	% alive
Combined			
Treatment A	860	140	86
Treatment B	700	300	70
DPB$<$ 50			
Treatment A	50	50	50
Treatment B	250	250	50
DPB$\geq$ 50			
Treatment A	810	90	90
Treatment B	450	50	90

Exercises

4.6.1 The conversion rate of a webpage is the proportion of visitors who take action on a webpage, i.e. the user does not just view the page and leave, he interacts with the page. A website landing page may have many sources, e.g. paid search, unpaid search, paid ads, etc. You have conducted an A/B for two versions of a landing page, and you wish to determine which version has more conversions. Your programmer misinterpreted your instructions, and rather than show different versions to each successive customer, he ran version A on one day and version B the next day. Each page got 14 000 visits. Version A has 830 conversions, while version B has 970 conversions. The difference is statistically significant. Your boss tells you to implement version B.
Before you can do this, your boss (who is computer literate if not statistically literate) calls you back to his office. He has checked the source of the referrals to landing page. He shows you the following:

	Version A			Version B		
Source	**Visits**	**Conversions**	**Rate**	**Visits**	**Conversions**	**Rate**
Paid search	10 000	500	0.05	2 000	80	0.04
Free search	2 000	130	0.065	2 000	90	0.045
Paid ads	2 000	200	0.10	10 000	800	0.08
Overall	14 000	830	0.059	14 000	970	0.069

"So version A does better on each source, but does worse overall? What is the meaning of this? Should we implement version A or version B?" How do you answer your boss?

4.7 Test and Roll

In Exercise 2.1.10 we found that the usual approach to calculating sample sizes can sometimes produce sample sizes that are too large to be useful. What can be done when the necessary sample size is far too large?

Traditional statistical testing is designed to avoid type I errors. In the context of A/B testing, traditional statistical testing is designed to avoid reaching the mistaken conclusion that A and B are different when in fact they are the same. For some situations, and online advertising is such a situation, there is no harm in committing a type I error. If I run an A/B test and mistakenly conclude that A is better than B when, in fact, they are the same, then I will run ad A instead of ad B. Would I have made more money if I had run ad B instead? No! In this situation minimizing the probability of making a costless error is not necessarily a good criterion.

An alternative course of action is to focus instead on a specific decision problem. Feit and Berman (2019) focus on the "test and roll" problem, where the data scientist plans to test treatments on a fraction of the population and then deploy (or "roll out") the better treatment on the remainder of the population. Let the population be N. In the test stage, n_1 customers are exposed to treatment 1, and n_2 different customers are exposed to treatment 2, where $n_1 + n_2 < N$, the population size. After the test, the winning treatment is rolled out to the remaining $N - n_1 - n_2$ customers. The goal is to maximize profit over both stages. The maximization strategy by which the sample size formula is derived explicitly incorporates the trade-off between learning in the test phase and earning in the roll phase. Sometimes N is known precisely, as in marketing from a mailing list of N clients. In other cases, N can be derived from a marketing budget: If each ad costs X, how many ads can we show? In still other cases, N can be the expected traffic at a website over some period of time: we are developing a campaign for the summer months, during which we expect our website will get N visitors.

The profit-maximizing sample size for each sample is

$$n_1 = n_2 = n^* = \sqrt{\frac{N}{4}\left(\frac{s}{\sigma}\right)^2 + \left(\frac{3}{4}\left(\frac{s}{\sigma}\right)^2\right)^2} - \frac{3}{4}\left(\frac{s}{\sigma}\right)^2 \tag{4.9}$$

We already have discussed N and how it might be determined. We have two standard deviation symbols in this formula, s and σ! We must distinguish between them.

In Section 2.4 we introduced the beta-binomial model and its Bayesian approach that requires a prior distribution for the parameter of interest. Here again we appeal to the Bayesian methodology. Our parameters of interest are the means of the responses for the two treatments, m_1 and m_2. We will give them identical priors, so we really have only one prior distribution to worry about, that for the common prior mean m, $m = m_1 = m_2 \sim N(\mu, \sigma^2)$. These values μ and σ are not parameters of the model *per se*, but we do need to know these parameters to implement the model, so they are called *hyper-parameters*. These hyper-parameters represent our beliefs about how the two treatments perform. We can deduce them from previous experiments, but doing so is beyond the scope of this book. If we had data on one hundred different websites, we could estimate the mean and variance of the average sales across all the websites; this might be done using hierarchical modeling, for example.

By way of preface, suppose we are estimating the click rate, which is a proportion. If we need the standard deviation of this click rate, an easy way to do this is to recognize that the mean click rate ($\hat{\mu}$) is a proportion, and so it has a standard deviation equal to

$$s = \sqrt{\hat{\mu}(1 - \hat{\mu})} \tag{4.10}$$

If the mean click rate of the prior data is $\hat{\mu} = 0.55$, then $s = 0.4975$.

Suppose an analyst did some hierarchical modeling and computed $\mu = 0.68$ and $\sigma^2 = 0.03^2 = 0.0009$. This represents our best beliefs about the current mean and variance of the click rate. We now have μ and σ^2 for our prior distribution. Taking this distribution to be normal, we can write $N(0.68, 0.03^2)$ for the prior. We can use $\hat{\mu} = 0.68$ to estimate s as done above, so $s = 0.4665$. Let $N = 100\,000$. When we plug everything into Equation 4.9, we get $n^* = 2283.998$, which we round up to 2284. Remember, always round up for sample size calculations, so if we had computed 2283.1, we still would have rounded up to 2284.

A very nice feature of this formula is that $n^* \leq \sqrt{N}(s/s\sigma)$, and hence, the profit-maximizing sample size is always less than the population size.

Rather than use a calculator to apply Equation 4.9, there is a website that will do the calculations for you: www.testandroll.com.

Exercises

4.7.1 How many visitors can be expected at a website over a three-month period? Choose a website and estimate this number. There are various websites that will provide this information; one such is www.similarweb.com.

4.7.2 For the above problem in the text ("Suppose an analyst…"), calculate the sample size using classical methods. Assume you want to find a lift of 2%; the current mean is 0.68, and $0.02 \times 0.68 = 0.0136 = d$, the size of the difference you want to detect. Use the traditional values of $\alpha = 0.05$ and $\beta = 0.8$. Use $\hat{\mu}$ to estimate the standard deviation, as done in the text.

4.7.3 The total market basket at a particular type of website averages about 10 dollars, but an updated estimate is needed. An analyst used a hierarchical model to estimate $\mu = N(10.36, 4.40^2)$ for a class of similar websites. For this particular website, $s = 103.77$. The advertising budget will allow for $N = 1\,000\,000$ impressions. What is the profit-maximizing sample size $n^* = n_1 = n_2$?

4.7.4 For the above exercise, suppose you wanted to detect a difference a lift of 2%, i.e. $d = 0.20$? Assume typical values for α and β. What sample size do you need?

4.8 Chapter Exercises

4.1 In Section 4.3 we asserted that actually conducting an experiment for the car rental problem was not possible. Why not?

4.2 For a two-sample t-test, to detect an effect that equals 1/2 of the standard deviation with $\alpha = 0.05$ and $\beta = 0.80$, how many observations do you need? Remember, this depends on whether the test is paired or not. In particular, fill in the below table. What type of relationship can you establish between the number for a paired test and the number for an unpaired test?

Effect size	Paired	Unpaired
1 std dev		
1/2 std dev		
1/3 std dev		
1/4 std dev		

4.9 Learning More

We know how to decide whether version A or version B should be shown to a general audience. What if we want to target individuals? There is a method for targeting individuals that goes by many names: "uplift modeling," "differential response analysis," "incremental value modeling," "true lift models," or "heterogeneous treatment effects." It's definitely applicable in experimental settings, but it's slightly too advanced for this book. The technically inclined reader may find it worthwhile to learn this method.

Section 4.1 "Audio/Video Reprise"

• Our assertion that the direction of the alternative for a hypothesis test depends on the action to be taken as a result of the test will strike many as heresy, but it is not. Many academics have shed blood over this question, but they are academics, and they do not believe that a practical decision will be made on the basis of a single hypothesis test. Part of this flows from a justified academic belief that questions cannot be settled by a single experiment (Pillemer, 1991). Another part flows from a belief that the analyst is always somewhat interested in rejections from the other side. This sentiment is expressed by Dallal (2012): "What damns one-tailed tests in the eyes of most statisticians is the demand that all differences in the unexpected direction – large and small – be treated as simply nonsignificant. I have never seen a situation where researchers were willing to do this in practice."

Academic researchers might never be willing to do this, but decision makers in business do so regularly. As examples 4.1 and 4.2 of Section 4.1.1 make clear, the idea that the decision to be taken determines the direction of the alternative is very tenable, and not at all outlandish.

Section 4.2 "Typing Test"

• For a two-sample test of means, independent samples, the variance of the difference is given by

$$\frac{\sigma_1^2}{n_1} + \frac{\sigma_2^2}{n_2}$$

whereas the variance of the paired difference is given by

$$\frac{\sigma_1^2}{n} + \frac{\sigma_2^2}{n} - \frac{2\rho\sigma_1\sigma_2}{n}$$

• A more detailed analysis of the economics of the boys' shoes exericse can be found in Easterling (2015) (pp. 48-49).

Section 4.3 "A/B/n Tests"

• A/B/n tests are discussed in Kohavi et al. (2009b), with many examples given.

• Our recommendations for testing the homogeneity of variance in ANOVA comes from Vorapongsathorn et al. (2004). Our advice on verifying the normality assumption for ANOVA may conflict with what you learned in your first statistics class. Typically, such a course teaches students to either test all the residuals at once *or* test them for each level. Both approaches are incorrect: sample size and homoscedasticity must be taken into account; see Kutner et al. (2005, p. 781).

There actually is a fourth standard assumption of ANOVA, additivity, but verifying this assumption is beyond the level of this book, so we just assume it.

• Yes, you were asked to make a box plot of the CarRental data and also to make a box plot of the residuals, and the box plots were practically the same (except the residual box plots have mean zero). There really was no need to plot the residuals by type of car, but we had you do it anyhow because it's always good practice to plot the residuals. More generally, it's good practice to plot everything. To paraphrase John Tukey, who invented the concept of "exploratory data analysis," a graph is usually the best way to find what you weren't looking for. So make lots of graphs.

Section 4.4 "Minimum Detectable Effect"

• Some researchers neglect to do a power analysis before conducting an experiment, fail to reject the null, and then wonder if their test had enough power.

To answer this question, they take the sample mean, variance, etc. and compute power. This is called a "post hoc power analysis" or "observed power," and it is a Bad Idea™ – see Goodman and Berlin (1994), Lenth (2007), Senn (2002), and Hoenig and Heisey (2001). As mentioned earlier, there is a one-to-one correspondence between "observed power" and the *p*-value. If the *p*-value is small, then observed power will be high, and if the *p*-value is large, then observed power will be low. So there is no point in computing observed power because the *p*-value already conveys the same information.

Another approach to analyzing a design is given by Gelman and Carlin (2014), based on Type S (sign) and Type M (magnitude) errors. It's a bit too technical for this book, but if you're technically inclined we highly recommend it. Essentially, if your experiment doesn't have enough power, it is very easy to get the sign wrong and/or only be able to discover really large effects – you'll miss the most likely sized effects. They give an example where a statistically significant result has a 0.40 probability of being the wrong sign and any significant result overestimates the magnitude of the true effect by a factor of 25.

The minimum detectable effect was coined by Bloom (1995), who gives an extended treatment of the topic. Hoenig and Heisey (2001) point out that the minimum detectable effect is flawed if it is taken as the truth. We have pointed out that the MDE cannot be taken as the truth, and we advocate it only as a last-ditch effort when whoever designed the experiment was in too much of a hurry and did not perform proper power calculations. Skipping over much math, a quick way to think about why MDE is bad, but post hoc power is worse is because the former depends only on one noisy estimate (the standard error) while the latter depends on two noisy estimates (the coefficient and the standard error).

Nonetheless, there are many who ignore the solid maths of statistical theory and advocate the calculation of post hoc power. For example, Bababekov et al. (2018) published an article in the top-ranked surgery journal, advocating the calculating of post hoc power. Gelman (2019) and many others published rejoinders in that journal, but apparently to no effect. Most of the original authors doubled down on their advocacy and published a follow-up article (Bababekov et al. (2019)), which reported on the following "research": the authors searched several journals for medical studies on humans and discarded all the findings that were statistically significant. For the remaining studies that were not statistically significant, they computed observed power and found low observed power. If you think about it for a moment, the result of this "research" could have been predicted with absolute certainty before it was even begun. What is surprising is that four of the insignificant studies did achieve high power. This, of course, means they made a mistake in their calculations! This mistake would have been obvious to anyone who knows anything about post hoc power. Althouse (2020) points out this error and laments that "[S]tatisticians have been playing Whack-A-Mole with this particular myth for several years, and apparently must continue to do so." So someone may ask you to compute post hoc power and even point to seemingly prestigious journals in which post hoc power is advocated and used. Don't fall for it.

Section 4.5 "Subgroup Analysis"

• The results of two individual subgroup tests cannot be compared to each other. Suppose there is a significant effect in subgroup A, but there is no effect in subgroup B. An analyst may then be tempted to claim that there is a difference between subgroup A and subgroup B, but to do so is incorrect. A similar mistake often is made in multiple regression analysis. Suppose a multiple regression has four independent variables, X_1, X_2, X_3, and X_4. Let the coefficient on X_2 be significant and the coefficient on X_4 be insignificant. Even though many analysts make this mistake, it is absolutely incorrect to say that there is a difference between the coefficients of variables X_2 and X_4. This concept is well explained in the article by Gelman and Stern (2006). The article by Sleight (2000) makes the point that subgroup analyses are "fun to look at – but don't believe them" due to a "combination of reduced statistical power, increased variance and the play of chance," and gives several examples where subgroup analyses were accepted blindly and turned out to be wrong.

• An ignorant journal editor can demand subgroup analyses even when the authors know the editor is wrong. The most famous example of authors complying with such a demand is the study that showed daily aspirin helps patients recover from heart attacks (ISIS-2, 1988). The editors insisted that a subgroup analysis be done; the authors refused because the results wouldn't be valid; the editors threatened to not publish the article without a subgroup analysis. The authors, tongue-in-cheek, performed one on astrological signs, showing that aspirin was ineffective for patients born under Libra or Gemini. Peto, one of the authors of the original article, tells the story in Peto (2005). Another study found that Leos were more likely to suffer from a gastrointestinal hemorrhage than other signs, while Saggitarians were more likely to break a leg (Austin et al., 2006).

• The problem of testing multiple hypotheses on the same data set has many names – multiple comparisons, snooping, fishing, dredging, and p-hacking – and the result is a spuriously significant result, i.e. a type I error. In particular, when p-hacking occurs, one thinks that the probability of a type I error is some small number, e.g. 0.05, but in reality it is much higher. This problem can arise when applying the same test to many different data sets or applying many tests to one data set. As an example of the former, consider flipping a coin 10 times and concluding that it is biased if it comes up heads 10 times. If the coin is fair, this will happen only with probability 0.0107. Suppose we test 100 fair coins. The probability that each does not reject the null hypothesis is $0.9893^{100} = 0.341$. So if we use this test with $\alpha = 0.0107$ on each coin, the overall α for the cumulative test on 100 coins really equals 0.34. There is one-third chance that we'll find a coin that is "unfair" according to a statistical test (even though we know it's really fair).

The garden of forking paths is not the same thing as p-hacking. In particular, the garden of forking paths tells us why multiple comparisons can be a problem even when the hypothesis is specified in advance and there is no "fishing expedition." For an explanation, see http://www.stat.columbia.edu/~gelman/research/unpublished/p_hacking.pdf.

• An advanced form of "slice and dice" comes from the literature on "estimating heterogeneous treatment effects." One strand of this research employs the data mining technique called "random forests" to construct groups of similar observations and provide confidence intervals for hypothesis testing. These types of "forests" are called "causal forests." It will be some time before this method is available in software, but you should keep an eye out for it, as it will be a very powerful method. A key article in this literature is Wager and Athey (2018).

• The data set `ttvariables.csv` and the associated exercises are based on the chapter by McCullough (2017).

Section 4.6 "Simpson's Paradox"

• Section 6 of Kohavi et al. (2009a) discusses how Simpson's paradox can crop up in A/B testing on the web and ruin an experiment.

• The UCI Machine Learning Repository (https://archive.ics.uci.edu/ml/index.html) has the "Bank Marketing Dataset." To this data set we have appended a randomly generated variable "version" that takes on the values A and B with equal probability, as if the customer had been shown one of two versions of the loan offer.

• The original Berkeley graduate admissions article is Bickel et al. (1975). The kidney stones example is from Julious and Mullee (1994), and the septic shock example is from Norton and Divine (2015). For further examples of Simpson's paradox in the context of business analytics, see the article by Berman et al. (2012). The smokers example is from Appleton et al. (1996).

• Understanding and interpreting differences between groups when data are aggregated can be a difficult task. Simpson's paradox, Kelley's paradox, and Lord's paradox are the three primary paradoxes in this area, though Simpson's is by far the most common. Wainer and Brown (2007) examine all three paradoxes on a common data set.

Section 4.7 "Test & Roll"

• The actual setup of the test and roll strategy is analogous to a two-armed bandit problem. Usually bandit problems are multiarmed, and the original problem was like this. Consider a row of slot machines that accept coins as payment. In the old days you would put a coin in a slot and pull a lever (arm) to make it go; nowadays you insert a money card and push a button. Each machine has a different payoff rate, e.g. 91%, 80%, and 83%. You want to find the machine that has the highest payout rate so that you can sink most of your money into it. You have to allocate some portion of your coins to each machine to estimate the various payoff rates. Once you find the machine with what you believe is the highest payout rate, you spend the rest of your money on that machine.

How should you go about figuring out which machine has the highest payout rate? One way is to put some small proportion of the total coins into each machine in equal proportion and get preliminary payoff rates. Then take another small proportion of the total coins remaining, but not distribute them in equal proportion. Instead, give more coins to machines with higher payoff rates and fewer coins to machines with lower payoff rates. We give more coins to machines with higher payoff rates because that's where we need precision. We don't need a good estimate of the low payout rates, so we don't waste data (coins) on them. Repeat this process until you achieve statistical significance for one machine. How, precisely, to determine statistical significance in this situation is very complicated, and we won't get into it, but you get the general idea.

You may wonder why we compute sample sizes for test and roll using `power.t.test` instead of `power.prop.test`. After all, isn't the number of clicks in N impressions really a binomial? Yes, it is. However, when N is large, then the binomial can be approximated very well by a normal distribution that has the same mean and variance as the binomial distribution being approximated. The maths for the proofs of the test and roll strategy are all based on the normal distribution (being an approximation to the binomial) because the normal distribution is much easier to work with when executing these types of proofs. And these proofs weren't easy to being with: look at the appendix of Feit and Berman (2019) if you're interested.

4.10 Appendix on One-Sided CIs, Tests, and Sample Sizes

Suppose we are only interested in detecting whether μ is negative; we do not care whether it is zero or positive. Then the null and alternative hypotheses are

$$\begin{aligned} H_0 &: \mu \geq 0 \\ H_A &: \mu < 0 \end{aligned} \tag{4.11}$$

where H_0 asserts that μ is zero or positive and H_A asserts that μ is negative. When first learning about one-sided tests, it can be difficult to write the null and alternative hypotheses correctly. A good way to approach this is to write the alternative first. What are you looking for evidence of? That is the alternative. After writing the alternative, the null is very easy to write. In the present case, we are looking for evidence that μ is negative, so we write that first. Then the null all but writes itself. Since there is only one tail, we make the tail size α, not $\alpha/2$ as would be the case for a two-tailed test. Which tail gets the α? The alternative tells us. If $\bar{x}$ falls in the tail, then H_0 is rejected; if $\bar{x}$ is very negative, then we disbelieve the null. So α goes in the left tail, on the negative side of the null distribution as in Figure 4.10. It is called the "null distribution" because it assumes that the null hypothesis is true. It is very important to put α in the proper tail, or else the answer will be wrong even if all the calculations are otherwise correct.

Statistics textbooks often say that a mere expectation is sufficient reason to write a one-sided hypothesis: if you expect the result to be negative, then use a one-sided hypothesis. This is incorrect. As pointed out in the text (see Section 4.1.1), only use a one-sided test if a "significant" result in the unexpected direction leads to the same decision as a finding of no effect.

Just as a two-sided CI can be used to test a two-sided hypothesis, so a one-sided CI can be used to test a one-sided hypothesis. In both cases, if the value of the parameter under the null hypothesis, e.g. $\mu_0 = 0$, falls in the interval, do not reject the null; if it falls outside the interval, reject the null in favor of the alternative. The difficulty with using one-sided intervals to test one-sided hypotheses is that there are two one-sided intervals and it is not always obvious which is the correct one to use.

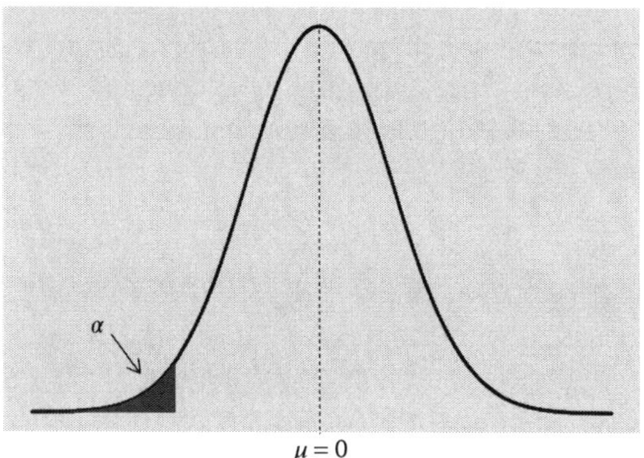

Figure 4.10 Null distribution for a one-sided, one-sample test.

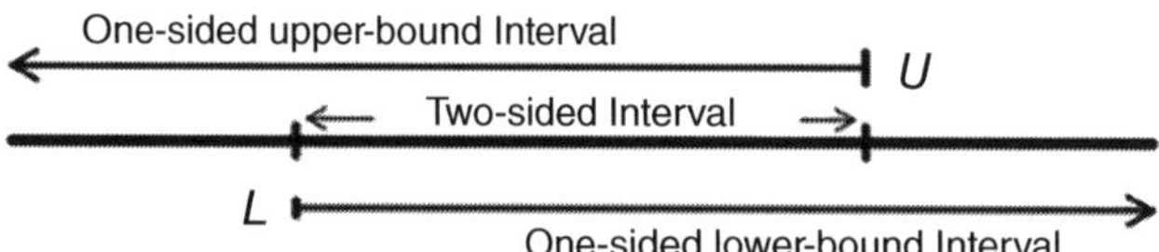

Figure 4.11 A two-sided interval is the intersection of two one-sided intervals.

Recall that a two-sided interval (L, U) is really just the intersection of two one-sided intervals (see Figure 4.11). These two intervals are a one-sided upper-bound interval $(-\infty, U)$, sometimes also called a "lower interval" because it resides on the left side of the real line and has a tail on the left, and a one-sided lower-bound interval (L, ∞), sometimes called an "upper interval" because it resides on the right side of the real line and has a tail on the right. A two-sided 95% interval that has 2.5% in each tail is just the intersection of two one-sided intervals, each of which has 2.5% in each tail. Therefore, if we want a 95% one-sided CI, all we have to do is get a 90% two-sided CI and break it up into a pair of one-sided intervals.

A one-sided interval is used to bound the range of likely values for the population parameter. The generic forms of the one-sided interval are given below:

Some call it	Description	Interval	Others call it
Upper-bound interval	Parameter no larger than U	$(-\infty, U)$	Lower interval
Lower-bound interval	Parameter at least as large as L	(L, ∞)	Upper interval

where, for example, in the case of one-sample with σ unknown, $L = \bar{x} - t_\alpha s/\sqrt{n}$ and $U = \bar{x} + t_\alpha s/\sqrt{n}$. Some texts refer to an upper-bound interval as a "lower interval," but we find this imprecise: Does it mean that the bound is on the lower side, or that the bound extends infinitely in the lower direction? It can be hard to remember.

The question arises, when do we use which of the above intervals?

For picking a one-sided CI, simply try to phrase the problem in one of the following two formats:

1. We are 95% confident that the population parameter is at least $L \Rightarrow (L, \infty)$.
2. We are 95% confident that the population parameter is no larger than $U \Rightarrow (-\infty, U)$.

The "population parameter" could be a mean, or a proportion, or a difference between means or proportions.

Using the one-sided interval to test an hypothesis is a little more complicated than the two-sided case. Consider

$$H_0 : \mu \leq \mu_0 \tag{4.12}$$

$$H_A : \mu > \mu_0 \tag{4.13}$$

To find evidence in favor of the alternative that will make us reject the null, it is important to remember that the CI gives us likely values of μ. If we want to reject the null, then we want the likely values of μ (the CI) to be *above* μ_0. To accomplish this, we want to put a lower bound, L, on μ. Hence, we want an interval of the form (L, ∞). Then, if $\mu_0 \leq L$, we will reject the null hypothesis.

Similarly, for the converse case

$$H_0 : \mu \geq \mu_0 \tag{4.14}$$

$$H_A : \mu < \mu_0 \tag{4.15}$$

to find evidence in favor of the alternative, we want to put an upper bound, U, on μ. To make us believe that the alternative is true, we want likely values of μ to be much less than μ_0; we want the CI to be below μ_0. Hence, we want an interval of the form $(-\infty, U)$ so that if $\mu_0 \geq U$, then null is rejected in favor of the alternative.

Let us consider four examples from Verhoeven and Wakeling (2009) to make the ideas more concrete.

Example 4.7 Liquid chlorine used in swimming pools has a shelf life of 90 days (2160 hours). A company sells an additive that is claimed to extend the life of the liquid chlorine. A statistical test doesn't tell us how long the shelf life is extended; we need a CI for this. We are only going to buy the additive if it makes the chlorine last longer. We do not care if the test statistic winds up in the opposite tail; if the additive shortens the life makes no difference to us. (It may make a difference to the company that sells the additive–the company would *not* want to conduct a one-sided test to determine whether the additive works.)

An additive to extend the life of the chlorine was added to nine jugs of chlorine. The shelf life of these jugs was 2159, 2170, 2180, 2160, 2167, 2171, 2181, and 2195. Testing $H_0 : \mu \leq 2160; H_A : \mu > 2160$ yields $t = 3.98$, p-value < 0.005. We want to find a 90% one-sided interval to estimate how long is the life of the chlorine with the additive.

We want to say we are 90% confident that the life is at least x. Therefore we want an interval of the form (L, ∞), and all we have to do is calculate $L = 2172.4 - 1.397(9.382/\sqrt{9}) = 2168$ so the one-sided 90% CI is $(2168, \infty)$. We are 90% confident that the chlorine with additive lasts at least 2168 hours (about 90 and 1/3 days). Using the 90% CI to test the null hypothesis, we see that $2160 < 2168$ so the null is rejected.

Example 4.8 A pizza parlor wants to change its delivery process to reduce delivery time from the current 25 minutes. The parlor is only going to adopt the new delivery process if it reduces the mean delivery time. If the process increases the delivery time or the delivery time remains the same, the decision will be the same: the pizza parlor will adopt the new process only if it reduces the delivery time. Therefore a one-sided test is in order.

A sample of 36 orders has a mean time of 22.4 minutes and a standard deviation of 6 minutes. Testing $H_0 : \mu \geq 25; H_A : \mu < 25$ yields $t = -2.60$ with a p-value < 0.01, which is strong evidence that the mean time is reduced, but it doesn't tell the pizza parlor by how much the mean time is reduced.

We want to say that the mean time is no more than x; we want an interval of the form $(-\infty, U)$. The one-sided 95% CI is $(-\infty, 24.1)$, but since time can't be negative, we rewrite this as $(0, 24.1)$. With 95% confidence the new mean waiting time is no more than $U = 24.1$ minutes. Using the CI to test the null hypothesis, $25 > 24.1$, the null-hypothesized value does not fall in the CI, so the null is rejected.

One-sided intervals for proportions are handled in a directly analogous fashion, as a pair of examples will make clear. For testing $H_0 : p \leq p_0; H_A : p > p_0$, we want a lower-bound interval $(L, 1)$. For testing $H_0 : p \geq p_0; H_A : p < p_0$ we want an upper-bound interval $(0, U)$.

Example 4.9 An insurance company pays 85.1% of its claims on time. The company has developed a new process designed to increase this percentage. The process will only be adopted if it really increases the percentage; if the test statistic falls in the wrong tail (it really decreases the percentage) does not matter. A one-sided CI is called for. A sample of 200 claims has 180 paid on time. Testing $H_0 : p \leq 0.851; H_A : p > 0.851$ yields $z = 1.95$, p-value ≈ 0.025.

We want to be 95% confident that true on time payments *is at least* some amount, so we calculate L:

$$L = 0.90 - 1.645\sqrt{\frac{0.9 \cdot 0.1}{200}} = 0.865$$

The one-sided 95% CI is $(0.865, 1)$. We are 95% confident that the new system pays at least 86.5% of its claims on time. An increase from 85.1 to 86.1% is appreciable for a high-volume company. To test the null hypothesis, observe that $p_0 < L$ (the null-hypothesized value does not fall in the CI) so the null is rejected.

Example 4.10 A company makes 32% of its deliveries late. The company wants to test a new process to reduce this number. The new process will only be adopted if the new process really reduces the number of late deliveries, so a one-sided test is appropriate. A statistical test finds that 22 of 118 deliveries are late. Testing $H_0 : p \geq 0.32; H_A : p < 0.32$ yields $z = 3.12$, p-val ≈ 0.001.

Wondering how small this new proportion is, we would like to be 95% confident that the new proportion *is no larger than* some proportion. Accordingly, we calculate U,

$$U = 0.186 + 1.645\sqrt{\frac{0.186 \cdot 0.814}{118}} = 0.24$$

so the 95% one-sided CI is $(0, 0.24)$. We think that the new system will reduce late deliveries by at least 8 percentage points from 0.32 to 0.24. This is substantial.

The above examples make very clear that an hypothesis test alone does not provide sufficient information to make a practical business decision and that CIs are better than hypothesis tests.

Moreover, considering all these examples at once allows us to deduce an easy way to remember how to write one-sided intervals. As in the case of choosing a tail for a one-sided test, it is the alternative that informs the decision. Consider Table 4.12.

Whichever side of the test alternative is unbounded gets the unbounded limit of the CI (or the zero or one of the CI in the case of proportions). The alternative hypothesis $\mu > 216$ implies only a lower limit for μ, so there is no upper limit, so the interval will be of the form (L, ∞). Similarly for the other cases. The situation is depicted graphically in Figure 4.12, where the null is rejected because μ_0 does *not* fall in the CI. If the null were not to be rejected, then L would be on the left side of the distribution with the arrow extending all the way to the right, and μ_0 would then fall in the CI.

Sample Sizes

The usual sample size formulae are for two-sided tests and will be excessively conservative (require too many observations) when applied to a one-sided test. Let

Table 4.12 One-sided alternatives and intervals.

	Alternative	Interval
Example I	$\mu > 216$	$(2168, \infty)$
Example II	$\mu < 25$	$(-\infty, 24.1)$
Example III	$p > 0.851$	$(0.8561, 1)$
Example IV	$p < 0.32$	$(0, 0.24)$

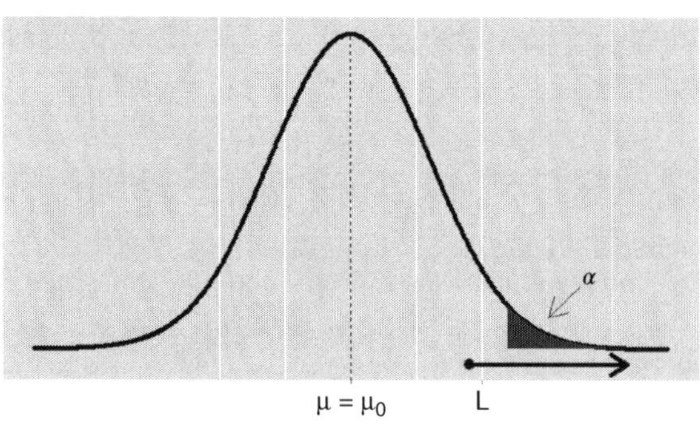

Figure 4.12 Lower bound interval (L, ∞) where $L = \bar{x} - z_\alpha s/\sqrt{n}$.

us consider a one-sided upper-tail test without loss of generality. The subscript n on σ reminds us that the standard error depends on the sample size and H_A is the alternative hypothesis:

$$\begin{aligned}
\text{Power} &= P(X \geq \mu_0 + z_{1-\alpha}\sigma_n | H_A) \\
&= 1 - P(X \leq \mu_0 + z_{1-\alpha}\sigma_n | H_A) \\
&= 1 - P\left(\frac{X-\mu}{\sigma_n} \leq \frac{\mu_0 + z_{1-\alpha}\sigma_n - \mu}{\sigma_n} | H_A\right) \\
&= 1 - N\left(\frac{\mu_0 - \mu}{\sigma_n} + z_{1-\alpha}\right) \\
&= N\left(\frac{\mu - \mu_0}{\sigma_n} - z_{1-\alpha}\right)
\end{aligned}$$

Now, for some probability of a type II error, β, power is $1 - \beta$. Substituting this into the left-hand side of the above and then applying the inverse normal transformation to both sides and then solving for $1/\sigma_n$ on the left-hand side yields

$$\begin{aligned}
1 - \beta &= N\left(\frac{\mu - \mu_0}{\sigma_n} - z_{1-\alpha}\right) \\
z_{1-\beta} &= \frac{\mu - \mu_0}{\sigma_n} - z_{1-\alpha} \\
\frac{1}{\sigma_n} &= \frac{z_{1-\beta} + z_{1-\alpha}}{\mu - \mu_0}
\end{aligned}$$

The above general formula for $1/\sigma_n$ allows us to solve for the sample size for both normal and binomial problems. For the normal problem, simply substitute $\sigma_n = \sigma/\sqrt{n}$ and solve for n. For the binomial, use the replacement $\sigma_n = \sqrt{\pi(1-\pi)/n}$ and replace μ with π. Then the sample size formulae are

$$n = \left(\sigma \frac{z_{1-\beta} + z_{1-\alpha}}{\mu - \mu_0}\right)^2 \tag{4.16}$$

and

$$n = \pi(1-\pi)\left(\frac{z_{1-\beta} + z_{1-\alpha}}{\pi - \pi_0}\right)^2 \tag{4.17}$$

5

Designing Tests with Small Samples

So far we have had large samples, which makes experimentation relatively easy. There are many reasons, though, that a large sample might not be available. An experiment might be very complex and only able to be repeated a few times. Observations might be very expensive, and the budget is limited. Some experiments are time consuming, and to collect a large sample would take more time than can be spared.

Small samples pose serious problems for the statistician, and detecting small differences in small samples is particularly difficult. Ziliak (2008) discusses a case from the Guinness Brewery in 1898, in which the brewery's first scientist, Thomas Case, was analyzing the percentage of resins in hops. One sample of size $n = 11$ had 8.1% soft resin content, while another sample of size $n = 13$ had 8.4%. This difference of 0.3% was nontrivial from a business perspective, and the pressing question was this: Is the difference a real effect or is it just random noise? The quality of the brew turned on the answer to this question. Yet, the question could not be answered. Because the standard deviation was not known and had to be estimated, the small sample size meant that the normal table could not be used. Shortly after this, Case hired William Sealy Gosset, who as a result invented "Student's" t-distribution to handle situations like this.

By the end of this chapter, readers should:

- Know how to block an experimental design.
- Implement a Latin square design.
- Understand that careful matching can lead to a dramatic reduction in residual variance, thus increasing the signal to noise ratio and making it easier to detect effects.
- Compare rerandomization, propensity score, and optimal matching.

Business Experiments with R, First Edition. B. D. McCullough.

Companion Website: www.wiley.com/go/mccullough/businessexperimentswithr

5.1 Case: Call Center Scripts (ANOVA)

A call center for a large car rental firm handles customer requests and complaints and uses scripts to help its employees respond to common issues. As the call center is constantly striving to improve efficiency, it often creates new scripts. Three new scripts have been written, and the call center wants to know which one allows the agents to resolve the problem fastest. The current problem of interest is flat tires.

When a customer calls in to report a flat tire, the agent has to make sure the customer is out of harm's way, collect enough information to dispatch a truck, and attend to many other minor details. Of course, the agent may well be dealing with irate or distressed customers. We wish to employ the script that accomplishes all the goals in the least amount of time. The response measure is "number of seconds per flat tire call." Naturally, we will choose the script that has the smallest average time per flat tire call.

Incoming calls are randomly assigned to agents. When an agent gets a flat tire call, he enters this information into his computer terminal, and one of the four flat tire scripts is randomly assigned to him. At the end of the day, there have been 24 flat tire calls (which, we admit, is a convenient number, as will be seen later). The data are in file `CallCenter.csv`. We will use the completely randomized design (CRD) introduced in Section 4.3 in the context of A/B/n tests. First, of course, we plot the data as shown in Figure 5.1. It is not obvious that there is a difference between the scripts, though it appears that script A could be better than the others.

The median of script A is lower than the other medians, but our sample size is small, and there is much overlap between the box plots. It is not clear that any one script is better than the others. Our ambivalence concerning a significant effect is confirmed when we analyze the data using traditional one-way ANOVA and get the below ANOVA table.

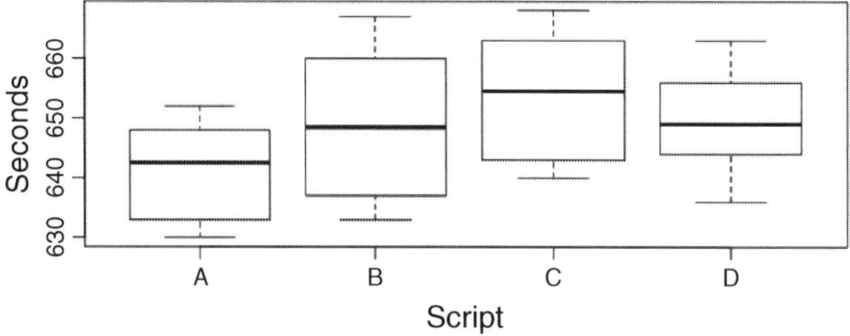

Figure 5.1 Boxplots of call center data.

Table 5.1 One-way ANOVA on call center scripts.

	df	SS	MS	*F*	*p*-Value
Script	3	486.17	162.06	1.39	0.275
Residuals	25	2331.67	116.58		

Try it!

Reproduce the results in Table 5.1.

There is no point in looking at the Tukey HSD intervals; they will all include the origin. Apparently, no one script is better than any of the others. But only apparently. In point of fact there are differences between the scripts, but we can't see them because our sample size is too small. It's too expensive to increase the sample size because agent time is valuable. And it's really not necessary. The existing sample will suffice if only we know how to analyze it properly. Our goal is to reduce the residual error (the unexplained variation) and to decrease the noise so that we can observe the signal. This is particularly important in small sample situations.

5.1.1 Blocking

A blocking variable is simply a categorical variable that explains some of the variation in the response variable. The blocking variable is not part of the experiment, and the experimenter may or may not have control over the blocking variable. The blocking variable separates the sample into blocks. A block is a group of observations that are relatively homogeneous compared to the entire sample. Since the observations in a block are homogeneous, their variance will be lower than the entire sample. The treatments are applied within each block. This gives cleaner estimates of the treatment effects, because the variation from block to block does not contaminate the estimates of the treatment effects. Some of the hitherto unexplained variance (residual error) is now explained by the differences between the blocks. Recall how a paired t-test can reduce the variance compared with a simple t-test. What pairing is to the t-test, blocking is to ANOVA. It makes the groups more homogeneous and thus reduces variances so that treatment effects can be more easily uncovered.

It turns out we didn't really have a CRD, because we didn't use a different agent for each call. We reused agents. Only six agents received the test scripts; each agent (conveniently) got exactly four calls, and (again, conveniently) each agent got each script at exactly the same time. A crosstab of the variables agent and script counts the number of cases in each cell, as shown in Table 5.2. In passing, we remark

Table 5.2 Crosstab of counts for scripts A, B, C, and D and agents 1, 2, 3, 4, 5, and 6.

	1	2	3	4	5	6
A	1	1	1	1	1	1
B	1	1	1	1	1	1
C	1	1	1	1	1	1
D	1	1	1	1	1	1

that this is a *balanced design* because it has the same number of observations in each cell. By contrast, if we didn't have the same number of observations in each cell, the design would be *unbalanced*. A crosstab is a convenient way to check for balance. Balanced designs are much easier to analyze than unbalanced designs, so we have a strong preference that our designs be balanced whenever possible.

We happen to know that some agents are better than others in the sense that they usually complete calls more quickly than other agents. The response variable is affected by the variable agent. We can use this variable as a blocking variable to reduce unexplained variation.

As always, we do not proceed immediately to calculation: First, we display the data. This was easy in two-sample and CRD analyses. Now, with blocking, it will be more difficult. It may take some time to find a display that reveals the information, but, as one of the pioneers of experimental design wrote long ago (Shewart, 1939, p. 88), "Original data should be displayed in a way that will preserve the evidence in the original data."

Honestly, we didn't do a good job looking at the data in Figure 5.1. We have three relevant variables, and we only looked at two of them. There are many ways to incorporate a third variable into a plot: some will be illuminating, and others will not; we will explore this idea in Exercise 5.1.2. One way is a traditional dot

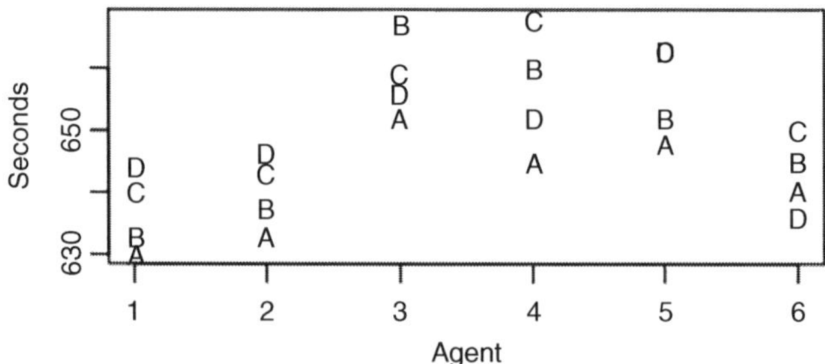

Figure 5.2 Plot of call center data.

plot for two of the variables and then use the third variable to label each point. See Figure 5.2 for an example.

Looking at Figure 5.2, we can see clearly that script A is the best. It has the lowest cumulative time for agents 1 through 5 and second lowest for 6. Only by considering agent can we see that script really matters; this is true for both the graphical and statistical analyses. Agents are not all equally fast – seconds is affected by agent. Therefore, we must use agent as a blocking variable.

Software Details

Reproduce Figure 5.2...

Technically, a scatterplot is for two continuous variables. We only have one continuous variable. But we will make use of the fact that the categorical variable agent is represented and use the scatterplot command to make a dot plot.

```
# agent is a factor, must convert it to numeric for
plotting
df <- read.csv("CallCenter.csv",header=TRUE)
head(df) # check to make sure the data are read
correctly
plot(as.numeric(df$agent),df$seconds,type="p",pch="")
text(as.numeric(df$agent),df$seconds,df$script)
```

This experimental setup, blocking with ANOVA, is called the randomized complete block design (RCBD). By *complete* we mean the all treatments are observed in each block. The hallmark of the RCBD is that each block gets each treatment exactly once. It works because the observations in each block are more homogeneous than randomly selected observations; the variance of the observations in a block is less than the variance of an equal number of randomly selected observations. This reduction in variance makes it easier to see the treatment effect.

We further assume there are no interactions between the factors, nor is there an interaction between the treatments and the blocks. As you may recall from your previous studies of regression analysis, an interaction occurs when the effect of one variable, such as the script, depends on another variable, such as the agent. In this case, an interaction would mean that some agents do better with one script and other agents with another. We will cover interactions more in Chapter 6, but for now we just note that the analysis here assumes no interactions.

Now we turn to the formal statistical analysis including blocking, and we get the ANOVA table shown in Table 5.3 where agent is the blocking variable.

Table 5.3 One-way ANOVA with blocking on call center data.

	df	SS	MS	*F*	*p*-Value
Script	3	486.2	162.1	5.46	0.0097
Agent	5	1886.8	377.4	12.73	0.0001
Residuals	20	444.8	29.7		

In contrast to the previous ANOVA for which script has a *p*-value of 0.275, we now see that script is significant. Comparing the two ANOVA tables, we see that the mean squared error of the residuals (the noise) in the model drops from 116.6 to 29.7, a 75% reduction. One of the assumptions of the CRD is that the observations are relatively homogeneous, but this assumption is violated, so the CRD doesn't work. Observe also that agent is significant. We expected this, and, indeed, we want the blocking variable to be significant; otherwise we wouldn't be blocking with it in the first place.

Let's look at the Tukey HSD intervals in Table 5.4. At the 12% level (i.e. $\alpha = 0.12$), script A is significantly faster than either B, C, or D. We also have the graphical evidence from Figure 5.2. Even if we set $\alpha = 0.05$ so B-A and D-A were not significant, we'd still have to pick a script, and it's clear that A would be the best choice. This was not at all obvious when the data were analyzed with plain ANOVA. Notice that we only produced Tukey intervals for scripts, but we could have produced them for agents, too, if we were interested in knowing which agents were fastest and which were slowest. This is explored in Exercise 5.1.6.

Blocking is a method of controlling for *nuisance factors*. A nuisance factor has some effect on the response, but is not of direct interest to the researcher. The nuisance factor has an effect on the response, including the fact that the nuisance factor in the model reduces the error, thus better enabling the researcher to observe the effects of the variables of interest. Below we give two examples of blocking.

Table 5.4 Tukey HSD intervals for call center data with blocking.

Scripts	Difference	Lower	Upper	*p*-Value
B-A	7.7	−1.4	16.7	0.112
C-A	12.5	3.4	21.6	0.006
D-A	8.2	−0.9	17.2	0.085
C-B	4.8	−4.20	13.9	0.441
D-B	0.5	−8.6	9.6	0.999
D-C	−4.3	−13.4	4.7	0.531

Software Details

To perform ANOVA with blocking (and get the Tukey intervals also)...
Convert agent from numeric to factor before proceeding:

```
df$agent <- as.factor(df$agent)
summary(aov(seconds~script+agent,data=df))
TukeyHSD(aov(seconds~script+agent,data=df),
which="script")
```

Example 5.1 A real estate company that employs appraisers suspects that different appraisers result in markedly different valuations to the same property. To test this, the manager randomly draws five properties and sends each appraiser out to appraise all five. A completely randomized design might not be appropriate in this case, because the properties are not identical. One might have more square footage, another might be in an undesirable neighborhood, etc. There is variation in the valuation not only due to the appraiser but also due to the property itself. The analyst should assign the properties numbers and then use this as a blocking variable.

Example 5.2 A pharmaceutical company wishes to compare three methods for assaying the purity of a drug. Twelve doses of the drug are available for testing. A technician can only perform three analyses per day. There are two options for conducting the test: (i) the study can be performed by four analysts in one day, or (ii) one analyst can complete the study in four days. In the first option block on analyst, in the second option block on day.

There are two primary disadvantages of the RCBD. First, when the number of treatments is large, the blocks become too large. Second, if there is an interaction between block and treatment effects, the error can increase instead of decrease.

Exercises

5.1.1 For both of the two examples given at the end of Section 5.1.1, identify the treatment, the response, and the blocking variable.

5.1.2 Reproduce Figure 5.2, except have script on the X-axis and label the points by agent. Does this graph show that script matters? If not, what does this graph show?

5.1.3 Reproduce the results in Tables 5.3 and 5.4.

5.1.4 Perform model checking on the ANOVA in Tables 5.1 and 5.3. See the beginning of Section 4.3 for the steps in model checking ANOVA.

5.1.5 Look again at Figure 5.2. For agent 5, there is no script C. Why not?

5.1.6 For the call center data, obtain the Tukey HSD intervals for `agent` and determine which agents are faster and which are slower. Why must we take these intervals with a grain of salt?

5.1.7 Obtain the Tukey HSD intervals for Table 5.1 and verify that they all include the origin.

5.1.8 Use the data in `yxz.csv`. These data were generated according to the model $Y = a + bX + cZ + \epsilon$ where $a = 10$, $b = 1$, $c = 2$, and ϵ is normally distributed. X and Z are uncorrelated, so regressing Y on X alone should give a good estimate of b. Do this. What is the standard error of the estimate (standard deviation of the residuals)? Plot fitted values of Y on the y-axis and actual values of Y on the x-axis. On the plot put a line with slope $= 1$ and y-intercept $= 0$. Now repeat this exercise, this time regressing Y on X and Z. Compare the two sets of results. How does the reduction in variance manifest itself?

5.1.9 Is it even worth running the script experiment? Comparing the best script with the worst script, how much time is saved every year? (Find an estimate of how many flat tires occur in the United States each year. Find an estimate of how many cars are in the United States. Estimate the probability that a car gets a flat in the course of a year. Choose an auto rental firm, e.g. Hertz or Enterprise. Find out how many cars it has. Assume half the cars are driven each day. Estimate how many flat tire calls to the service center each day.)

5.2 Case: Facebook Geo-Testing (Latin Square Design)

"DigiPuppets" is a start-up company making finger puppets that work with mobile apps, as shown in Figure 5.3. DigiPuppets needed a cost-effective marketing strategy to show results to potential investors in order to get funding.

The business question facing the company was this: Which advertising themes would be best at attracting consumer attention for DigiPuppets using paid

Figure 5.3 DigiPuppet on a child's finger. Source: DigiPuppets.

Table 5.5 Initial experimental design for DigiPuppets.

Treatment	Story	Education	Tech	Lessons	Cute	Shared time
City	Philly	Orlando	KC	Houston	Denver	LA

advertising on Facebook? They chose six themes, each having different images and different text: Cute, Education, Lessons, Shared Time, Story Time, and Tech. The target audience was Facebook users who were parents of preschoolers. The response measure was clickthroughs on a button that says, "Learn More." The original design focused on six medium-sized US cities, as shown in Table 5.5.

What is wrong with this design? If the Story treatment has a significant effect, you can't tell whether the effect is because the Story treatment is better than the other themes or because people in Philadelphia just happen to like DigiPuppets more than people in the other cities. We say that the theme and city variables are *confounded* or *aliased*; we cannot separate the theme effect from the city effect. We will have much more to say about aliasing in Section 7.7.4.

A solution would be to assign several cities to each treatment as in a CRD and run the experiment for several days, but that would have broken the company's budget. They only had enough money to run ads in a few cities for a few days. What to do?

The solution was to use the Latin square design. DigiPuppets implemented the experimental design presented in Table 5.6.

Inspection of Table 5.6 shows that theme and city are not confounded. With 6 themes and 6 days, the experiment is run a total of 36 times. Observe, however,

Table 5.6 Latin square design for DigiPuppets.

	Philly	**Orlando**	**KC**	**Houston**	**Denver**	**LA**
Tuesday	Story	Education	Tech	Lessons	Cute	Shared
Wednesday	Education	Story	Shared	Cute	Tech	Lessons
Thursday	Tech	Shared	Education	Story	Lessons	Cute
Friday	Lessons	Tech	Cute	Education	Shared	Story
Saturday	Cute	Lessons	Story	Shared	Education	Tech
Sunday	Shared	Cute	Lessons	Tech	Story	Education

that not all combinations are tested; therefore, this is an *incomplete design*. For example, Philly and Story on a Thursday is not tested. In fact, most of the combinations are not tested, but this is part of the trade-off when deciding to use a Latin square.

A major virtue of the Latin square design is its efficiency. Much smaller sample sizes lead to lower costs, faster data acquisition, etc. Suppose we had three factors (e.g. one treatment and two blocking variables), each with four levels; then there are $4^3 = 64$ combinations. Conceptually, running 64 different experiments is not difficult, but as a practical matter it might be impossible, especially if each experiment is expensive to run or difficult to set up. How might one investigate these three factors without running 64 experiments? The answer is a Latin square design, in which only $4^2 = 16$ experiments are necessary, a reduction of nearly 80%.

First, though, we have to determine the response variable. The data file `DigiPuppets.csv` contains two possible dependent variables, clickthroughs and impressions. The same amount of money was spent on each city, but Facebook charges slightly different prices (depending on supply and demand), so the number of impressions in each city differs. To measure consumer response, it is probably best to use the clickthrough rate (CTR), which is defined as 100*clickthroughs/impressions, which is a percentage. To add the variable ctr to the existing data frame, use the command `df$ctr <- 100*df$clickthroughs/df$impressions`

Analyzing a Latin square design via ANOVA is straightforward. There is one dependent variable, in this case ctr; two blocking factors that are usually listed first, city and day; and one treatment factor that is usually placed after the blocking factors, theme. As can be seen in Table 5.7, both the day of the week and the city had significant effects at the 10% level (i.e. $\alpha = 0.10$). The p-value on theme is small, too. While it is not less than 0.05, we have to take a decision on the question: Are the themes all equally effective? The evidence suggests it is more likely there is a difference than not. The next step is to examine the pairwise differences to

Table 5.7 ANOVA on Latin square for DigiPuppets.

	df	SS	MS	*F*	*p*-Value
City	5	0.6624	0.132	2.196	0.0952
Day	5	0.9287	0.186	3.080	0.0320
Theme	5	0.7831	0.157	2.597	0.0576
Residuals	20	1.2063	0.060		

determine which theme is most effective. None of the pairwise differences is significant; you are asked to confirm this result in Exercise 5.2.4. This is one of those unfortunate results that sometimes occurs when working with data. We know that the themes are not all equally effective, but the data don't speak loudly enough for us to hear which theme is the best. In order to figure out which theme is best, we would have to run another experiment.

Try it!

Reproduce Table 5.7. Be sure to perform model checking.

Before moving on we need to discuss the degrees of freedom. One way to think about this is that you have n data degrees of freedom, and these are used to estimate various quantities. For a k-parameter model, the model degrees of freedom are k, and these are used to estimate the parameters of the model. In Table 5.7, these are 5 each for mean squares associated with city, day, and theme. The remaining $n - k$ degrees of freedom are left over – residual – and are used to estimate the variability in the dependent variable, which is the residual mean square. If $n - k$ is small, you're not going to get a good estimate of the unexplained variance. For the Latin square design with one observation per cell, residual $df = (n - 1)(n - 2)$, which equals 20 in Table 5.7; this is more than enough to get a good estimate of the residual mean square. Essentially, it's as if you're trying to compute the variance using 20 observations. For some Latin square designs, if $n = 3$ or $n = 4$, there may not be enough residual degrees of freedom to get a good estimate of the residual mean square, which makes it difficult to estimate effects. If you have only two residual degrees of freedom, it's as if you're trying to estimate the variation of the residuals using only two observations; your estimate probably isn't going to be very good.

5.2.1 More on Latin Square Designs

A Latin square is a mathematical artifact, an $n \times n$ matrix containing n different symbols arranged so that each symbol appears exactly once in each row and once in each column. When the symbols represent factors in an experiment, the Latin square is a very useful experimental design for handling an experiment with exactly two factors, each of which has n levels, and one treatment that also has n levels. Suppose that the four treatments are A, B, C, and D, while the factors are day (Wednesday, Thursday, Friday, Saturday) and time (morning, afternoon, evening, and night). One possible Latin square arrangement is shown in Table 5.8.

Notice this does not test all possible combinations, so it is another example of an incomplete design. For example, it does not test treatment A on Wednesday afternoon, nor does it test D on Thursday morning. In fact, it does not test most of the combinations; it tests only 16 of the 64 ($= 4^3$) possible combinations. If we choose the 16 combinations carefully, we can nevertheless still conduct a good test because the design will be orthogonal. (We will discuss orthogonality in Chapter 6; if you're not familiar with this concept, don't worry.) Orthogonality requires that in each column and in each row, every treatment is used once and only once. When this happens, we can get good (i.e. unbiased) estimates of the effects. The representation above is called the *standard form*: all treatments are listed in natural order (in this case, A, B, C, and D) in the first row and first column, and each subsequent row/column shifts the order by one. It is easy to confirm that the standard form is an orthogonal design. An example of a nonstandard form that is orthogonal is given in Table 5.9.

A poor choice for filling the squares of a Latin square leads to confounding. A very bad choice for a Latin square arrangement leads to perfect confounding, as in Table 5.10.

Table 5.8 Standard form for Latin square design.

	Wednesday	**Thursday**	**Friday**	**Saturday**
Morning	A	B	C	D
Afternoon	B	C	D	A
Evening	C	D	A	B
Night	D	A	B	C

Table 5.9 Another orthogonal Latin square design.

	Wednesday	**Thursday**	**Friday**	**Saturday**
Morning	C	D	A	B
Afternoon	D	A	B	C
Evening	A	B	C	D
Night	B	C	D	A

Table 5.10 A bad choice for a Latin square design.

	Wednesday	Thursday	Friday	Saturday
Morning	A	B	C	D
Afternoon	A	B	C	D
Evening	A	B	C	D
Night	A	B	C	D

Table 5.11 Randomizing a Latin square design: assign random numbers to rows and columns and arrange the rows in order and then the columns.

(a)					(b)					(c)				
	4	2	3	1		4	2	3	1		1	2	3	4
2	A	B	C	D	1	B	C	D	A	1	A	C	D	B
1	B	C	D	A	2	A	B	C	D	2	D	B	C	A
4	C	D	A	B	3	D	A	B	C	3	C	A	B	D
3	D	A	B	C	4	C	D	A	B	4	B	D	A	C

If we use the design in Table 5.10 and if the results of the test show that A is the best treatment, we cannot rule out the possibility what we have seen is not the superiority of A, but that Wednesday is the best day to run the experiment. Moreover, we cannot distinguish between the day of the week and the treatment. In this design, we say that the days and the treatments are perfectly *confounded* or *aliased*. We will discuss confounding more in Chapter 8. Observe also that the above design is *not* orthogonal. While it is true that each treatment is used once and only once in each row, it is not true of the columns. Hence, the design is not orthogonal.

It is, as always, important to randomize the execution of the experiment. You can't just randomize the treatments in a row, because that might not preserve orthogonality. We need to randomize the order in which the treatments are assigned. The usual way to do this for a Latin square design is to generate random numbers 1 through n for the rows and again for the columns, e.g. if $n = 4$, obtain 4, 2, 3, 1 for the rows and 2, 1, 4, 3 for the columns, as shown in Table 5.11. Assign these numbers to the rows and columns as shown below in (a), then reorder the rows as shown in (b), and then reorder the columns as shown in (c).

Using panel (c) to randomize Table 5.8, we would proceed as follows: Wednesday morning will get treatment A, Thursday morning will get treatment C, Friday morning will get treatment D, etc.

Table 5.12 Data for new product example.

Message	City			
	1	2	3	4
I	A 42	B 41	C 45	D 46
II	B 39	C 34	D 48	A 40
III	C 30	D 32	A 36	B 38
IV	D 32	A 30	B 33	C 33

To make these ideas concrete, let us consider a couple more examples. Be sure you can identify the dependent and independent variables.

Example 5.3 A company wishes to test the market for a new product. Two factors are of interest, the type of advertising campaign (A: TV commercial, B: direct mail, C: Facebook, and D: search ads) and the message of the campaign (I: "easy and fast," II: "low cost," III: "guaranteed or your money back," and IV: "free delivery"). Test campaigns were run in four cities (1, 2, 3, and 4). Sales in thousands during the campaign period are presented in Table 5.12.

Example 5.4 Suppose the manager at a factory wonders whether the type of background music played affects the workers' productivity. A different type of music is played each day of the week, and the day's output is recorded. Since there are five days to the workweek, we'll consider five types of music (A, B, C, D, and E). Since this is to be a Latin square, we'll run the experiment for five weeks. At the end of the experiment we have the data given in Table 5.13.

5.2.2 Latin Squares and Degrees of Freedom

When engineers use Latin squares, they frequently use one observation per cell because they have well-defined models with small errors. In this situation, the degrees of freedom for the error (dfE) is $(n-1)(n-2)$ where n is the number of levels per factor. If $n = 2$, then dfE = 0. If $n = 3$, dfE = 2. To have enough degrees of freedom to obtain a decent estimate of the error variance, we may well need $n \geq 4$ so that $(n-1)(n-2) \geq 6$. If increasing the number of levels of the factors is not an option, then replication (more than one observation in each cell) may be necessary. If r is the number of replicates per cell, then dfE $= rn^2 - 3n + 2$, greatly increasing the error degrees of freedom. Next we illustrate the effect of too few residual degrees of freedom and the benefit of replication.

Table 5.13 Latin square for background music exposure.

	Day				
Week	**M**	**T**	**W**	**T**	**F**
1	A	B	C	D	E
	113	119	120	120	125
2	B	C	D	E	A
	116	121	123	136	119
3	C	A	E	B	D
	120	118	122	119	119
4	D	E	A	C	B
	119	112	117	116	140
5	E	D	B	A	C
	112	124	123	122	122

An auto company wonders which of its three models of car, 1, 2, or 3, gets better gas mileage. Gas mileage is affected by the fuel type and the speed at which the car is driven, so the three common fuel types, A, B, and C, and three typical speeds, 25, 50, and 75 miles per hour, are used as blocking variables. For a specified quantity of fuel, each car is run around a racing track until it runs out of gas; total number of miles traveled is the response variable. The data are in `CarData.csv`. Note that car and speed are coded as numeric data but are really factors; so, after reading in the data, these two variables will have to be converted to factors. An easy way to do this is `df$car <- as.factor(df$car)`. To increase the degrees of freedom, each combination is run twice, so the data file has two distance variables, distance1 and distance2.

Let us first pretend that these were two separate experiments and analyze distance1 and distance2 separately. The box plots are presented in Figure 5.4.

There's some overlap, not too much, and it looks like car 1 might get the best mileage. But the ANOVA tables in Table 5.14 say otherwise. Observe that dfE in each table conforms to the formula given above.

When the data are treated as one experiment with replication so that distance1 and distance2 are combined into a single variable distance, the box plot is even less compelling than before, as shown in Figure 5.5; yet, the ANOVA table tells a completely different story as can be seen in Table 5.15. Of course, the box plot does not show all the data; it omits both speed and fuel. This idea is explored in Exercise 5.2.

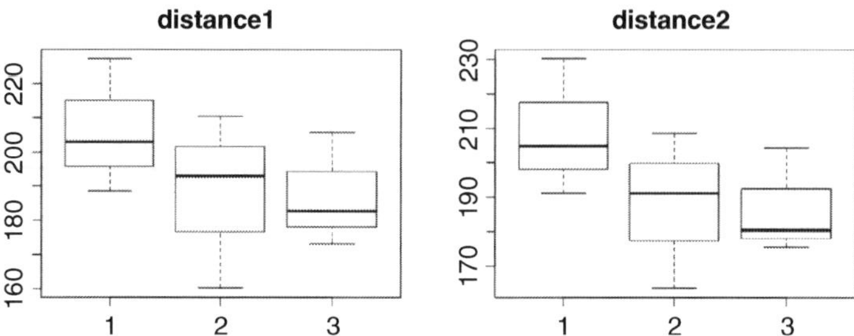

Figure 5.4 Box plots for separate distances.

Table 5.14 ANOVA results for distance1 (left) and distance2 (right).

		distance1				distance2			
	df	**SS**	**MS**	***F***	***p*-Value**	**SS**	**MS**	***F***	***p*-Value**
Speed	2	356.9	178.4	0.3	0.74	325.8	162.9	0.4	0.70
Fuel	2	1257.5	628.7	1.3	0.44	1184.7	592.3	1.5	0.39
Car	2	697.4	348.7	0.7	0.59	926.0	463.0	1.2	0.46
Residuals	2	1002.6	501.3			779.1	386.6		

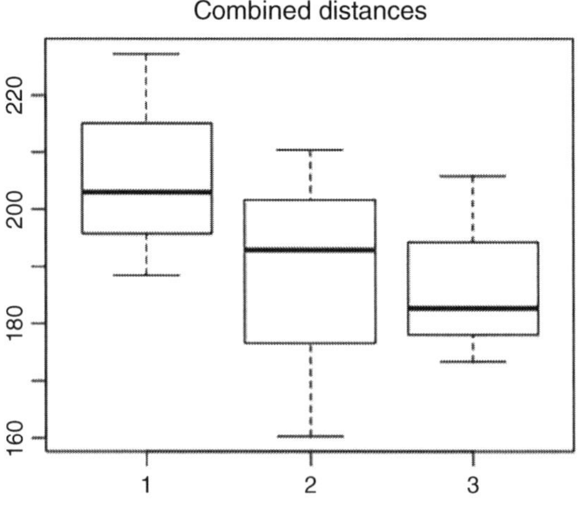

Figure 5.5 Box plots for combined distances.

In the ANOVA table, observe that dfE has increased in accordance with the formula given above. The mean squared error, which estimates the residual variance, has dropped from the 400–500 range (the separate tests) to 163 (the combined test); it has been reduced by more than half!

Software Details

To analyze a Latin square with replications by ANOVA...

The data have to be arranged properly. Suppose there are $n = 3$ factor variables, f1, f2, and f3, and $k = 2$ replicates of the response variable, r1, r2, all arranged in a $m \times 5$ matrix: f1, f2, f3, r1, r2 where each variable has m observations; in the CarData problem $m = 9$. The data will have to be restructured so that the factor variables are repeated completely for each response, as a $2m \times (n + k - 1)$ matrix. For the CarData problem this is 18×4: f1, f2, f3, r1 f1, f2, f3, r2 with only four variable names, the first three of which may remain the same but the fourth must be created anew: f1, f2, f3, r. To be clear, the first column of the new matrix is the first column of the old matrix repeated twice (because there are two replicates), and similarly for columns two and three. The fourth column of the new matrix is the last two columns of the first matrix, one on top of the other.

This may require some data munging.

Once the data are formatted correctly, Latin squares with replicates is just a straightforward ANOVA with blocking.

```
df <- read.csv("CarData.csv") df$speed <- as.factor
(df$speed) df$car <- as.factor(df$car) df1 <- df[,-5]
df2 <- df[,-4] colnames(df1)[4] <- "distance" boxplot
(distance car,data=df1,main="distance1") colnames(df2)
[4] <- "distance" boxplot(distance car,data=df2,main=
"distance2") df3 <- rbind(df1,df2) boxplot(distance car,
data=df3,main="distance") lm1 <- lm(distance~speed+
fuel+car,data=df1) anova(lm1) lm2 <- lm(distance~speed+
fuel+car,data=df2) anova(lm2) lm3 <- lm(distance~speed
+fuel+car,data=df3) anova(lm3)
```

Exercises

5.2.1 Suppose each observation costs \$10 000. How much did the Latin square without replication cost? How much did the Latin square with replication cost? How much would a CRD cost? Suppose all three cars can run

Table 5.15 ANOVA on combined distances.

	df	SS	MS	*F*	*p*-Value
Speed	2	677.95	338.97	2.08	0.1713
Fuel	2	2439.32	1219.66	7.48	0.0088
Car	2	1615.28	807.64	4.95	0.0292
Residuals	11	1793.10	163.01		

on the track at the same time, and a workday is approximately 8 hours (the experiment is not run continuously with 24 hour workdays). Approximately how many days would it take to execute the Latin square with and without replication? How many days to execute the CRD?

5.2.2 Consider the standard form in Table 5.8. Use a six-sided die to randomize the treatments in each row. Is the resulting design orthogonal?

5.2.3 Reproduce Table 5.14. Be sure to check the models.

5.2.4 Compute the Tukey intervals for the DigiPuppets example. Which theme is most effective? Least effective?

5.2.5 Let the random numbers for rows be 1, 3, 4, 2 and let the random numbers for columns be 4, 2, 1, 3. Randomize Table 5.9.

5.2.6 Reproduce Table 5.15. Perform model checking. Also get the Tukey intervals and interpret them. See the beginning of Section 4.3 for the steps in model checking ANOVA.

5.2.7 Analyze the DigiPuppets data with impressions as the response variable. What does this tell us?

5.2.8 Make graphs that show all the data as in the car rental problem. We can't show car, speed, and fuel all in one graph, but we can make a pair of graphs.

5.2.9 Analyze the new product data in Table 5.12.

5.2.10 Analyze the background music data in Table 5.13.

5.3 Dealing with Covariate Imbalance

Large samples make statistical analysis easy, but we do not always have the luxury of a large sample, and small samples have particular difficulties of their own. Suppose a financial firm offers free consultations to select customers as part of a cross-sell program. Each consultation costs $300 in resources: employee time, marketing materials, follow-up efforts, etc. The budget for the project is $12 000, which means forty customers can get the free consultation. Forty customers are randomly selected from the customer list. Twenty are randomly assigned to receive treatment A and 20 to receive treatment B. The experiment shows that customers preferred treatment A. Subsequent analysis of the 40 customers shows that treatment A was given to 14 longtime customers and 6 newer customers, while treatment B was given to 9 longtime customers and 11 newer customers. Question: Did the statistical test show that A is preferred because A is better, or because longtime customers are more likely to choose A than B? Answer: We can't tell.

Imagine that you have to decide whether remodeling will increase sales at a chain of convenience stores. Each remodel costs $100 000 and your budget for the test is $800 000. Sixteen stores are chosen randomly, eight of which are remodeled. The resulting test shows that the remodel generates increased sales. A subsequent analysis shows that of the eight remodeled stores, six are close to competitors. Question: Should the remodel be rolled out to all stores or only to stores that are near competitors? Answer: We don't know.

In both of the above cases, we have encountered what is called *covariate imbalance*. If the covariate has an effect on the response, and the covariate is not (nearly) equally represented in the treatment and the control groups, then the experiment is said to suffer from covariate imbalance. This increases the error variance and makes it harder to detect effects. The requirement of near-equal (but not absolutely equal) representation is applied, because small imbalances don't have much of an effect and equal representation (i.e. perfect covariate balance) will practically never occur in a large samples.

To see that perfect covariate balance will rarely occur, imagine flipping a coin 10 times. The probability of exactly five heads is 0.25. Flipping a coin 200 times, the probability of exactly 100 heads is 0.056. Flipping five coins, the probability of being more than 0.1 away from 0.5 is 1 – 0.25 = 0.75. Meanwhile, the similar probability for 200 flips is 1 – 0.995 = 0.005 because the probability of getting between 80 and 120 heads in 200 flips is 0.995. In large samples, the probability of getting comparable treatment and control groups is quite high. When the sample size is small, you're practically guaranteed to have covariate imbalance. "Large" and "small," however, are relative terms. Consider a binary variable that is present in 30% of a sample of size 50 that must be divided into treatment and control. The probability that the two group proportions will differ by more than 10% equals 0.38. If $n = 100$, then the probability equals 0.27. If $n = 200$, then the probability that they differ by more than 10% equals 0.09. Finally, if $n = 400$, then the probability equals 0.02 (Kernan et al., 1999).

To get a feel for the imbalance problem, we treat the banking data in the file `BankData.csv` as a population (remember that it has $N = 45\ 211$ observations). To keep things simple, we'll focus only on three variables: amount (continuous), housing (binary), and loan (binary). The population mean for amount is 1362.27; the population proportion for housing (=yes) is 0.556; the population proportion for loan (=yes) is 0.160.

We take three samples of size 10 and calculate the sample statistics and do the same for samples of size 30, 100, 250, 1000, and 5000. The results are presented in Table 5.16. We would like the control and treatment samples to be similar and for both of them to resemble the population – yet we often don't know the population, so we settle for having the control and the treatment look alike. Look at the first row. In the control sample, amount has a mean nearly half again as much as in the treatment sample. The second $n = 100$ row has the same trait, and the proportion for loan differs by 0.10. The proportions for the binary variables settle down quickly, but even for the third $n = 250$, there is a difference of 0.09 for loan.

Should we be unfortunate enough to get a random sample with severe covariate imbalance, it can adversely affect our results. This is *not* a failure of the random number generator. In fact, imbalance is going to happen if the number of covariates is large enough. Suppose we were to conduct a two-sample test of means to make sure the control and treatment means are approximately the same. Even if the random number generator worked flawlessly, we'd still get a type I error 5% of the time (if $\alpha = 0.05$). With 10 covariates, the probability of having at least one type I error is $1 - 0.95^{10} = 0.40$. Some people advocate testing for covariate imbalance and then trying to correct for it. This is a mistake; see Pemantle et al. (2019). A severe covariate imbalance on a prognostic variable cannot be corrected after the fact, because that's multiple testing.

It might seem that blocking would be useful, but it's really of limited use in this situation, and there are two problems. First, blocking as a strategy for inducing covariate balance requires knowing in advance that covariate might cause the imbalance. Second, you can't block on every covariate or even many covariates (every time you introduce a new blocking variable, you lose some degrees of freedom, and eventually you wind up with too few degrees of freedom). The way to avoid covariate imbalance, then, is to take preventive action and use a matched design.

In sum, if the sample size is large, a two-sample test of means between control and treatment will reject the null for some variables if the number of variables is

Table 5.16 Control and treatment means for various sample sizes.

		Amount	Housing	Loan		
$N = 45\ 211$	Mean	1362.27	0.160	0.556		
	St dev	3044.8	0.367	0.497		
		Control			Treatment	
	Amount	Housing	Loan	Amount	Housing	Loan
$n = 10$	1189.80	0.10	0.5	733.60	0.00	0.50
$n = 10$	1160.4	0.10	0.40	1003.5	0.30	0.60
$n = 10$	1228.5	0.10	0.40	2031.2	0.10	0.50
$n = 30$	1564.07	0.20	0.63	1421.50	0.23	0.53
$n = 30$	1038.67	0.10	0.67	1872.03	0.13	0.63
$n = 30$	1300.00	0.07	0.60	1025.13	0.13	0.53
$n = 100$	1412.17	0.19	0.59	1280.83	0.17	0.50
$n = 100$	1914.46	0.18	0.53	1277.54	0.18	0.43
$n = 100$	1467.31	0.17	0.48	1112.07	0.21	0.54
$n = 250$	1345.16	0.19	0.60	1251.45	0.15	0.55
$n = 250$	1169.79	0.13	0.57	1392.13	0.13	0.52
$n = 250$	1279.03	0.18	0.63	1328.76	0.16	0.54
$n = 1000$	1468.63	0.14	0.54	1305.66	0.15	0.56
$n = 1000$	1470.60	0.16	0.57	1627.38	0.16	0.54
$n = 1000$	1278.24	0.16	0.56	1465.15	0.15	0.56
$n = 5000$	1336.87	0.16	0.55	1459.60	0.16	0.56
$n = 5000$	1411.47	0.16	0.55	1369.40	0.17	0.56
$n = 5000$	1372.71	0.16	0.55	1392.45	0.17	0.55

large. This is just life – type I errors cannot be eliminated; you have to live with them. You can expect that these differences, while statistically significant, will not be practically significant. When the sample size is small, then the probability of a large difference that is practically significant is large, and preventative measures such as matching should be undertaken in advance.

5.3.1 Matching

We have previously seen the ability of matching to reduce variance in Section 4.2, and matching will do the same here. To see this, let τ be the treatment effect and let the variable t indicate treatment if $t = 1$ and control if $t = -1$. Using $+1$ and -1

instead of 0 and 1 makes the math easier. Let Y be the outcome and X a baseline covariate. A *baseline covariate* is a variable that is measured or observed before the experiment begins. Measure X in deviations so that $x = X - \overline{X}$. Then some calculation shows that the variance of the treatment effect is

$$V(\tau) = \frac{2\sigma^2}{n} \frac{\sum x^2}{\sum z^2 - 2(\sum tx)^2/n} \tag{5.1}$$

The first factor on the right-hand side is due to X affecting Y. The second factor depends on the actual values assumed by X. In the event of perfect covariate balance, $\sum tx = 0$, and the treatment variance is minimized. Having covariate balance reduces the variance of the estimate of the treatment effect. Matching is just a special case of blocking when the size of the block is 2.

In the above examples, we could choose one long-term customer and one short-term customer and then flip a coin to assign treatment. We could match stores, too: pick one close to a competitor, one far from a competitor, and then flip a coin to assign treatment. This form of matching can only accommodate one or two relevant covariates due to the curse of dimensionality. With one binary covariate, there are 2 strata, with four there are 16 strata, and with eight there are 256. Then, if we need so many observations in each stratum, we're going to have to run a lot of experiments. And what if the covariates are not binary, but continuous? How can we match on those? The more variables you match on, the harder it is to get a match. See Exercise 5.3 for a demonstration of this fact.

Matching requires subject matter expertise. In the event that matching cannot be exact, and this is usually the case for continuous covariates, the limits of an approximate match must be specified in advance. For example, performing an experiment on women's fashion, a 2-year age difference might be significant for 18-year-old women, but not for 30-year-old women.

There are sophisticated methods of matching that handle all these difficulties, but they are beyond the scope of this course. We briefly mention three of them. At the level of this book, you are not expected to know how to implement these strategies, but you should know of their existence and perhaps how to recognize a situation in which they might prove useful. In such a case you can either read the books and articles in the references to gain the requisite expertise or call an expert.

5.3.2 Rerandomization

The first method is rerandomization, which, strictly speaking, isn't a form of matching but does seek to achieve the same goal. Athey and Imbens (2017, §7.2) give an example. Suppose we have a population of $N = 100$, consisting of 50 men and 50 women. We draw a sample of $n = 60$ randomly. If the sample does not have exactly 30 men and 30 women, we reject the sample and redraw, continuing until we get a sample with exactly 30 men and 30 women. Actually, we don't have to have exactly 30/30; we could settle for 29/31 or even 28/32. The point is that we keep drawing new samples (rerandomizing) until we get a sample that is approximately balanced. The usual *p*-value computed by the software will be too big and will have to be adjusted downward by a sophisticated statistical procedure. As the number of covariates increases, it takes longer to find a rerandomized sample that has balance, but the idea stays the same. On a historical note, the eminent statistician Donald Rubin, one of the founders of modern causal statistics, of course, had a dissertation adviser; his name was Cochran. Cochran actually knew R. A. Fisher, the godfather of modern experimental statistics. Rubin once had a conversation with Cochran about rerandomization, which went like this (Rubin, 2008, p. 1351):

> Rubin: What if, in a randomized experiment, the chosen randomized allocation exhibited substantial imbalance on a prognostically important baseline covariate?
> Cochran: Why didn't you block on that variable?
> Rubin: Well, there were many baseline covariates, and the correct blocking wasn't obvious; and I was lazy at that time.
> Cochran: This is a question that I once asked Fisher, and his reply was unequivocal. Fisher said, "Of course, if the experiment had not been started, I would rerandomize."

Figure 5.6 gives horizontal box plots for the differences between treatment mean and control mean (covariate imbalance) for 1000 regular (pure) randomizations and 1000 rerandomizations for several variables. As can be seen, in *all* cases, the spread of the former is much greater than the spread of the latter, demonstrating the virtue of rerandomization as a method for improving covariate balance.

5.3.3 Propensity Score

Harry's sells men's grooming supplies online and in retail stores like Target. In their subscription business, customers sign up to receive razor blades and foaming shave gel every few months. Because this is a new mode of purchasing for many customers, Harry's advertises on many channels from online display advertisements to Internet radio commercials. When they spend money on advertising, they hope it will produce positive returns – that is, that the advertising brings in new customers who eventually drive more profit to Harry's than Harry's spends on advertising.

The marketing team at Harry's was considering launching a new advertising campaign focused on advertising within Facebook's mobile app. This is a tricky

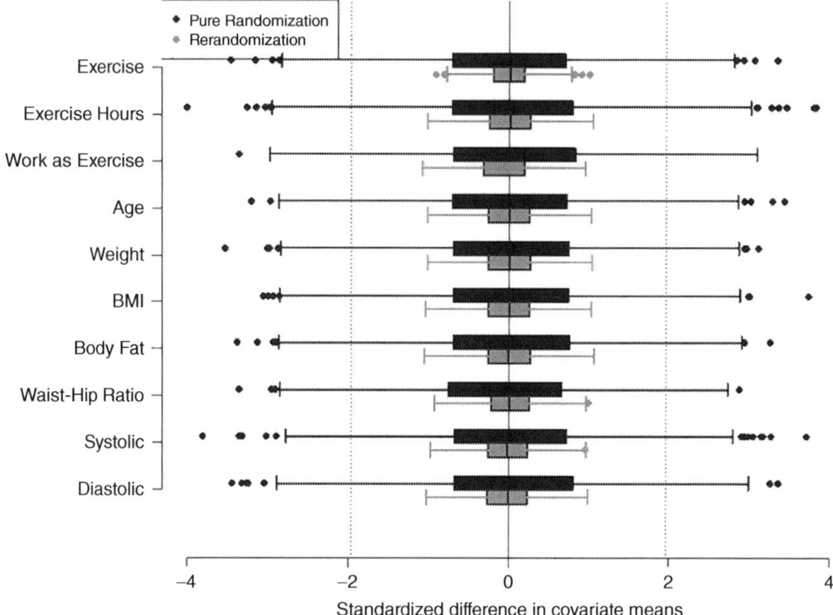

Figure 5.6 Randomization vs. rerandomization. Source: courtesy of Kari Lock Morgan.

proposition for Harry's, because it's widely recognized in the industry that shopping behavior on mobile platforms is very different from shopping behavior on desktop platforms. Namely, many customers use their mobile devices for browsing and exploration and delay purchasing until they're back at a desktop where entering credit card information is easier. Harry's was concerned that their usual approach to evaluating advertising effectiveness, counting how many users purchase immediately after seeing the advertisement, wouldn't work for mobile advertising. If they did a user-level test and randomly showed the ads to some users and not to others, but they were unable to track the sales for those users as they moved from mobile to desktop, then Harry's might conclude that the mobile ads don't work when they really do. (Those in marketing might refer to mobile advertising as a "top of funnel" channel, because it generally reaches consumers who are relatively early in the purchase process.)

Instead, Harry's analytics team planned to run the ads in four distinct geographic regions. The sales for those four regions could be compared with the sales for four control regions. At $n = 8$, this is a tiny sample size, and so it was important to maximize the amount of useful information generated in the experiment. If the analytics team were to choose each set of four cities at random and then measure sales for each city, there would be a substantial amount of variation in sales from city to city. If the control cities happen to be larger than the treatment cities, Harry's might mistakenly conclude that the ads don't work when they really do. For the Harry's team, the solution was *propensity score matching*. To implement this method, one computes *propensity scores* for the experimental units and then matches units that have similar scores. In essence, the matching is done on the basis of the estimated propensity of being assigned to the treatment. In the present case, Harry's analysts used fitted values from a regression model as propensity scores.

Propensity scoring is regularly used in the analysis of observational data (see Rosenbaum (2010)) and is becoming more widely used in experimental design. First, the Harry's team assembled all the data they could find about cities. They had historic sales data for each city, and they also used US census data, which describes the population of each city. Using this historical data, they built a regression model to predict weekly sales (the outcome they will use in the experiment) as a function of everything else they knew about each city. They then used the predicted weekly sales from the regression and found four pairs of cities that had approximately the same fitted value for sales. This is explored in Exercise 5.2. For the experiment, one city in each pair was randomly chosen to get mobile advertising (and not the usual desktop advertising), while the other city got its usual desktop advertising (but not mobile advertising). See Figure 5.7 for example advertisements.

Because this was a brand-new direction for the analytics team, they had no idea what effect sizes or variances might be, and they could not do a power analysis before conducting the experiment. Instead, they viewed this $n = 8$ experiment as an exploratory experiment to gather information so that they could do a power analysis in contemplation of launching a full-scale mobile marketing campaign. The paired t-test revealed no difference between the two methods. This is explored in Exercise 5.3.

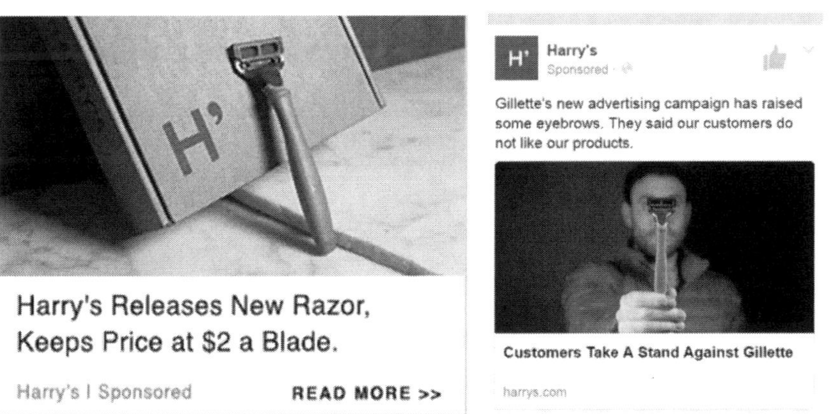

Figure 5.7 Harry's online banner (left) and mobile Facebook (right) ads. Source: courtesy Harry's.

5.3.4 Optimal Matching

The third method is optimal matching. To give a feel for why computer matching is better than matching by hand, consider the following example from Greevy et al. (2004). Suppose you had eight observations on age:

$$24, 35, 39, 40, 40, 41, 45, 56.$$

This distribution is symmetric and most dense near its center, like a normal distribution. It would be natural first to form a first pair $\{40, 40\}$ by picking the two closest and then pick the next two that are closest $\{39, 41\}$, followed by $\{35, 45\}$, and $\{24, 56\}$. The total absolute difference is $0 + 2 + 10 + 32 = 44$. This is a "greedy" algorithm that only looks at the current step (making a pair) without looking at the overall picture (whether this pair is the best choice for minimizing the total absolute difference). An optimal computer match would be $\{24, 35\}$, $\{39, 40\}$, $\{40, 41\}$, and $\{45, 56\}$, which has a total absolute difference of $11 + 1 + 1 + 11 = 24$. It would take a long time to find this nonobvious matching by hand, but a computer can do it easily. Matching on many covariates is technically more complicated but can be done easily. Matching on many covariates, say 20, is routinely done in the analysis of observational data. Matching should be done before the experiment is conducted, but should it for some reason be done after, never match on the outcomes!

Matching is not a solved problem. There are many ways to match, and we don't know which way is best. Nonetheless, even if the matching is not perfect, imperfect matching is often better than no matching at all.

The downside of matching is that, while managers understand a coin flip so they can buy into randomization, they often can't grasp how matching helps. With many different ways to match, a manager might ask, "Is this the best way to match?" And if the answer is "no" (because nobody knows the best way to match) and you can't tell him the best way, he might not want to match. Therefore, be prepared to demonstrate the deleterious effect of covariate imbalance and to suggest that in such a situation, some matching is probably better than no matching.

5.3.5 Sophisticated Matching: Selling Slushies

Mastercard's "Test & Learn" (formerly "Applied Predictive Technologies") provides business analytics software to large businesses to capture data and then analyze it, often for the purposes of performing experiments.

As an example of the benefits of sophisticated matching, Test & Learn had a client, a chain of convenience stores, that was interested in increasing the number of frozen slushy drinks sold. Test & Learn created a homogeneous sample of

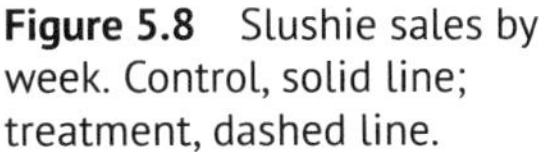

Figure 5.8 Slushie sales by week. Control, solid line; treatment, dashed line.

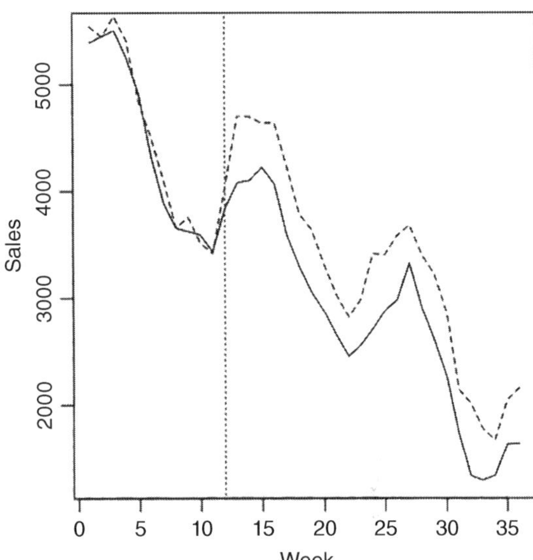

200 stores from which to create treatment and control groups. To verify the homogeneity of the sample, store sales were tracked for 12 weeks before the promotional campaign was begun. The promotional campaign began in week 13, with three campaigns denoted A (27 stores), B (39 stores), and C (32 stores). Figure 5.8 shows the sales over the 36-week period. Note how closely the treatment and control groups track each other during the initial 12 weeks. This is the strength of good matching. Absent matching, both would be jagged lines with the same trend, the jaggedness representing noise that would make detecting an effect more difficult. We have strong graphical evidence that the three campaigns have a positive effect on slushie sales. This will be pursued in Exercise 5.7

Exercises

5.3.1 Do some "balance testing." Use the data in `BankData.csv`. Draw a sample of size n where n is divisible by 2. Randomly divide the sample into equal-sized treatment and control groups. Test whether the means/proportions are the same in both treatment and control groups for the variables amount, housing, and loan. Report on your results: Is there balance or not? Do this for various choices of n.

5.3.2 Use the data in `HarrysDemographicData.csv` to predict monthly sales for each city (a linear regression with sales as the dependent variable will do this). Sort the cities according to predicted sales. Drop the three largest cities because their predicted sales are not at all close to each other. Match the next 16 cities into eight pairs: 1 and 2, 3 and 4, etc. We are interested in pairs for which the predicted sales are very close to each other.

5.3.3 Harry's actually ran the experiment on only eight cities, that is, four pairs. The data in `HarrysData.csv` contain the results of the experiment. Column 1 indicates the pair number, column 2 gives sales for the treatment cities, and column 3 gives sales for the control cities. Do a matched-pair analysis.

5.3.4 For the Harry's data, after doing the paired test, do an independent samples two-sample t-test and see the result. How much of a reduction in variance does the matched-pair method yield?

5.3.5 Welders are exposed to many toxic chemicals that can cause health problems, and these problems can manifest themselves in DNA protein cross-links, measured by percent of DNA protein cross-links in white blood cells (DNC). DNC is also thought to be affected by age, race, and smoking status. A group of 10 welders is available for analysis. A group of 20 potential controls is collected. Match each of the 10 welders to one of the potential controls. (You are doing this by hand. Different persons will have different ideas of what is a "match" and so will get different answers.)

5.3.6 Suppose you rerandomize several times to get good balance, but your software doesn't perform the necessary technical adjustment. Your p-value is 0.0237 and $\alpha = 0.05$. What should you do?

5.3.7 For the slushie campaign, test whether the campaigns were successful and, if so, which of the three campaigns was most successful. The actual experiment was conducted during weeks 13–36, so load the file `SlushieWks13-36Stacked.csv`. The data are already stacked to facilitate ANOVA. Make a box plot with Sales as the response and Campaign as the treatment. What do you see? Perform ANOVA with Sales as the response and Campaign as the treatment. If necessary, compute the Tukey HSD intervals.

Welders					Controls				
ID	**Age**	**Race**	**Smoker**	**DPC**	**ID**	**Age**	**Race**	**Smoker**	**DPC**
1	38	C	N	1.77	1	48	AA	N	1.08
2	44	C	N	1.02	2	63	C	N	1.09
3	39	C	Y	1.44	3	44	C	Y	1.10
4	33	AA	Y	0.65	4	40	C	N	1.10
5	35	C	Y	2.08	5	50	C	N	0.93
6	39	C	Y	0.61	6	52	C	N	1.11
7	27	C	N	2.86	7	56	C	N	0.98
8	43	C	Y	4.88	8	47	C	N	2.20
9	39	C	Y	4.88	9	38	C	N	0.88
10	43	AA	N	1.08	10	34	C	N	1.55
					11	42	C	N	0.55
					12	41	AA	Y	1.49
					13	36	C	Y	0.65
					14	42	C	N	1.78
					15	56	AA	Y	1.02

5.4 Chapter Exercises

5.1 A company wishes to test the market for a new snack. Two factors are of interest: the advertising campaign and the product description on the package. Four types of campaigns are considered, and four different descriptions are considered:

Campaign		Description	
A	TV commercials	I	"Contains vitamins and minerals"
B	Newspaper ads	II	"High energy"
C	Prize in the package	III	"Low cost"
D	Free sample	IV	"Low calorie"

The various test campaigns were conducted in four cities, 1, 2, 3, and 4. Sales, in thousands of dollars, were recorded as follows:

	1	2	3	4
I	A 42	B 61	C 65	D 66
II	B 60	C 55	D 59	A 61
III	C 49	D 51	A 47	B 49
IV	D 53	A 51	B 52	C 52

Analyze the data.

5.2 Refer to the call center example. The first time ANOVA was performed, no effect was found, and only script was included as a treatment. What was the mean squared error for this model? The second time, both script and agent were included, and both treatments were found to be significant. What was the mean squared error for this model?
What can you deduce about the relationship between mean squared error and the finding of significant effects?

5.3 This question shows how hard it is to match when the number of covariates gets large. This is often called "the curse of dimensionality." As the number of dimensions increases, the probability of getting a match decreases markedly.

a) Consider two binary factors, A and B, each of which is 50% ones and 50% zeros. What is the probability that a randomly drawn observation is positive for both A and B?
b) Consider four binary factors, each of which is 50% ones. What is the probability that a randomly drawn observation is positive for all four factors?
c) Consider ten binary factors, each of which is 50% ones. What is the probability that a randomly drawn observation is positive for all ten factors? What can you conclude about the effect of the number of factors on the probability of a match?
d) Redo the above three parts, where each factor is now 10% ones. What can you conclude about the effect of the proportion of ones on the probability of a match?

5.5 Learning More

Here it is apropos to introduce an instructive article (Young, 1996) from which we quote a paragraph. Perhaps we should have used this quote in Chapter 3 on "Designing Tests," but you wouldn't have been able to appreciate it.

> Even when there is the potential for the solution of a problem by use of a well designed experiment, the method has often failed or lead people in the wrong direction. Why is this? One reason is that people undertake to do a "DOE" or "Design of Experiment" without thinking about the meaning of the words they are using, especially the critical word "design." This word reflects the importance of careful up-front planning in the experimental process; too often this critical stage is given far too little attention. For example, critical decisions on what factors should be included in the experiment, what levels should be used and whether the levels of one factor (say line speed in a continuous heat soak operation) should depend on another (say temperature level) are often given far too little attention.

From the abstract of the article, "In addition to the careful choice of factors and levels, three essential components of a well-planned (designed) experiment are the techniques of blocking, replication and randomization. The impact of these techniques in improving both the effectiveness and efficiency of planned experiments is illustrated using a case study on the closing effort of a hatchback car." As the topics of blocking, replication, and randomization have been covered, you are now able to appreciate Young's remarks, whereas if the above had been presented in Chapter 3, you might not have seen the point.

Section 5.1 "Call Center Scripts"

- We have described the randomized complete block design. There is also a randomized incomplete block design when the number of treatments exceeds the number of units in each block. The interested reader should consult a standard text on experimental design for details.
- Some good advice, attributed to the pioneer George Box, is to "Block what you can, and randomize what you cannot." The reason this is such good advice is that "[r]andomization without blocking is incapable of achieving what blocking can, which is to eliminate one component of estimation error entirely, setting $\Delta T_X = 0$, rather than merely ensuring that $E(\Delta T_X) = 0$" (Imai et al., 2008, p. 493). Take another look at the denominator in Equation (5.1).
- Below is an extended discussion on why blocking works – because it reduces variance in the estimator. This may be beyond the statistics you learned in your last statistics course; if so, just skip it.

When the sample size is small, it becomes important to incorporate additional variables into the analysis to decrease the error variance (reduce the noise) so that effects can be better detected. Why does blocking work? More generally, why does reducing the residual variance (unexplained variation) enable us to see effects that we otherwise would miss? The theoretical reason for this is straightforward. We use a regression analysis to explain this, because it's easier than using the ANOVA framework; as we shall see in Chapter 6, there is an intimate relationship between ANOVA and regression that enables us to use regression to solve ANOVA problems. Therefore, if we explain this for regression, we are also explaining it for ANOVA.

Let Y be the response and let X be a dummy variable taking on the value 0 for control and 1 for treatment. The problem can be cast in a regression framework as

$$Y = \alpha + \beta X + \nu \tag{5.2}$$

where the error term ν includes all other factors affecting Y. β is the population parameter, and $\hat{\beta}$ is the estimator thereof. Because X is uncorrelated with the error term, $E[\hat{\beta}] = \beta$. This is the great advantage of experimental data over observational data: the experimental design guarantees that X and the error term are uncorrelated, whereas with observational data this is practically never true.

For sake of argument, suppose there is only one other factor that affects Y, and let us call this variable Z, so the true model is

$$Y = \alpha + \beta X + \gamma Z + \epsilon \tag{5.3}$$

We can see, then, that $\nu = \gamma Z + \epsilon$ because ν is an error term that includes everything affecting Y except X. Therefore, the error variance of Equation 5.2 is greater than the error variance of 5.3, i.e. $V(\nu) > V(\epsilon)$, which means that using Equation 5.3 will allow us to detect smaller effect sizes than using Equation 5.2. To see this more clearly, elementary calculations show that the variance of $\hat{\beta}$ computed from Equation 5.2 is

$$V_1(\hat{\beta}) = \frac{\sigma_\epsilon^2 + \gamma^2 V(Z)}{nV(X)} \tag{5.4}$$

whereas the variance of $\hat{\beta}$ computed from Equation 5.3 is

$$V_2(\hat{\beta}) = \frac{\sigma_\epsilon^2}{nV(X)(1 - r_{XZ}^2)} \tag{5.5}$$

where r_{XZ}^2 is the squared sample correlation between X and Z.

Comparing Equations 5.4 and 5.5, we see that the numerator of the former cannot be smaller than the latter, so whether V_2 can be larger than V_1 depends on r_{XZ}^2. Some algebra shows that $V_2 < V_1$ when $1 - r_{XZ}^2 > \sigma_\epsilon^2/(\sigma_\epsilon^2 + \gamma^2 S_Z^2)$. The experimental design virtually guarantees this condition because $E[r_{XZ}^2] = 0$ by construction. Due to sampling variation, r_{XZ}^2 will not exactly equal zero, but it should be quite close to zero in most cases.

The takeaway from this analysis is that adding variables that affect the response decreases the error variance. When the residual variance decreases, the signal to noise ratio increases, enabling us to detect effects that we otherwise would miss. This idea is explored in Exercise 5.1.8. One common way to incorporate an additional variable into the experiment is called "blocking." As mentioned already, but it is worth reiterating, this is particularly important when the sample size is small.

The above analysis is discussed in greater detail in Franklin (1991).

Section 5.2 "Geo-Testing"

Latin square is a fully confounded design, so a significant interaction effect usually swamps the main effects, rendering them null. Thus, if a significant main effect is found, then it is usually the case that there is no interaction. On the other hand, if no main effect is found, there may be a significant interaction (Hamlin, 2005, p. 338).

• Latin square is great for preliminary studies because it is quicker and cheaper than other methods.

• If three blocking variables are used in a Latin square, it is called a Graeco-Latin square design.

Kutner et al. (2005, p. 1185) offer the following advice for sizing Latin square designs:

> Because of the limitations of the degrees of freedom for experimental error… Latin Squares are rarely used when more than eight treatments are being investigated. For the same reason, when there are only a few treatments, say, four or less, additional replications are usually required.

Section 5.3 "Dealing with Covariate Imbalance"

Some persons think that covariates should always be balanced and, if imbalance is found, then it should be somehow corrected by some sort of *ad hoc* statistical adjustment. This is nonsense. Imagine you have 20 covariates and apply a balance test to teach one. Even if there is no imbalance, each test has a 5% chance of being a type I error, so if a test shows that there is imbalance, you wind up correcting for type I error and not any true imbalance. The article by Pemantle et al. (2019) clearly explains why balance testing is Bad Idea™, as does Senn (1994). Just because other people do balance testing, it doesn't mean that you should.

• An extended discussion of Equation 5.1 can be found on pages 89 *et seq* of Senn (2007).

• A decent optimal matching article is Lu et al. (2011).

• The "slushie" example is based on a real problem, but the data are faked. Thanks to Cornelius Kaestner for providing us with the example, and thanks to Mastercard for permission to use it.

- The “Harry’s” example is based on a real problem, but some details were changed to protect trade secrets. The data for Exercises 5.3.3, and 5.3.4 are fake for the same reason. Thanks to Michael Kaminsky for providing the example and data, and we thank Harry’s for permission to use this example. A more detailed account can be found at https://medium.com/harrys-engineering/matching-cities-for-small-sample-experiments-harrys-engineering-be4e88c2b112
- The matching of welders example is adapted from (Rosenbaum, 2017, p. 164).

6

Analyzing Designs via Regression

This section of the book begins our formal treatment of experiments involving many variables. For analyzing several variables the method of linear regression is a natural choice. True, the ANOVA method involves several variables, but, we shall see, it's really just a special case of regression, as are many of the methods that we've already used. Linear regression allows us to handle many different experimental methods in a consistent fashion, rather than having to learn a new way to analyze each method, as we have done in previous chapters.

By the end of this chapter, readers should:

- Understand that two-sample tests and ANOVA are special cases of regression.
- Be able to write out the prediction equation from regression software output.
- Explain why effect coding (−1, 1) for binary variables is necessary.
- Know the difference between dummy variables and effect coding.
- Define an interaction.
- Interpret an interaction.
- Analyze interaction plots.
- Realize that letting software determine coding for dummy variables and effect-coded variables is parlous.
- Know the difference between a design matrix and a model matrix.
- Be able to use pretreatment variables to decrease the error variance (and why it is important to decrease the error variance).

6.1 Experiments and Linear Regression

So far we have utilized many different methods for analyzing relatively simple experimental designs: one-sample test of proportions, two-sample test of means, paired tests, and ANOVA, among others. It turns out that all these can be viewed as special cases of linear regression. The linear regression framework is very

Business Experiments with R, First Edition. B. D. McCullough.

Companion Website: www.wiley.com/go/mccullough/businessexperimentswithr

useful because instead of needing more methods for the advanced techniques in the rest of this book, we can just use linear regression. However, we'll have to take care in properly choosing the independent variables so that linear regression will work properly to analyze these more sophisticated experimental designs. You may wish to review the regression chapter of your introductory statistics text before proceeding.

The theoretical linear regression model

$$Y = \beta_0 + \beta_1 X_1 + \beta_2 X_2 + \cdots + \beta_k X_k + \varepsilon \tag{6.1}$$

has as its empirical counterpart the estimated regression equation

$$y = b_0 + b_1 x_1 + b_2 x_2 + \cdots + b_k x_k + e \tag{6.2}$$

where y and x_i are realizations of Y and X_i, the sample statistics b_i estimate the unknown population parameters β_i, and the unknown population errors ε are estimated by the sample residuals e.

The *fitted values* are given by

$$\hat{y} = b_0 + b_1 x_1 + b_2 x_2 + \cdots + b_k x_k \tag{6.3}$$

Some people call these "predicted values," but nothing is being predicted here, since we already know the true values of y. We reserve the use of the word "predict" for situations where we really are trying to predict something that is unknown.

In Chapter 2 we performed a two-sample test of means on the audio/video data and reported the results thusly:

> The test showed that video calling significantly increased sales per customer by about \$10.44 per customer over audio sales call (95% CI = (\$2.62, \$18.26)).

Let's obtain the same result using regression.

Use the file `AudioVideo.csv`. Let y = `sales_one_week` and let d be a dummy variable for the categorical variable `call_type` that equals unity for "video" and zero for "audio." Formally, we define the *dummy variable* d as

$$d = \begin{cases} 1, & \text{if video} \\ 0, & \text{if audio} \end{cases} \tag{6.4}$$

In this context, audio is the control (or "baseline") and video is the treatment. The regression equation is

$$y = b_0 + b_1 d \tag{6.5}$$

The intercept is the baseline, and the slope is the (possibly negative) addition to the baseline if the dummy variable equals one. Put another way, b_0 is the mean sales for audio, and b_1 is the amount added to that to obtain mean video sales, which equals $b_0 + b_1$. Run the regression of y on d and get the 95% confidence for the slope coefficient. The intercept equals 110.292, which is the control mean, the baseline. The slope coefficient equals 10.442, which estimates the effect of the treatment. So the regression estimate of mean audio sales is 110.292, while the estimate of mean video sales is $110.292 + 10.442 = 120.734$. The 95% confidence interval for the slope is (2.62, 18.26), which is exactly what we found in Chapter 2.

Try it!

For the email response case results in Section 2.2.1, we found that the 95% CI was $-0.025 \pm 0.0653 = (-0.0903, 0.0403)$. Let's reproduce this result using regression.

Don't let the software automatically create the dummy variable. Create a dummy variable dummy so that $A = 1$ and $B = 0$ for email_type. Create y so that yes $= 1$ and no $= 0$ for response. Regress y on d and get the 95% confidence interval for the slope. Interpret your results.

What would happen if d had been created so that $A = 1$ and $B = 0$ or y so that yes $= 0$ and no $= 1$?

```
df <- read.csv("EmailResponse.csv")
df$dummy <- ifelse(df$email_type=="A",1,0)
df$y <- ifelse(df$response=="yes",1,0)
lm1 <- lm(y~dummy,data=df)
summary(lm1)
confint(lm1)
```

Software Details

Don't let the software create dummy variables for you...

When software packages automatically create dummies for a categorical variable, they often do strange things. In particular, the software package might not code the dummy variables the way that you would. Therefore, as a general rule, it is always best to code the dummy variables yourself, unless you know for a fact that your software package does it the way that you want.

Next consider the $A/B/n$ test from Chapter 4 that used the data set `CarRental.csv`, which tested the null hypothesis that various types of cars (minivan, sedan, etc.) have the same rental time. It may be useful to obtain descriptive statistics on this data set before proceeding (calculate the means for each category). The ANOVA table is shown in Table 6.1.

Table 6.1 ANOVA table for car rental data.

Analysis of variance					
Source	**df**	**Adj SS**	**Adj MS**	***F*-Value**	***p*-Value**
Label	3	2295	764.92	54.49	0.000
Error	108	1516	14.04		
Total	111	3811			

Let us first analyze the coefficients given in Table 6.2. The intercept represents compact cars. The coefficient on luxury represents the additional days that a luxury car is rented compared with a compact car, $9.994 + 6.285 = 16.279$, which happens to be the mean number of days that a luxury car is rented. Similarly, the rest of the coefficients can be interpreted.

We can see that the model estimated by R is

$$\text{days} = b_0 + b_1\text{luxury} + b_2\text{minivan} + b_3\text{sedan} \tag{6.6}$$

where days is the number of days each vehicle is rented and luxury is a dummy variable that equals unity when the vehicle is a luxury car and zero otherwise, and similarly for minivan and sedan. Thus, when luxury = minivan = sedan = 0, the vehicle is a compact.

Try it!

Don't let the software create the dummy variables for you! Create dummy variables so that you can run a regression and get the same results as Table 6.2.

Software Details

To use regression to perform an ANOVA...

```
newdf <- stack(df)
# make the dummy variablese
```

Table 6.2 Regression results on car rental data.

	Coef	**Std err**
Intercept	9.994	0.59
Compact	6.285	0.96
Luxury	14.52	1.16
Minivan	4.684	0.88

```
compact <- ifelse(newdf$ind=="compact",1,0)
sedan <- ifelse(newdf$ind=="sedan",1,0)
luxury <- ifelse(newdf$ind=="luxury",1,0)
minivan <- ifelse(newdf$ind=="minivan",1,0)
lm1 <- lm(newdf$values~luxury+minivan+sedan)
summary(lm1)
```

Why is compact not in the regression?

As noted previously, regression and ANOVA give the same F-statistic. The same hypothesis has been tested, and the same answer has been produced; both methods are doing the same thing. As far as the regression coefficients are concerned, notice that "compact" is missing; it is the baseline, represented by the intercept. Also notice that the same information in the Tukey intervals in Figure 4.6 is contained in the regression coefficients. For example, Figure 4.6 says that the difference between luxury and sedan is 1.603, which happens to equal the difference of the above regression coefficients: 6.285 − 4.682 = 1.603. We could also use the regression output to form all the confidence intervals for pairwise differences, but that would take us too far afield. The point is that once again, regression and ANOVA are doing the same thing and getting the same answer.

When data are stacked – when one variable is categorical and has k levels, most statistical packages will automatically create $k-1$ dummy variables when the variable is used in a regression. It is important that the package recognize the variable as categorical. If the k categories are represented by numbers, e.g. $1, 2, \dots k$, the package might treat them as numbers and not create the necessary dummy variables. Regardless, unless you're sure that your statistical packages will create dummy variables the way that you want, you should create them yourself.

Categorical information can be incorporated via the use of dummy variables. Suppose it may matter whether an observation is male or female. Then this may be captured by the dummy variable

$$D = \begin{cases} 1, & \text{if male} \\ 0, & \text{if female} \end{cases} \tag{6.7}$$

In the simple regression (Y and X) case with one dummy variable, the response function is

$$E[Y] = \beta_0 + \beta_1 X + \beta_2 D \tag{6.8}$$

whence

$$E[Y] = (\beta_0 + \beta_2) + \beta_1 X_1 \quad \text{if male} \tag{6.9}$$

$$E[Y] = \beta_0 + \beta_1 X_1 \quad \text{if female} \tag{6.10}$$

and we can see clearly that the dummy variable amounts to a shift in the y-intercept when the condition is satisfied. A test of whether the two regression lines in Equations 6.9 and 6.10 are the same (whether there is no difference in the intercept between men and women) is a test of $H_0 : \beta_2 = 0$ against $H_A : \beta_2 \neq 0$ in Equation 6.8. The parameter β_0 is the baseline, which, in this case, is the mean effect for female, while β_2 is the (possibly negative) additional effect for male.

Exercises

6.1.1 The data file `malefemaleincome.csv` contains three variables on 200 persons: years, the years of employment; income, the amount of money the person is paid; and sex, whether the person is male or female. Assume that income increases with the number of years of service. Assume that this rate of increase is the same for both sexes (so that there is only one slope in the regression). Test whether there is a difference in the intercept for men and women. Interpret the regression coefficients.

6.1.2 This is an exercise in working with regression equations to re-express the same result in different ways. Cast the landing pages example into regression. Create a dummy variable d so that version $A = 0$ and version $B = 1$. Create y equal to sales. Regress y on d. The 95% CI for the slope is $(-0.47, 1.30)$. Reconcile this interval with the one given in Chapter 2:

> The test showed that Landing Page A and Landing Page B produced similar sales per website visit (95% CI = $(-1.30, 0.47)$).

Specifically, how could you make the intervals the same?

6.1.3 The two-sample test of proportions is also a special case of regression. Cast the email response problem into a regression framework using the data set `EmailResponse.csv`. Define numeric variables x = 1 for version A and 0 for version B and y = 1 for a response of yes and 0 for no. Regress y on x. Get a 95% CI for the slope and compare it with Equation 2.9.

6.2 Dummies, Effect Coding, and Orthogonality

For simple $0 - 1$ dummy coding, the coefficients are relative to the intercept, which is the baseline. By contrast, effect coding – a "dummy variable" with -1 and $+1$ instead of 0 and 1 – calculates coefficients relative to the grand mean. There is a much more important difference between these two methods: when there are only two levels, effect coding produces orthogonal variables, whereas dummy coding produces variables that are not orthogonal. Orthogonality of regressors is a very desirable property, as we shall see.

Consider a simple experiment with two factors, A and B, each of which has two levels: a low level and a high level. The experiment is run once at each combination, and the results are presented in Table 6.3. Because it is an experiment with factors, it is called a *factorial experiment*. Since there are two factors and each factor has two levels, it is more precisely called a 2^2 factorial experiment, where the base indicates the number of levels of each factor and the exponent indicates the number of factors. $2^2 = 4$ is, of course, the number of unique treatment combinations of the factors. A conventional way to represent these data graphically is using a *square diagram* as in Figure 6.1.

We define the *main effect of a factor* as the average change in the response due to a change in the factor. We can calculate the main effect of factor A as

$$\text{Main effect factor } A = (30 + 50)/2 - (20 + 40)/2 = 40 - 30 = 10. \tag{6.11}$$

Table 6.3 Simple experiment.

Response	Factor *A*	Factor *B*
50	hi	hi
20	lo	lo
40	lo	hi
30	hi	lo

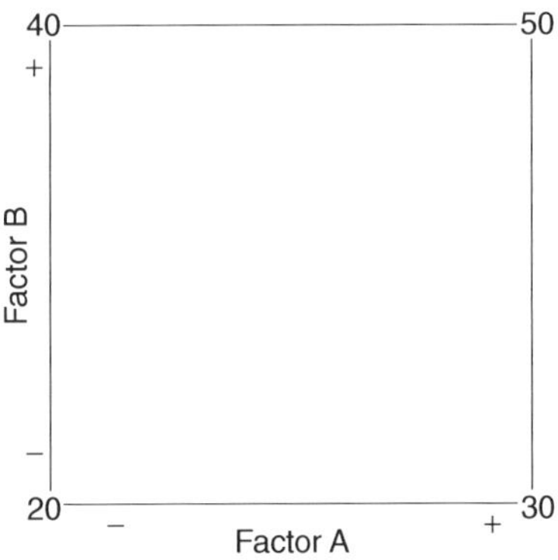

Figure 6.1 Square diagram for simple experiment.

When factor A increases from low to high, on average, the response increases by 10. Similarly, the main effect of factor B is

$$\text{Main effect factor } B = (40 + 50)/2 - (20 + 30)/2 = 45 - 25 = 20. \tag{6.12}$$

When factor B increases from low to high, on average, the response increases by 20.

We do not have to calculate these by hand. We can use regression with dummy variables to compute these effects using the variables in Table 6.3, after factor A and factor B have been converted to dummy variables. To be clear, the data in numerical form for this regression are on the left side of Table 6.4. Typical regression output is given in Table 6.5.

Observe that the regression results exactly reproduce the main effects calculated in Equations 6.11 and 6.12. Further observe that the R-squared = 1.0 because the regression is a perfect fit. Since all the residuals equal zero, the standard error of the coefficient is also zero, and the t-stat will equal the coefficient divided by zero.

Now let's do this with effect-coded variables, which are given in the right side of Table 6.4. Typical regression output is given in Table 6.6.

Observe again that the R-squared = 1.0 because the regression is a perfect fit. Observe this time that the regression slope coefficients (not the intercept) are one-half the main effects calculated in Equations 6.11 and 6.12. This makes sense

Table 6.4 Numeric data for Table 6.3.

Response	dummyA	dummyB	effectA	effectB
50	1	1	1	1
20	0	0	−1	−1
40	0	1	−1	1
30	1	0	1	−1

Table 6.5 Regression results for simple experiment with dummy variables.

	Coeff	Std err	t-Stat	p-Value
Intercept	20	0	inf	0
dummyA	10	0	inf	0
dummyB	20	0	inf	0
		R-squared = 1.0		

Table 6.6 Regression results for simple experiment with effect-coded variables.

	Coeff	Std err	t-Stat	p-Value
Intercept	35	0	inf	0
effectA	5	0	inf	0
effectB	10	0	inf	0
		R-squared = 1.0		

because the dummy regression uses a difference of one (1 – 0) to estimate the effect, while the effect-coded regression uses a difference of two (1 – (–1)). To get an estimate of the main effect from the effect-coded regression, therefore, simply double the appropriate regression coefficient (and double the standard error, too).

Try it!

Reproduce the results in Tables 6.5 and 6.6.

```
response <- c(50,20,40,30)
dummyA <- c(1,0,0,1)
dummyB <- c(1,0,1,0)
effectA <- c(1,-1,-1,1)
effectB <- c(1,-1,1,-1)

lm1 <- lm(response~dummyA+dummyB)
summary(lm1)
lm2 <- lm(response~effectA+effectB)
summary(lm2)
```

When predictors are orthogonal, we get clean estimates of the parameters: a change in x_1 implies a change in y, and nothing else is changing that would change y. But when predictors are not orthogonal, then x_1 and x_2 move together, and it's hard to say that a change in x_1 and not in x_2 produced the change in y. We need something like dummy variables to represent the levels of the factors, but we need these "somethings" to be orthogonal. What we need are *contrasts*.

A *contrast* is a linear combination of the parameters where the weights sum to zero. If μ_i is the mean effect of the ith category, each category gets a weight c_i, so the linear combination is $\sum_{i=1}^{k} c_i\mu_i$ with the requirement that the weights sum to zero given by $\sum_{i=1}^{k} c_i = 0$. Contrast coding ensures that the coefficients are uncorrelated with each other (i.e. orthogonal), which is a necessary condition for ANOVA, and ANOVA routines implicitly incorporate contrasts. Contrast coding is not a necessary condition of regression, so if we want to use regression to perform ANOVA, we have to explicitly incorporate contrasts. For now we use a special case of contrast where "hi" = +1 and "lo" = −1; this special case is also called *effect coding*. This produces orthogonal variables in the case that the categorical variable has exactly two levels.

By definition, two vectors are orthogonal if their dot product equals zero. Consider two vectors, each of length three: $a = a_1, a_2, a_3$ and $b = b_1, b_2, b_3$. Their dot product is $a \cdot b = a_1 b_1 + a_2 b_2 + a_3 b_3$. In the case of effect coding for factors with two levels, this is trivially true: the effect coding for the factors in Table 6.3 are effectA $= [1, -1, -1, 1]$ and effectB $= [1, -1, 1, -1]$; the dot product of these two variables is $(1)(1) + (-1)(-1) + (-1)(1) + (1)(-1) = 1 + 1 + (-1) + (-1) = 0$.

Using regression with dummies is an easy way to estimate main effects when there are no *interactions*, which we will define later. When interactions are present, regression with dummies does not work. We'll see why regression with dummies doesn't work in such a situation and how to make regression work when interactions are present (because calculating effects by hand is not something we want to do). First, however, we must dispense with a common experimental approach called "one factor at a time" (OFAAT) experimentation.

Some persons are of the opinion that it is best to change only OFAAT so as to isolate the effect of that particular factor. By contrast, in experimental design, we typically change more than one factor at a time. The prototypical experimental design method is called *factorial design*, about which we'll learn more in the next chapters. There are two big problems with OFAAT. First, it is inefficient; experimental design estimates of effects are more precise than estimates produced by OFAAT. Factorial designs use all the observations to estimate the effect of each factor, while OFAAT uses only two observations to estimate the effect of each factor. Second, OFAAT is unable to detect interactions between variables. Czitrom (1999) gives more advantages of designed experiments over OFAAT and illustrates the advantages with several examples. The article by Box (1990) gives an example where a dramatic improvement to ball-bearing manufacture – worth tens of millions of dollars – was achieved by a very simple experiment that had been overlooked for decades because the engineers were too focused on OFAAT and completely missed an easy interaction.

Box (1990) also gives an amusing example in which the target measurement is the number of rabbits in the hutch and the factors are does and bucks, each with a low level of zero and a high level of one. The target measurement takes on the value of zero when the control situation obtains: there is neither doe nor buck in the hutch. Following the OFAAT methodology, place a doe in the hutch and wait for a period of time. The total number of rabbits in the hutch is one. Remove the doe and insert the buck. Wait a period of time; the total number of rabbits is still one. The experiment is over with the conclusion that neither does nor bucks affect the number of rabbits in the hutch. Modern experimental design will vary both factors at once by placing a buck and a doe in the hutch simultaneously, waiting a period of time, and then observing that there now are several rabbits in the hutch! The OFAAT methodology would never discover this interaction. Analyzing OFAAT experiments is statistically simple, and analyzing interactions statistically is difficult. Thus, when we abandon the OFAAT methodology, we necessarily admit the possibility of interactions, and we must be able to handle them statistically.

Exercises

6.2.1 Consider the data in Table 6.4. Verify that the variables dummyA and dummyB are not orthogonal. Verify that the variables effectA and effectB are orthogonal.

6.2.2 Change the responses of Table 6.3 to $30, 35, 40, 25$ and calculate the main effects by hand. Then compute the main effects by regression for both dummy variables and effect coding. For each of the two regressions, drop one of the variables and see what happens to the coefficients.

6.3 Case: Loan Experiment Revisited (Interactions)

Before analyzing the loan experiment data, we have to understand what interactions are, how to model them, and how to analyze them.

6.3.1 Interactions

To motivate the idea of an interaction, consider Table 6.7 representing output due to two factors, A and B, each at a low level and a high level.

The effect of increasing B from low to high, holding A constant, is 3. The effect of increasing A, holding B constant, is 5. If there is no interaction, then $X = 18$. If the interaction is positive, then $X > 18$, and if negative, $X < 18$. Another way to think about it is this: the effect of one factor depends on the level of the other factor.

More formally, two variables are said to interact if the effect of one depends on the level of the other. An example will help to clarify this concept. Suppose a credit card company is interested in determining the effect of interest rates and annual fees on credit card offers. The company will offer a high or low interest rate and

Table 6.7 2 × 2 table of responses to factors A and B.

	B_L	B_H
A_L	10	13
A_H	15	X

a high or low annual fee; there are a total of four possible combinations. A large number of offers are sent to each combination, and the response rate is observed as a percentage. The results are shown in Table 6.8.

A more informative way to view this information is via an *interaction plot*, as shown in Figure 6.2.

As can be seen, when the interest rate is low, there is not much difference between the high and low fees. However, when the interest rate is high, there is a marked difference between the low and high fees. Thus, the effect of the fee depends on whether the interest rate is high or low. The interest rate and the annual fee interact. If there were no interaction, then the lines would be parallel. (Of course, having uncovered this relationship graphically, our next step would be to confirm it statistically by running the appropriate test.) When an interaction

Table 6.8 Response to interest rate and annual fee.

	Low fee	**High fee**
Low rate	2.25	2.03
High rate	2.35	1.72

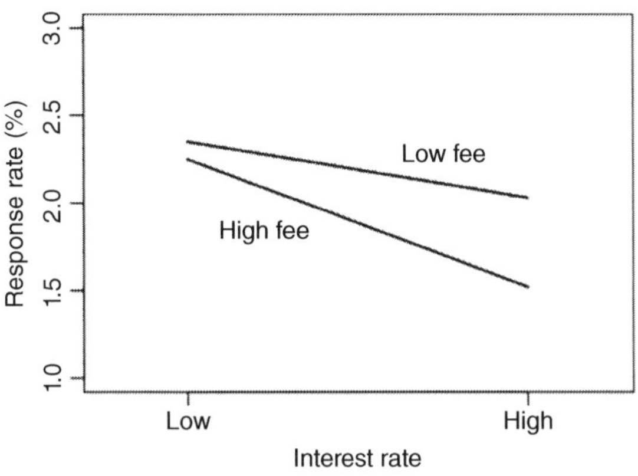

Figure 6.2 Interaction plot for interest rate and annual fee.

is present, it really doesn't make sense to speak of the "main effect of a factor" because the effect of one factor depends on the level of another factor.

There are two equivalent ways of describing this (or any) interaction: (1) the change in the response rate due to changing the interest rate depends on the fee, or (2) the change in the response rate due to changing the fee depends on the interest rate. Sometimes you may find that one way just makes more sense than the other or is easier for the client to comprehend.

Dummies fail to work in the presence of interactions because dummy variables are not orthogonal (think "are not uncorrelated" if you're not familiar with orthogonality). Let's see how to incorporate effect coding into our regression analyses.

Try it!

Take the data in Table 6.8 and turn it into three variables: response, rate, and fee. Be sure to code rate and fee with effect coding; use −1 for low and +1 for high.

Regress response on only rate. Next regress response on rate and fee; then add an interaction term and run the new regression. Compare the coefficients from the two regressions. Notice that the slope coefficients do not change when the interaction term is added. Now do likewise for the dummy regression. What is the difference?

```
response <- c(2.25,2.35,2.03,1.72)
rate <- c(-1,1,-1,1)
fee <- c(-1,-1,1,1)

lm1 <- lm(response~rate+fee)
summary(lm1)
lm2 <- lm(response~rate+fee+rate:fee)
summary(lm2)

rate <- c(0,1,0,1)
fee <- c(0,0,1,1)

lm3 <- lm(response~rate+fee)
summary(lm3)
lm4 <- lm(response~rate+fee+rate:fee)
summary(lm4)
```

One of the virtues of effect coding is that the effect variables are orthogonal (uncorrelated), so adding or dropping an effect variables from the regression does not change the estimates of the remaining coefficients when interactions are present. This is not true for dummy regression. The property of orthogonal variables that coefficients don't change when variables are dropped or added will be especially useful when we begin dropping variables from our regression analyses.

Consider the experimental results presented in Table 6.9, in which an interaction effect is present. The interaction plot presented in Figure 6.3 clearly shows the interaction between the two factors.

The three effect variables are orthogonal, since their dot products equal zero: $\text{effectA} \cdot \text{effectB} = (-1)(+1) + (+1)(+1) + (-1)(-1) + (+1)(-1) = -1 + 1 + 1 - 1 = 0$. Similarly, $\text{effectA} \cdot \text{effectAB} = 0$ and $\text{effectB} \cdot \text{effectAB} = 0$. All the effect variables are pairwise orthogonal. The same is not true of the dummy variables; this is explored in Excercise 6.2.1.

Recall that *main effect* of a factor A is the change in the response variable when the level of the factor increases from low to high. This, of course, implies taking the average of factor A across the levels of factor B. Examining Table 6.9 we can calculate the main effect of A as $(35 + 31)/2 - (6 + 20)/2 = 33 - 13 = 20$. (Remember, it really makes no sense to speak of a "main effect" in the presence of an interaction, we are really just saying "main effect" to refer to the calculation; it's not really a main effect.)

Table 6.9 Effect coding vs. dummy coding.

Response	effectA	effectB	effectAB	dummyA	dummyB	dummyAB
20	−1	+1	−1	0	1	0
31	+1	+1	+1	1	1	1
6	−1	−1	+1	0	0	0
35	+1	−1	−1	1	0	0

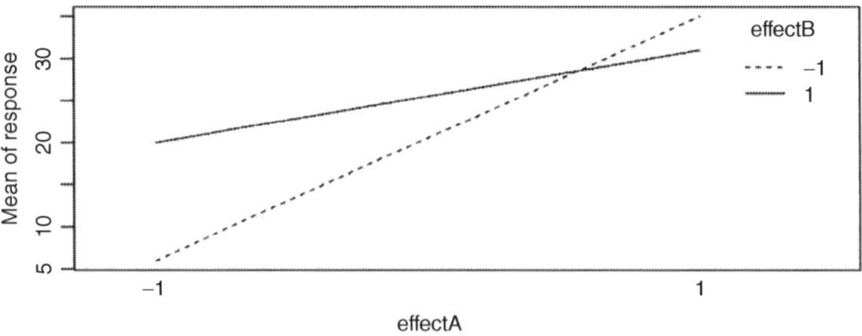

Figure 6.3 Interaction plot for data in Table 6.9.

The *interaction effect* between two factors A and B is the average difference between the effect of A at the high level of B and the effect of A at the low level of B: $(31 - 20)/2 - (35 - 6)/2 = -9$. (The interaction effect can equivalently be defined as the average difference between the effect of B at the high level of A and the effect of B at the low level of A.) The interaction is a *second-order* effect, the product of two factors. Main effects are *first-order* effects. Taking the square of a factor, i.e. multiplying a factor by itself, is also a second-order effect (being the product of two factors), but we do not employ squared factors in this book.

The above definitions are perhaps more easily visualized using the square diagram in Figure 6.4 compared with computing them from Table 6.9. An excellent discussion of how and why these effects are computed can be found in Chapter 2 of Goupy and Creighton (2007). For our purposes, it suffices that the computer will calculate them for us as long as we use effect coding properly.

Let us apply regression to these data using both dummy variables and effect-coded variables. First regress response on effectA and effectB and then regress response on dummyA and dummyB. Now form the interaction variables effectAB as the product of effectA and effectB, and similarly for dummyAB. Then add these variables to their respective regressions. Results are presented in Table 6.10. (When you run these regressions, you'll get point estimates but no standard errors for the interaction regressions – why not?) Notice that in the no-interaction case, both the effect regression and the dummy regression give equivalent answers and correctly estimate the main effects – remembering to double the regression coefficients in the case of the effect regression.

Since effect coding makes A, B, and AB orthogonal, we can add AB without changing the estimated coefficients in the effect regression. This is not true for the dummy coding, because dummy variables are not orthogonal. When the interaction term is added to the model, the dummy regression no longer estimates main effects. Indeed, this time all the coefficients change. Therefore, if you need to run a

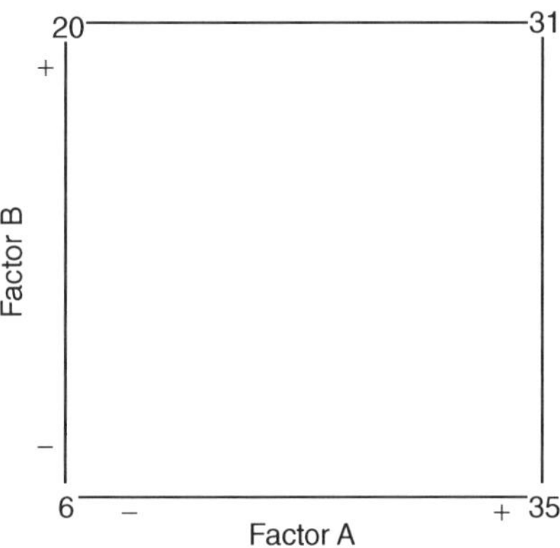

Figure 6.4 Square diagram for main and interaction effects for data in Table 6.9.

Table 6.10 Results of regressions on effect coding, dummies, and interactions.

	No interaction		Interaction	
	Effect coding	**Dummies**	**Effect coding**	**Dummies**
Intercept	23	10.5	23	6
Slope A	10	20	10	29
Slope B	2.5	5	2.5	14
Slope AB	—	—	−4.5	−18

model with an interaction term, you should use effect coding rather than dummy coding. Since you almost always have to test for interactions, you should always use effect coding rather than dummy coding.

6.3.2 Loan Experiment

Let us analyze Example II from Section 1.4; reread it before proceeding. The data are `LoanExperiment.csv`. We have taken the liberty of creating effect coding for these data, see `LoanExperiment2.csv`. Be sure to compare these two files and understand the relationship between them. When creating effect coding, it is useful to have some mnemonic device for remembering which condition is +1 and which condition is −1; that way we don't have to refer constantly to a sheet of paper on which we've defined the effects. We like to use "good is +1" and "bad is −1." For example, the regions are Northeast and Midwest. We code Northeast as +1 and Midwest as −1 because the author lives in the Northeast, so of course the Northeast is good. (If you live in the Midwest, you can switch the numbers yourself.) NegativeExample has levels Yes and None: yes is good and gets +1, while none is bad and gets −1. Complete mnemonics are provided in Table 6.11. These, of course, are our mnemonics. If you don't like them and think others would be easier to remember, then do so. We also introduce the column "factor" with levels A, B, C, D, and E, because writing individual letters is often easier than writing out variable names.

Since we have five factors, each at two levels, this is a 2^5 factorial experiment. Let us write the *design matrix* for this experiment. In Table 6.9 we wrote "+1" and "−1" in a haphazard order. We should be systematic about the way in which we order the "+1" and "−1" in an experiment, and this systematic way is called the *standard order*: the first row is all "−1," and the first column alternates "−1" and "+1"; the second column alternates pairs, "−1" and "−1," followed by "+1" and

Table 6.11 Mnemonics for remembering level effect coding.

Variable	Factor	−1 level	+1 level
Type	*A*	Lease	Loan
Region	*B*	Midwest	Northeast
Description	*C*	Current	Enhanced
PositiveExample	*D*	current	Enhanced
NegativeExample	*E*	No	Yes

"+1"; the third column alternates quadruplets; and so on. See Table 6.12, where the first column gives the "standard order." The standard order is distinguished from the "run order," which is the random order in which the experiments are executed.

Software Details

To create an effect-coded numeric variable from a categorical variable...

```
df$RegionNumeric <- ifelse(df$Region=="Northeast",1,-1)
```

Let us graphically explore the data. First, plot levels of each independent variable against the response variable, as shown in Figure 6.5. We clearly see that Document, Type, Region, and NegativeExample have no effect on the completion rate, while Description has some effect and PositiveExample has a very strong effect.

Next we examine possible interactions, of which there will be $C(5, 2) = 10$; to conserve space, we show only two, displayed in Figure 6.6, though in practice you should always plot all of them. (If the $C(5, 2)$ notation is unfamiliar to you, refer to the Learning More section of this chapter.) We see that there is a strong interaction between NegativeExample and Region: using a negative example decreased the response rate in the Northeast but increased it in the Midwest. There is no interaction between PositiveExample and NegativeExample.

Run the regression with PercComplete as the dependent variable, with all five independent variables and all possible two-way interactions, of which there will be $C(5, 2) = 10$. Results are presented in Table 6.13. When looking at this table, notice how we almost run out of room defining the interaction terms. For this reason we'll soon begin using letters A, B, C, etc. to name the factors.

Table 6.12 Design matrix for loan experiment.

Std order	Type	Region	Description	PositiveExample	NegativeExample
1	−1	−1	−1	−1	−1
2	1	−1	−1	−1	−1
3	−1	1	−1	−1	−1
4	1	1	−1	−1	−1
5	−1	−1	1	−1	−1
6	1	−1	1	−1	−1
7	−1	1	1	−1	−1
8	1	1	1	−1	−1
9	−1	−1	−1	1	−1
10	1	−1	−1	1	−1
11	−1	1	−1	1	−1
12	1	1	−1	1	−1
13	−1	−1	1	1	−1
14	1	−1	1	1	−1
15	−1	1	1	1	−1
16	1	1	1	1	−1
17	−1	−1	−1	−1	1
18	1	−1	−1	−1	1
19	−1	1	−1	−1	1
20	1	1	−1	−1	1
21	−1	−1	1	−1	1
22	1	−1	1	−1	1
23	−1	1	1	−1	1
24	1	1	1	−1	1
25	−1	−1	−1	1	1
26	1	−1	−1	1	1
27	−1	1	−1	1	1
28	1	1	−1	1	1
29	−1	−1	1	1	1
30	1	−1	1	1	1
31	−1	1	1	1	1
32	1	1	1	1	1

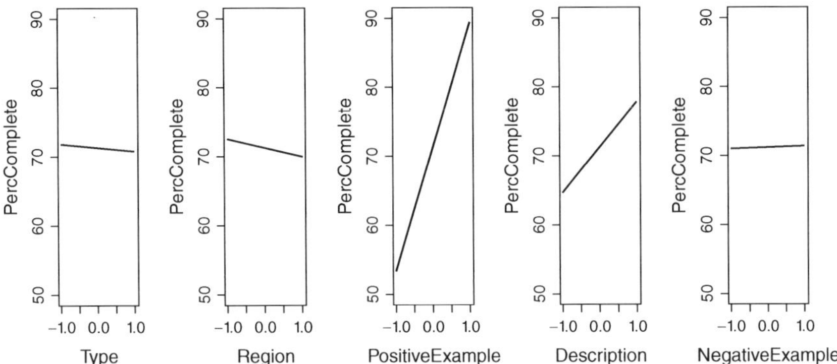

Figure 6.5 Main effect plots for loan experiment.

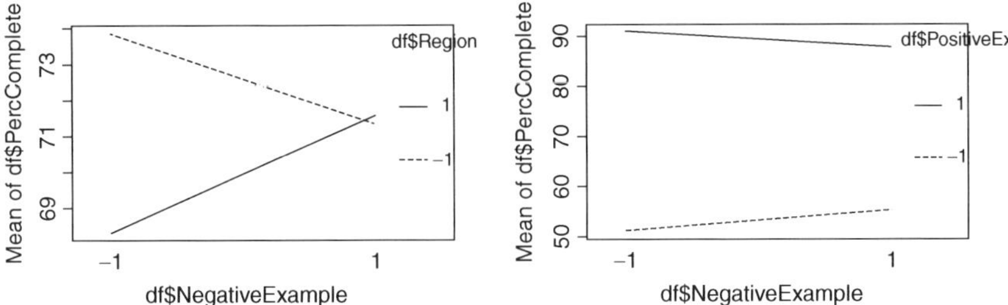

Figure 6.6 Two interaction plots for loan experiment.

Software Details

About creating interaction variables…

With even a few variables, creating and entering all the interaction terms is tedious, so many software packages have shortcuts to do this. Some packages automatically create plots for main effects and interactions; for other packages these plots have to be created individually. Since we have created our own effect-coded variables in `LoanExperiment2.csv`, we are going to leave them as numeric and *not* change them to nominal or factors.

To run a regression with main effects and two-way interaction effects and plot main and interaction effects…

We show code for one plot of each type. You'll have to create the rest.

```
lm1 <- lm(PercComplete~.*.,data=df)
summary(lm1)
# main effect plot
```

(Continued)

(Continued)

```
plot(aggregate(PercComplete~Type,df,mean),type="l")
# interaction plot
interaction.plot(df$Type,df$Region,df$PercComplete)
```

Hint: In Figure 6.5 we put five plots in a row. To do this, before making your plots, issue the command: `par(mfrow=c(1,5))`. If you do this, for the rest of the session, all plots will be put five in a row. To stop this behavior, issue the command `par(mfrow=c(1,1))` after making the plot with five graphs in it.

Note also that all the plots have the same y-axis scale, which was chosen to accommodate the largest effect, that of PositiveExample. This was effected by adding the option `ylim=c(50,90)` to all the plots. For example, the first plot was made by issuing this command:

```
plot(aggregate(PercComplete~Type,df,mean),type="l",
   ylim=c(50,90))
```

If you let the software choose the y limits for each graph, they will be hard to interpret.

Table 6.13 Regression with all interactions, loan experiment.

	Estimate
(Intercept)	71.2750
Type	−0.5000
Region	−1.3125
Description	6.6125
PositiveExample	18.0625
NegativeExample	0.1875
Type:Region	0.3375
Type:Description	−0.7375
Type:PositiveExample	1.2875
Type:NegativeExample	−0.8375
Region:Description	−0.0250
Region:PositiveExample	0.4000
Region:NegativeExample	1.4500
Description:PositiveExample	−0.6000
Description:NegativeExample	−0.5500
PositiveExample:NegativeExample	−1.7500

Including the intercept, we have 16 parameters to estimate and 16 observations, so there are no degrees of freedom left to estimate the residual variance, and we cannot conduct statistical tests. Observe the coefficient for the interaction between Region and NegativeExample; it equals 1.45, which must be doubled to estimate the interaction effect, which is 2.90. This and the interaction between Type and PositiveExample might be the only interactions worth examining. The rest of the interactions are so small compared with the main effects on PositiveExample and Description that they can be ignored. (It is a general rule to ignore interactions that are small compared with significant main effects.)

After much consideration and analysis, the team decided to drop NegativeExample. In this case, the analysts used judgement to drop a statistically significant variable; they were not slaves to p-values. There were three reasons for dropping this significant variable:

1. If NegativeExample is dropped, then there is no interaction effect to worry about.
2. The effect of NegativeExample is small, and it can be safely ignored.
3. As a practical (not statistical) matter, the form would be easier to comprehend and fill out if it didn't include a negative example.

Run the regression again, this time with no interactions. Description and PositiveExample are the only significant variables, so drop the insignificant variables. Run the regression yet again, this time with just Description and PositiveExample. This appears to be our *reduced model*, the model that is left after having dropped irrelevant variables. Now, model checking is in order, which we leave to the reader.

The model checking procedure for regressions recommended by Mee (2009, §2.6) is:

1. Plot the residuals against $\hat{y}$
2. Plot the residuals against run order
3. Create a normal probability plot of the residuals
4. Identify outliers

We will be quite concerned with the R^2 of our regressions also with the estimate of the error variance (the two are kissing cousins). The estimate of the error variance is the sum of the squared residuals divided by the degrees of freedom:

$$\hat{\sigma}^2 = \frac{\sum_{i=1}^{n} e_i^2}{n-k-1} = \frac{\sum_{i=1}^{n} (y - \hat{y}_i)^2}{n-k-1} \tag{6.13}$$

where there are n observations, k independent variables (not counting the constant, hence the "−1" in the denominator), and $\hat{y}$ is the fitted value of the dependent variable. Obviously, the closer $\hat{y}$ is to y, the smaller will be the error variance and the higher will be the R^2. There is another way to make the error variance smaller, and that is to increase the degrees of freedom (which is the term in the denominator).

The square root of the error variance is the standard deviation of the error, which has many names, e.g. "root mean square error" or "S" or "standard error of the regression," while R calls it "residual standard error." Whatever the name, it can be useful to think of it as the standard deviation of the residuals. We will often pay attention to whether this quantity goes down or up as we pursue a modeling strategy. If RSE decreases, we are better able to see effects that are really there. If RSE increases, we are more likely to commit a type II error and miss effects that are truly present in the data.

Model checking is in order, and there are no red flags.

Having validated the model, we can now use it. In particular, we will use the estimated regression model to predict the response rate when the description is used along with the positive example:

$$71.2750 + 6.6125 * (+1) + 18.0625(+1) = 95.95 \tag{6.14}$$

As it turned out, the redesigned forms increased application completeness from 60% to more than 95%! Resources dedicated to reprocessing incomplete applications dropped significantly, resulting in great savings. Further, the cycle time for the loan process dropped markedly, resulting in increased customer satisfaction and increased market share. This experiment was a resounding success.

It is easy enough to do the math for Equation 6.14 by hand, but it can be useful to have the software do it for us, especially when there are many variables or we want many predictions.

Software Details

To create a prediction for a single observation...

The "predict" command requires as input a data frame with the values of the independent variables for which the prediction is wanted. So we will have to make a data frame with one row and two columns and the appropriate column names. First we create a matrix with one row and two columns with the correct numbers, then we assign the correct column names for the numbers, and finally we give the new data frame to the predict command.

```
lm1 <- lm(PercComplete~Description+PositiveExample,data=df)
newdf <-   data.frame(Description=+1,PositiveExample=+1)
predict(lm1,newdata=newdf)
```

Exercises

6.3.1 For the data in the right panel of Table 6.9, compute the pairwise dot products for the three dummy variables. Are the variables orthogonal? Why or why not?

6.3.2 Using the file `Houses.csv`, regress price (in thousands of dollars) on square footage, age (in years), and dummies for fireplace and type of house (colonial, rancher, or split level). Write an equation to predict the price of a colonial house with a fireplace that is 30 years old and has 1500 square feet. (This is sometimes also called a "response function.")

6.3.3 Sometimes, looking at a graph one way is not very illuminating, while looking at it another way is illuminating. Consider the right plot in Figure 6.6. Graph this, but switch it around so that PositiveExample is on the x-axis. How illuminating is this? Do the same for the left plot. Be prepared to make all interaction plots both ways, and then choose the one that works best.

6.3.4 Dummy coding and effect coding give the same predictions. Use the data from Table 6.9. Using the effects, run the regression and predict the response when factorA is high and factorB is low. Do the same for the regression with dummies.

6.3.5 Make an interaction plot for the data in Table 6.3. Before you make the plot, describe what it will look like.

6.3.6 Consider Table 6.9. Change 31 to 49. Now there is no interaction effect (to within computer rounding error). Verify this by making an interaction plot. Next run the regressions for effects and dummies (without interactions, of course). Verify that they give the same answer. What can you conclude?

6.3.7 Revisit Excercise 6.2.2. There is no interaction for this model. How do you know? (Create an interaction plot or run a regression; you should be able to do it both ways.)

6.4 Case: Direct Mail (Three-Way Interactions)

As we saw in the last section, to have a two-way interaction required two variables to interact. For a three-way interaction, we will need three variables, and this is a *third-order effect*. The cube of a factor also is a third-order effect, but we do not use cubed factors in this book. Products such as A^2B and BC^2 also are third-order effects, but we don't use these, either. A credit card company has sent out a direct mail solicitation with three factors:

- price (A) – The interest rate charged – either 4.99 or 10.99% (the usual rate exceeds 17%)
- duration (B) – Of the special low interest rate – either 6 months or 1 year
- promo (C) – A promotional bonus of frequent flyer miles, either 1000 or 5000

The design matrix is given in Table 6.14, where "std order" is an abbreviation for "standard order." It gives the row numbers moving from all low to all high levels. That the numbers are all in order indicates that the design has not been randomized yet; always run the experiments in a random order.

Just as there were two equivalent interpretations of a two-way interaction, there are three equivalent interpretations of a three-way interaction:

- It measures the effect on the response rate of the price*duration interaction when promo is changed from lo (4.99%) to hi (10.99%).
- It measures the effect on the response rate of the price*promo interaction when duration is changed from lo (six months) to hi (one year).
- It measures the effect on the response rate of the duration*promo interaction when price is changed from lo (no miles) to hi (1000 miles).

We can also write the data in the form used to estimate the model, which is called the *model matrix*, which shows the model that will be estimated. It will include a column of ones to represent the intercept term, as well as any factors and interactions for which we wish to compute effects. In the present case we have all the interactions, so the model matrix is given by Table 6.15. It is a good thing

Table 6.14 Design matrix for credit card case.

Std order	*A*	*B*	*C*
1	−1	−1	−1
2	1	−1	−1
3	−1	1	−1
4	1	1	−1
5	−1	−1	1
6	1	−1	1
7	−1	1	1
8	1	1	1

Table 6.15 Model matrix for credit card case.

Std order	Int	*A*	*B*	*C*	*A* : *B*	*A* : *C*	*B* : *C*	*A* : *B* : *C*
1	1	−1	−1	−1	1	1	1	−1
2	1	1	−1	−1	−1	−1	1	1
3	1	−1	1	−1	−1	1	−1	1
4	1	1	1	−1	1	−1	−1	−1
5	1	−1	−1	1	1	−1	−1	1
6	1	1	−1	1	−1	1	−1	−1
7	1	−1	1	1	−1	−1	1	−1
8	1	1	1	1	1	1	1	1

we can reference the factors by letters rather than variable names or we'd have trouble finding room to write "price*promotion*duration" instead of "ABC" for the three-way interaction.

The data are contained in `CreditCard1.csv`. The effects, rounded to four decimals, are presented in Table 6.16. Remember, the effects are twice the regression coefficients.

It appears that promo and duration have large main effects, while the main effect of price is small. Relative to the main effects, the interaction effect for price*duration is large, while the other interaction effects are small.

Try it!

Reproduce the results in Table 6.16. Do this, of course, by running a regression and doubling the coefficients.

Table 6.16 Effects for credit card data set.

price	(*A*)	−0.0775
promo	(*B*)	0.5775
duration	(*C*)	0.3725
price*promo	(*AB*)	−0.0775
price*duration	(*AC*)	−0.2725
promo*duration	(*BC*)	−0.0275
price*promo*duration	(*ABC*)	−0.0725

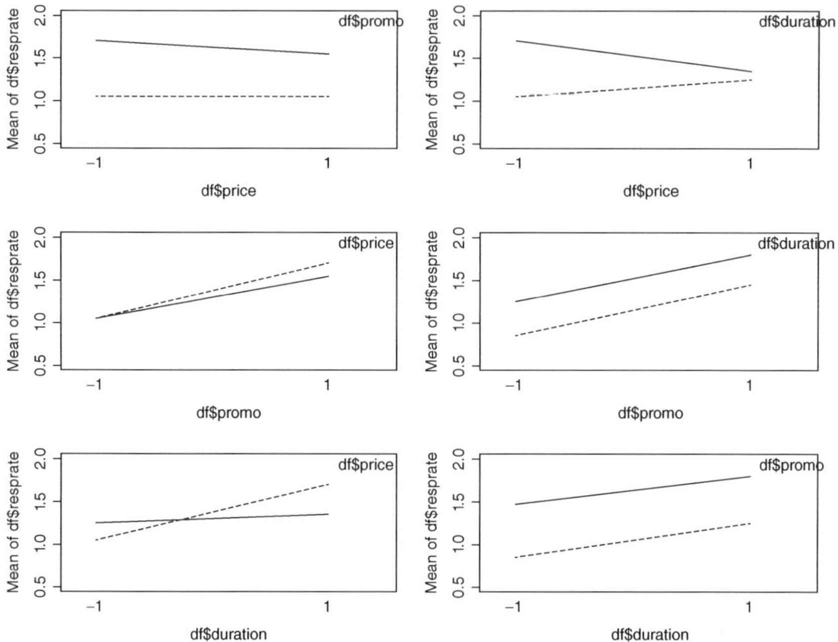

Figure 6.7 Two-way interaction plots for credit card example.

Let us interpret the two-way interactions, profiles of which are in Figure 6.7. Realize that in these six plots, there are three pairs, each pair showing the same information in perhaps a different way. For example, the top right and bottom left are a pair, each showing price and duration against the response.

In the first row, the first graph says that there is slight interaction between price and promo, but should we include it? Realize that the first graph on the second row represents the same information; both graphs depict the interaction between price and promo. The latter graph shows the almost parallel nature of the lines more clearly. The second graph on the first line shows a slightly greater interaction, but it is equivalent to the first graph on the third line, which more starkly shows the interaction between price and promo. The second graph on the second line and the second graph on the third line both depict the lack of interaction between promo and duration.

The regression estimates eight parameters (the intercept and seven slopes) with eight observations. This is an example of a *saturated model*, which is a model that has as many parameters as data points. It has no degrees of freedom left over to estimate the error variance. Since we have no residual degrees of freedom, we cannot conduct traditional hypothesis tests. How are we to determine which of the coefficients are significant and which are not?

Now, there does exist a population standard error, but we can't estimate it because we don't have enough observations; indeed, we don't have *any* degrees of freedom with which to estimate the error. Lenth proposed a method for computing a pseudo-standard error (PSE) (the precise details of which concern us not) to use in place of the standard error that we could compute if only we had enough observations. We make use of Lenth's PSE and construct a normal plot of the effects (normal probability plots were employed in Section 4.3). Such a normal plot is shown in Figure 6.8, and it works just like a normal probability plot, except that instead of applying the normal probability plot to data, we apply it to the estimated effects. The underlying theory is that insignificant effects are random noise and thus normally distributed. Hence, points that fall on the line are normal and thus insignificant, while points that fall off the line are not normal and thus significant. (There is also a "half-normal plot" that is applied to the absolute values of the effects, and it works the same way. For our purposes, it doesn't matter whether you use the normal effects plot or the half-normal effects plot.)

We treat half-normal plots as an aid to a modeling strategy rather than as a substitute for a modeling strategy. In particular, we treat these plots as suggestive and not conclusive because of the great variation in such plots when there are only a few coefficients; see Appendix 3a of Daniel and Woods (1980) for more on this point. For example, points that are close to the line may nonetheless prove to be significant. On the other hand, we wouldn't want to dismiss a point that is far from the line without using a statistical test to verify that such a point really was insignificant or for some other good reason. Recall that in the previous section, the variable NegativeExample, though statistically significant, was dropped from the model for good reasons.

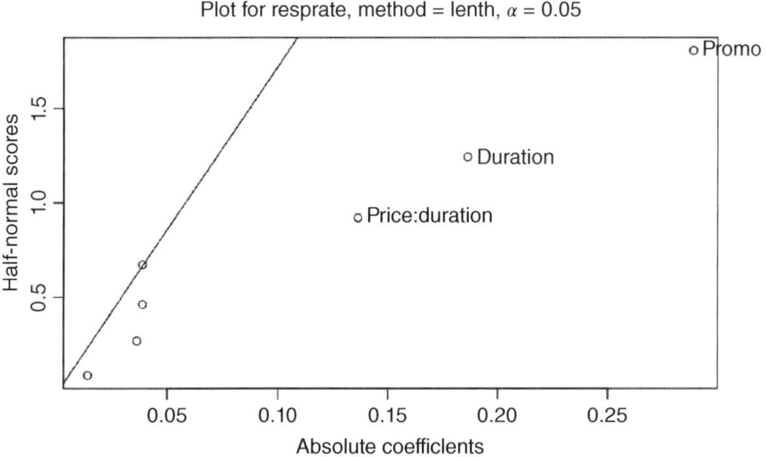

Figure 6.8 Half-normal plot for the fully saturated credit card experiment.

The half-normal plot in Figure 6.8 has identified three points that are far off a line drawn through all the points (from left to right: price*duration, duration, and promo). This result is consistent with our general intuition that small higher-order effects (small relative to the main effects) can be dropped. We're not going to worry about formal significance right now. If the point is far off the line, we're going to include that variable. If that variable shouldn't be included, our modeling efforts will tell us later. This will turn out to be a smart decision and suggests that we shouldn't use statistical tests mechanistically when we are modeling with real data.

Software Details

To create a half-normal plot...

The package "DoE.base" must be installed. This only has to be done once.

```
install.packages("DoE.base")
```

After it has been installed, it must be loaded into memory; this has to be done every time you want to use the package.

```
library(DoE.base)
```

Run a regression to create a linear model object.

```
lm1 <- lm(resprate~price*promo*duration,data=df)
```

We like to have a line drawn on this probability plot to make it easier to interpret the points, and it is often easier to produce the plot and draw the line all at once. To do this, execute the following commands.

```
hncoeff <- halfnormal(lm1, half=TRUE)$coef
pse1 <- ME.Lenth(hncoeff)$PSE
abline(a=0, b=1/pse1)
```

By default, only "significant" effects are labeled in the plot. If you want to see what all the labels are, just make alpha large.

```
halfnormal(lm1, half=TRUE, alpha = 0.95)
```

Our general modeling strategy for a saturated design is as follows. If the highest-order interaction coefficient is small relative to the coefficient of the factor variables themselves, drop the highest-order interaction from the model (if it's large, choose some other variable with a small coefficient to drop). This frees up a degree of freedom that can be used to estimate the error so that hypothesis testing on the coefficients can be conducted. (Realize that the more variables are dropped, the more degrees of freedom will be available to estimate the error, so the estimate of the error will improve.) Then drop the variable that has the highest p-value and rerun the regression. Repeat this process until all the remaining variables are significant (or are to be kept in the model for some reason even if they aren't significant, for example, due to the *marginality principle*). The marginality principle asserts that if an interaction between two variables is significant, both of the individual variables should be included in the model (and similarly for three-way interactions). For example, if the coefficient on the three-way interaction term ABC is significant, and the individual variables A and B are significant, too, but the coefficient on C is insignificant, the variable C should still be included in the model.

There may be statistically significant variables that are practically insignificant; drop these, too. If you are unsure whether a statistically significant variable is practically significant, compute the fitted values ($\hat{y}$) both with and without the variable. If the fitted values ($\hat{y}$) are practically the same, the variable is practically insignificant.

Let us apply our modeling strategy to the present case:

1. Create new interaction variables, for sake of space let us call them pripro, pridur, produr, and priprodur. Regress resprate on price, promo, and duration and the four interaction variables. The normal plot did not indicate the priprodur is important, and its coefficient is an order of magnitude smaller than the coefficients for promo and duration and the same order as the coefficient on price.
2. Let's just make sure that priprodur is insignificant. Though this can't always be done, we can do so in this case because a lower-order interaction term (produr) has a miniscule coefficient. To do this, drop produr, which has the smallest coefficient by far, and rerun the regression: the resulting *p*-value on priprodur is insignificant, so let's add produr back to the model and drop priprodur, and run the regression again.
3. Notice that the standard error of the regression is 0.102 and the standard error of the coefficients is 0.0362. The largest *p*-value is on produr, so drop it and run the regression again.
4. Notice that the standard error of the regression is now 0.0775, a large reduction, and the standard error of the coefficients is now 0.0274; the coefficients are being estimated more precisely with more degrees of freedom. The largest *p*-value now is a tie between price and pripro, so drop the interaction term pripro rather than the main variable price and run the regression again.
5. The standard error of the regression and the standard error of the coefficients both have increased slightly, but that's ok; we can expect that they are better estimates than those of the previous regression because of the increased degrees of freedom. All the coefficients except for price are significant, so we keep them, and we keep the variable price because of the marginality principle.

After the model has been estimated and both statistically and practically insignificant variables have been removed, the result is called the *reduced model*.

Table 6.17 Regression results for credit card model.

Parameter	Estimate	Std err	*t*-Stat	*p*-Value
Intercept	1.339	0.032	42.30	<.0001
Price	−0.039	0.032	−1.22	0.3082
Promo	0.289	0.032	9.12	0.0028
Duration	0.186	0.032	5.89	0.0098
Price*duration	−0.136	0.032	−4.31	0.0231

Our *reduced model* for this experiment will be

$$\text{resprate} = b_0 + b_1 * \text{price} + b_2 * \text{promo} + b_4 * \text{duration} + b_5 * \text{pridur} \tag{6.15}$$

The results of this regression are presented in Table 6.17.

Try it!

Reproduce the results in Table 6.17.

The experimental goal is to increase the response rate. The effect of price is negative, so if price is set at the low level (−1), then the effect of the term price * (−1) will be positive. Similarly, we want to set the levels of promo and duration at high levels (+1).

The coefficient of the interaction term price*duration is negative. To have an overall positive influence on response, we need price*duration to be negative. Can this be done? Yes, if price (−1) is low and duration (+1) is high. Looking at main effects, we saw that we want to set price to low and duration to high. Looking at the interaction effect, we saw again that we want to set price to low and duration to high. Both the main effects and the interaction effects are in agreement. This is not always the case, as will be seen in the next chapter in Section 7.2. In such a situation, the easiest way to decide what the levels should be is to compute the fitted values for all possible combinations as in Table 6.18 and pick the combination that has the highest fitted value for the response.

Try it!

What is the fitted value of the response rate when price is set at the low level and both promo and duration are set at the high level?

Table 6.18 Predictions (fitted values) for credit card model.

Price	Promo	Duration	Resprate	Prediction
1	1	1	1.55	1.63875
−1	1	1	2.05	1.98875
1	−1	1	1.15	1.06125
−1	−1	1	1.35	1.41125
1	1	−1	1.55	1.53875
−1	1	−1	1.36	1.34375
1	−1	−1	0.95	0.96125
−1	−1	−1	0.75	0.76625

Notice that as our intuition and the half-normal plot indicated, price is insignificant and the rest of the variables are significant. It was a good thing we ignored the significance test on the points in the normal effects plot. If we hadn't included the two "insignificant" points, we'd have never found out that they really are significant! All the standard errors are the same, which is to be expected because the independent variables are effect-coded. Model checking is in order, which we leave as an exercise.

Since the data set contains all the variables at all the levels, we can simply apply the estimated model to the data set and get fitted values for all possible combinations of the factors. This is shown in Table 6.18.

Immediately we can see that the highest predicted response rate is 1.988 75 when price is 4.99%, promo is 1000 frequent flyer miles, and duration is one year.

Software Details

To get fitted values from the regression model...

The lm object, say, `lm1`, already has the fitted values attached to it. These may be accessed by `lm1$fitted.values`. The fitted values can be added to the data frame by issuing the command

```
df$predictions <- lm1$fitted.values.
```

The data frame now looks just like Table 6.18.

When the effect-coded variables are arranged in a matrix, e.g. three leftmost columns of Table 6.18, the matrix is called a *design matrix*; it shows the specific experiments that will be run with which factors at high and low levels. The rows of the design matrix are sometimes called *design points*. We will see in the next chapter that a large class of experiments can be analyzed by regressing a response variable on a properly constructed design matrix. For a clear distinction between the design matrix and the model matrix, refer to Tables 6.14 and 6.15.

Exercises

6.4.1 Perform model checking on the credit card regression model.

6.4.2 Drop the interaction term from the credit card model. What happens to R^2? What happens to the predictions? What can you conclude?

6.5 Pretreatment Covariates in Regression

Modeling pretreatment covariates in a regression model sometimes is called ANCOVA. ANCOVA stands for "analysis of covariance" and is an extension of ANOVA that allows the researcher to control for variables (covariates) that are not part of the experiment. In order for these covariates to be unrelated to the treatment, they have to be measured before the experiment takes place, so they have to be pretreatment covariates. Whether you call this method ANCOVA or "pretreatment covariates in regression" does not matter, but you should be able to recognize both names.

Adding variables to a regression is useful if the additional variables explain the dependent variable. Recall the standard variance decomposition (refer to your introductory statistics book if necessary):

$$\begin{array}{ccccc} \text{SST} & = & \text{SSR} & + & \text{SSE} \\ \text{Total variation} & = & \text{Explained variation} & + & \text{Unexplained variation} \\ \text{in } y & & \text{due to } X & & \text{due to error} \end{array}$$

where SST is total sum of squares of the dependent variable, SSR is regression sum of squares (the "explained" sum of squares due to the independent variables), and SSE is error sum of squares (the "unexplained" sum of squares due to the residual). No matter how many variables we add to the regression equation, SST doesn't change because the dependent variable stays the same. When we add meaningful independent variables, SSR goes up, and necessarily, SSE goes down. (SSE going down is the same thing as the standard error of the regression going down.)

In order for an effect to be significant, it must be roughly twice the standard error of the regression. We have no control over the size of an effect, but we can influence the standard error of the regression. The smaller we can make the standard error, the more effects we can "see." Here we give a simple example.

Suppose we have the current method of university math instruction, Program A, and a new method, Program B. The outcome is the test score. A group of students are randomly assigned to either program, and then they take the test. The data are in `mathtest.csv` where progB $= -1$ for Program A and $+1$ for Program B. This

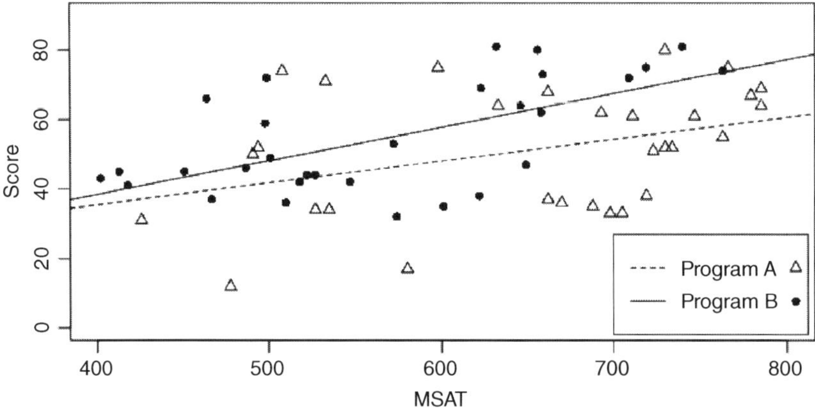

Figure 6.9 Score vs. MSAT.

is really a two-sample test of means, but we will handle it using regression. The coefficient on progB is insignificant, and the *R*-squared is less than 2%.

Maybe there is an effect, and we just can't see it because there is too much noise in the model. Perhaps we can introduce another variable that is correlated with the response that could soak up some of the unexplained variation. Indeed, the students' SAT math scores are just such a variable. Students with high SAT math scores are likely to do better on a math test than students with low SAT math scores. Let's add the variable MSAT to the model as a covariate. First, we'll check for interactions and find none, so we'll drop the interaction term. The resulting reduced form shows that both progB and MSAT are highly significant. The *R*-squared has increased to over 20%, and the residual standard error has declined; MSAT has soaked up some of the noise and allowed the effect of progB to be observed. Figure 6.9 confirms the positive relationship between score and MSAT, and the fitted lines for Program A and Program B show that, on average, Program B has a higher score than Program A.

In real life, we would not proceed as above. Even before running the experiment, we would decide that we want to use MSAT as a covariate. If we were to get an insignificant result and then start shopping for "useful" covariates, we would get hopelessly lost in the garden of forking paths.

Exercises

6.5.1 Run the regressions reported in this section and confirm the R^2s of less than 2% and more than 20%.

6.5.2 Reproduce Figure 6.9 as best you can; it was made with ggplot2, which you are not expected to know. But try using the `plot` command in R. At least, plot the points with different shapes. You can put lines on a plot using the `abline()` command, and change the type of line by using the `lty=` option.

6.5.3 What is the effect of using Program B instead of Program A? (Remember to double the regression coefficient.) Express this as a 95% CI (i.e. get a 95% CI for the coefficient on Program B and then double it).

6.5.4 What is the effect of MSAT? (Do *not* double the regression coefficient, as this is a covariate not an effect-coded variable.) Express this as a 95% CI.

6.5.5 The above CI for MSAT is for a one-point change in MSAT score and might be hard to interpret because of the leading zeros in the decimals. It might be better expressed as a CI for a 10-point change in MSAT score. Do so. Which CI do you think would be more easily communicated to an audience, the CI for a 1-point change or the CI for a 10-point change?

6.5.6 Revisit the boys' shoes example from Exercise 4.2.5. The data file `BoyShoes.csv` includes the variable Afoot, which records on which foot (right or left) the type A material was placed. Does which foot gets the material affect the rate of wear? First run a regression to perform the paired t-test. Then add Afoot to the regression as a covariate. Comment on your results.

6.6 Chapter Exercises

6.1 Your family runs a small web business and wants to improve sales and also wants to collect contact information from customers in order to send marketing emails about promotions and sales. To these ends, free shipping is offered with purchases totaling more than \$30 (low) or \$50 (high), and customers are given the option to create a profile before checking out; the profile is either required (high) or optional (low). Sales vary predictably throughout the week, so to eliminate this source of variation, the experiment is only run on Tuesdays for eight weeks. Data are presented below.

Date	Free shipping over	Profile required	Total sales
05 January 2010	\$30	Yes	\$3275
12 January 2010	\$50	No	\$3594
19 January 2010	\$50	No	\$3626
26 January 2010	\$30	No	\$3438
02 February 2010	\$50	Yes	\$2439
09 February 2010	\$30	No	\$3562
16 February 2010	\$30	Yes	\$2965
23 February 2010	\$50	Yes	\$2571

Create effect-coded variables and run the regression, being sure to include an interaction term. Is there a main effect? Is there an interaction effect? Should they offer free shipping at \$30 or \$50 dollars? (This is a trick question – remember the interaction!) What combination produces the highest revenue?

What is the effect of including the interaction? Run the regression with no interaction term and note the R^2. Compare it with the R^2 from the regression that includes the interaction.

6.2 This exercise explores a useful way to interpret interactions. Suppose you have estimated the model $y = a + bx_1 + cx_2 + dx_1x_2$ and obtained $a = 10, b = 5, c = 8$ and $d = 2$. The effect of x_1 is $b + dx_2 = 5 + 2x_2$, which cannot be interpreted without knowing something about x_2. Suppose the range of x_2 is $[1, 5]$. The low end of the effect of x_1 is $5 + 2(1) = 7$, and the high end of the effect of x_1 is $5 + 2(5) = 15$. This is *not* a confidence interval, and it does not represent uncertainty. It represents the range in the estimated treatment effect of x_1 on y. Suppose the range of x_1 is $[6, 10]$. What is the range of the effect of x_2 on y?

6.3 Distinguish between a design matrix and a model matrix.

6.7 Learning More

Section 6.1 "Experiments and Linear Regression"

• Why do we need regression in experimental design when we already have ANOVA? There are four reasons. First, ANOVA can't handle unbalanced designs (remember, a balanced design is when each cell has the same number of observations). Second, ANOVA can't handle continuous independent variables. Third, we can add additional variables to the analysis to soak up residual variation and decrease the noise; this is much easier to do with regression (this method is called ANCOVA and is discussed in Section 6.5). Fourth, very often we are concerned with the magnitude of effects, not just their existence, and magnitudes are front and center in regression (in the form of regression coefficients), whereas they are hidden away in ANOVA.

• ANOVA is very useful for observational data, when we wonder whether some variable affects another variable. When we design an experiment, we're usually pretty sure that one variable affects another – why else include the variable in the analysis? – and we want to estimate the effect of one variable upon another. For this latter purpose, regression is much more useful than ANOVA.

• For those familiar with matrix notation (if you're unfamiliar, skip this paragraph and don't worry about it), we have

$$y = X\beta + \varepsilon \tag{6.16}$$

where X is a matrix of n rows, one for each observation, and $k + 1$ columns, where the first column is entirely ones (for the intercept) and the remaining columns are the k variables. The covariance matrix of the vector of coefficients b is given by

$$Var(b) = \sigma^2(X'X)^{-1} \tag{6.17}$$

so that the variance of a particular coefficient, say, b_i, is given by the product of the standard error of the estimate σ^2 and the ith element of the main diagonal of the matrix $(X'X)^{-1}$. When the variables are orthogonal, then the covariances (off-diagonal elements of $(X'X)^{-1}$) are all zero, which makes life nice because adding and dropping variables doesn't change the coefficients.

The usual interpretation of b_1, say, is the *estimated* expected change in Y for a one-unit change in X_1. (Remember, β_1 is an unknown population parameter that is estimated by its corresponding sample statistic b_1, the same way that $\overline{x}$ estimates μ.) However, we prefer to summarize regression results using descriptive terminology: comparing two observations that differ by one in x_1 and that are identical in all the other x's, we predict y to differ by b_1 on average. It's much more time consuming to say or to write, but it's much more intuitive and easier for nonspecialists to understand.

The above representation uses an intercept in the model. This is not necessary. We could use two dummies and drop the intercept term:

$$D_m = \begin{cases} 1, & \text{if male} \\ 0, & \text{otherwise} \end{cases} \quad D_f = \begin{cases} 1, & \text{if female} \\ 0, & \text{otherwise} \end{cases} \tag{6.18}$$

in which case the regression equation becomes

$$Y = \beta_1 X + \beta_2 D_m + \beta_3 D_f \tag{6.19}$$

for which the response functions are

$$E[Y] = \beta_2 + \beta_1 X_1 \text{ if male} \tag{6.20}$$

$$E[Y] = \beta_3 + \beta_1 X_1 \text{ if female} \tag{6.21}$$

A test for the regression lines being the same is $H_0 : \beta_2 = \beta_3$ against $H_A : \beta_2 \neq \beta_3$. Observe that the same hypothesis (that the regression lines are the same) can be tested in different ways depending on how the model is written.

Section 6.2 "Dummies, Effect Coding, and Orthogonality"

• There are many different types of orthogonal contrasts, and they can be used to analyze the results of experiments that have different designs. Sometimes, different contrasts can be applied to the results of the same design to examine different aspects of the results.

Effect coding only allows for testing simple contrasts, in which each level of a factor is compared with the mean of a baseline level, and in more complicated experiments it can be necessary to test complex contrasts. There are many coding schemes, depending on what types of comparisons you want to make. If you want to compare each coefficient with a baseline, just use traditional 0–1 dummy variables (This is called *simple coding or treatment coding*). To compare each variable with a grand mean, use effects contrasts (also called *deviation coding*). If the variables have a specific order, to compare the level of one variable with the mean of subsequent levels, use *Helmert contrasts.* To compare levels of a variable with the mean of previously levels, use *reverse Helmert contrasts.* To compare each level with the next level, use *forward difference coding*. To compare each level with the previous level, use *backward difference coding.* If none of those does what you want, there are always *user-defined contrasts.* The reader interested in knowing more about contrasts can consult Chapter 4 of the text by Oehlert (2010), which is available on the web in pdf form via a Creative Commons license. For our purposes, we'll use effect contrasts 90% of the time, and, if there are no interactions and our audience only knows dummy variables, then we'll use dummy variables.

• $C(n, k)$, also written $\binom{n}{k}$, denotes the number of ways to choose k items from a group of n items when order does not matter. The formula is $C(n, k) = n!/(k!(n - k)!)$ where ! denotes "factorial." $C(5, 2) = 5!/(2!3!) = (5 \times 4 \times 3 \times 2 \times 1)/(2 \times 1 \times$

$3 \times 2 \times 1) = (5 \times 4)/(2 \times 1) = 20/2 = 10$. If you need more of a refresher, consult your first statistics text.

• Some packages have procedures for conducting experiments, and they automatically create necessary dummy variables, etc. As we have indicated, they may not create dummy variables or effect-coded variables the way we would want. There is another complication from such packages: some software packages report effects, while others report coefficients (Fontdecaba et al., 2014) and do not make this clear to the user.

• This is not a linear algebra book, but we want to get across the idea that a dot product equal to zero implies orthogonality, which implies correlation equal to zero. Figure 6.10 shows six vectors: a, b, c, d, e, and f. A vector that begins at the origin is denoted by the coordinates of its ending point, e.g. $a = (1, 0)$ means the vector a begins at the origin and ends at the point $(1, 0)$. Correlation indicates the extent to which two vectors point in the same direction. For example, the vectors c and e have a correlation of 1.0. The vectors a and f have a correlation of -1.0. In general, if two vectors make an acute angle, their correlation is positive, and if they make an obtuse angle, their correlation is negative. At that point where the angle is neither acute nor obtuse (an orthogonal angle), their correlation is zero. Orthogonal means "at a right angle to," so if a is orthogonal to b, then of course b is orthogonal to a, and they are at right angles to each other. Write out vector b (we have already written out vector a), and calculate the dot product of a and b. Do similarly for c and d.

Another consequence of orthogonality is that when the predictors are orthogonal, the standard errors of the coefficients are all the same – provided that there are enough residual degrees of freedom to estimate the standard errors. When the design matrix is orthogonal, the standard error for each estimated regression coefficient is the same and equal to the square root of (MSE/n) where MSE is mean square error and n is the number of observations.

When all the variables are uncorrelated, the covariance matrix is diagonal.

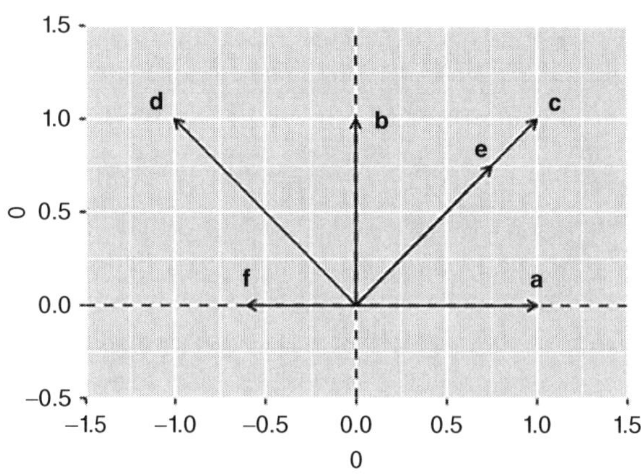

Figure 6.10 Pairs of orthogonal vectors.

Section 6.3 "Interactions"

• An interaction term in a regression is a product of two or more independent variables. For example, adding an interaction term to $Y = \beta_0 + \beta_1 X_1 + \beta_2 X_2 + \varepsilon$ yields $Y = \beta_0 + \beta_1 X_1 + \beta_2 X_2 + \beta_3 X_1 X_2 + \varepsilon$. In practice, we create a third variable $X_3 = X_1 X_2$ and run the regression $Y = \beta_0 + \beta_1 X_1 + \beta_2 X_2 + \beta_3 X_3 + \varepsilon$. Alternatively, we could call the interaction X_{12} to represent the fact that it's the product of X_1 and X_1, i.e. $X_{12} = X_1 X_2$, and then estimate $Y = \beta_0 + \beta_1 X_1 + \beta_2 X_2 + \beta_{12} X_{12} + \varepsilon$.

The slope in a regression with an interaction term no longer has a simple interpretation. The expected change in Y from a one-unit change in X_1 is $\beta_1 + \beta_3 X_2$, so the effect on Y of changing X_1 depends on the value of X_2; this is a nonlinear relationship.

Suppose we have estimated a model with two interactions terms, say,

$$\hat{y} = b_0 + b_1 x_1 + b_2 x_2 + b_3 x_3 + b_{13} x_1 x_3 + b_{23} x_2 x_3 \tag{6.22}$$

and we want to know the effect of X_1 and X_2 when $x_3 = 5$. Simply substitute $x_3 = 5$ into the equation to get

$$\hat{y} = b_0 + b_1 x_1 + b_2 x_2 + b_3 \cdot 5 + b_{13} x_1 \cdot 5 + b_{23} x_2 \cdot 5 \tag{6.23}$$

whence

$$\hat{y} = (b_0 + 5b_3) + (b_1 + 5b_{13}) x_1 + (b_2 + 5b_{23}) x_2 \tag{6.24}$$

and the effect of changing X_1 is seen to be $(b_1 + 5b_{13})$, while the effect of changing X_2 is $(b_2 + 5b_{23})$. If the value of x_3 had been something other than 5, all these effects would have taken on different values. Thus, the effect of changing X_1 or X_2 depends on the value of X_3.

A bit of terminology is in order here. The main effects, in this case X_1, X_2, and X_3, are collectively referred to as "first-order effects." The two-way interaction term is a "second-order effect." A three-way interaction, which we present in the next section, is a "third-order effect."

• The marginality principle, advocated by Nelder (1977), holds that in a model with main and interaction effects, the main effects are marginal to the interaction effects. This principle implies that in a model with interactions, a lower-order term is not tested if a higher-order term is significant. The reason is that the lower-order term is difficult to interpret if a higher-order term is significant. Moreover, when using such a model to make predictions, if the interaction is significant, then the main effects of the interacting variables should be included even if they are insignificant.

There is another important concept called the principle of effect hierarchy. This principle says that most of the variation in the response can be attributed to the main effects and second-order effects account for more of the variation in the response than third-order effects. What this means is that first-order effects are more important than second-order effects, which, in turn, are more important than third-order effects. This has implications for our modeling strategy. If we had a model with first-, second-, and third-order effects and we had to guess which coefficients would matter most, we would say that the first-order coefficients matter more than the second-order coefficients.

• One has to be careful to assess the probable effects of interactions before conducting an experiment. Suppose you were a web designer and you wanted to know the best combination of font color and background. You might run experiments every hour for several days, testing all manner of combinations. What would happen when you tested a white font against a white background?

• Look at the difference between Figures 6.5 and 6.6. The former has been "prettified," while the latter is standard R output. It's okay to use standard R graphs for yourself while conducting an analysis, but never show them to other people. Always prettify your graphs before displaying them to anyone else.

Section 6.4 "Three-Way Interactions"

• Some might suggest that our modeling strategy is naive and equivalent to backward stepwise regression and will be problematic for full factorial designs with several factors. We plead guilty with extenuating circumstances: this is a second course in statistics for persons whose first course didn't use calculus. Persons desiring a better modeling strategy can consult the superb text by Harrell (2015), appropriately entitled *Regression Modeling Strategies, 2e.*

• A finding of no difference is not meaningless; it gives us the opportunity to select the least costly option.

• Do not use *p*-values mechanistically. In the case of the credit card data, we ignored the formal test of significance and found that, in our final model, those variables were significant. It was a good thing we ignored the formal test. Suppose, though, that the (near) final model had shown that the insignificant variables really were insignificant. Then we'd have just dropped them from the model and wind up where we would have wound up if we had been slaves to *p*-values. No harm, no foul. (Box et al., 2005, p. 188) summed it up nicely:

> In the proper use of statistical methods *information about the size of an effect and its possible error* must be allowed to interact with the experimenter's subject matter knowledge. (If there were a probability of $p = 0.50$ of finding a crock of gold behind the next tree, wouldn't you go and look?)

Section 6.5 "Pretreatment Covariates in Regression"

• "Pretreatment covariates in regression" is sometimes called "ANCOVA," which is short for "analysis of covariance."

• The test score example is taken from an example in chapter 10 of the book by Seltman (accessed 25 October, 2018). Chapter 10 is devoted to ANCOVA, so if you wish to learn more, it is a good place to start. The book is available on the web for free.

• When factors have three levels, we can detect curvature in the response function. In this book we use only two levels. It is also possible to investigate curvature by augmenting the two-level design with adding center points to the design; see Chapter 6 of Montgomery (2017).

• A good article on why one factor at a time is not good is Czitrom (1999).

• The exercise about the family with the webstore is taken from p. 301 of Dunn (2019).

6.8 Appendix: The Covariance Matrix of the Regression Coefficients

In order to read Chapter 9 on custom design, you need to know something about the covariance matrix of the regression coefficients. You don't need to be an expert; you just need to have some idea of what it is and what it does. We need to introduce some terms from matrix math, but it's okay if you don't know matrix math. There are no exercises for this appendix because you don't even have to know enough to solve problems. So just read through this appendix and do your best to understand, and if you don't understand everything, that's okay. Familiarity with the covariance matrix is crucial to understanding Chapter 9, but that's all you need: familiarity.

The covariance matrix is computed $\hat{\sigma}^2(X'X)^{-1}$, where $(X'X)^{-1}$ is the matrix inverse of $(X'X)$. By "matrix inverse" we mean that multiplying $X'X$ by $(X'X)^{-1}$ produces an "identity matrix" that has ones on the main diagonal (from top left to bottom right) and zeros everywhere else. An example of an identity matrix is

$$I = \begin{pmatrix} 1 & 0 & 0 \\ 0 & 1 & 0 \\ 0 & 0 & 1 \end{pmatrix}$$

An example of the matrix math follows. Don't worry about how these calculations are done; just follow along. X' is the transpose of X, which means that first column of X becomes the first row of X', the second column of X becomes the second row of X', etc.; so X' has $k+1$ rows and n columns. As an example,

$$X = \begin{pmatrix} 1 & 2 \\ 1 & 3 \\ 1 & 4 \end{pmatrix} \quad X' = \begin{pmatrix} 1 & 1 & 1 \\ 2 & 3 & 4 \end{pmatrix}$$

Furthermore, matrices can be multiplied, so X' can be multiplied by X, and then the inverse of the product can be computed:

$$X'X = \begin{pmatrix} 3 & 9 \\ 9 & 29 \end{pmatrix} \quad (X'X)^{-1} = \begin{pmatrix} 4.8333 & -1.5 \\ -1.5 & 0.5 \end{pmatrix}$$

with the end result that multiplying $X'X$ (called "x prime x") and $(X'X)^{-1}$ (called "x prime x inverse") yields

$$(X'X) * (X'X)^{1-} = \begin{pmatrix} 1 & 0 \\ 0 & 1 \end{pmatrix}$$

Standard regression output includes the coefficients and their standard errors. Where do the standard errors come from? They come from the covariance matrix and important behind-the-scenes computation that we need to make explicit.

Recall that the regression model (in matrix form) is $y = X\beta + \epsilon$, where the error term ϵ has mean zero and variance σ^2. The square root of the error variance is often called "the standard error of the estimate." Now, σ^2 is a population parameter that we cannot know and must estimate. We estimate it by $\hat{\sigma}^2 = \sum (y_i - \hat{y}_i)^2$ where, of course, $e_i = y_i - \hat{y}_i$ is the residual from the regression. The sample value of the standard error of the estimate is often denoted $\hat{\sigma}$.

Use the data in `SR5.csv` to run a regression of Y on X1, X2, and X3. Now we are going to calculate the covariance matrix of the regression coefficients, which is

	(Intercept)	X1	X2	X3
(Intercept)	2.2028	−0.3447	−0.0653	−0.3045
X1	−0.3447	0.0816	0.0021	0.0142
X2	−0.0653	0.0021	0.0092	0.0126
X3	−0.3045	0.0142	0.0126	0.1086

If you take the square root of the values in the main diagonal, you'll get the standard errors of the coefficients from the regression output. The off-diagonal elements tell us how the coefficients covary. For example, 0.0126 means that if the coefficient on X3 goes up, the coefficient on X2 goes up, or if the coefficient on X3 goes down, the coefficient on X2 goes down. The value 0.0142 has a similar interpretation. Conversely, −0.3045 is negative, so if the coefficient on X3 goes down, the intercept will go up.

The coefficient on X3 is 2.461. If we drop the variable X3 from the regression, its coefficient will go to zero, so to speak, which is a decrease. What do we expect will happen to the intercept and the coefficients on X1 and X2? From the above paragraph, we can deduce that the intercept will increase and the coefficients on X1 and X2 both will decrease.

For the problem at hand, let X be the matrix composed of four columns, the first column is all ones, and the second through fourth columns are the variables X1, X2, and X3. So X has $n = 50$ rows where n is the number of observations, and $k + 1 = 3 + 1 = 4$ columns where k is the number of independent variables.

Try it!

Regress Y on just X1 and X2, and confirm that the intercept increases and the two slopes both decrease when X3 is dropped.

Software Details

To get the covariance matrix of the coefficients...

Run a regression in the usual fashion, e.g. `mylm <- lm(Y~X1+X2+X3, data=df) X <- model.matrix(mylm) # create the design matrix xpxinv <- solve(t(X)%*%X) # create x prime x inverse see <- mylm$sigma # get the standard error of the estimate see^2 * xpxinv # print the covariance matrix`

To see what effect coding does to the covariance matrix, load the file `LoanExperiments2.csv` and regress PerComplete on the five effect-coded independent variables (don't worry about making interaction terms). Get the covariance matrix of the coefficients. Drop any variable and see what happens to the remaining coefficients: They don't change! Why not? Because the X variables are orthogonal.

The covariance matrix allows us to assess the extent of the optimality of an experimental design. An optimal covariance matrix is one that has zeros off the main diagonal. There will come a time in the next two chapters when we will not be able to have all zeros off the main diagonal. In such a situation, we will have many possible covariance matrices from which to choose, and we will have to choose carefully if we want good experiments. When we have less-than-optimal designs, we can figure out how far from optimal they are. A moment's reflection suggests that, if we must have nonzero covariances off the main diagonal, we would want few of them and we would want their magnitudes to be small. We'll leave further reflections on this matter to a later chapter.

7

Two-Level Full Factorial Experiments

Having learned how to analyze multivariable experiments in the previous chapter, in this chapter, we introduce the workhorse of the experimental methods, the 2^k factorial design, in which there are k factors, each having two levels. In the previous chapter we really learned all we need to know about setting up and analyzing 2^k factorial designs, so part of this chapter will really be nothing new. What will be new are our discussions in Section 7.3 on the determinant of a matrix and in Section 7.4 on aliasing. Knowledge of the determinant and aliasing helps us understand full factorial designs. Moreover, we need to know about aliasing before we get to Chapter 8 on fractional factorial designs, because aliasing is the big problem with fractional designs. We need to know about the determinant before we get to Chapter 9 on custom designs, because custom designs are based on the determinant of the design matrix.

The 2^k factorial design is a workhorse for four reasons:

1) It is efficient in the use of information; factorial designs are optimal.
2) It is easy to set up, as it requires only effect coding. Different types of coding are necessary when there are more than two levels, and they can be very difficult to set up.
3) It is easy to analyze. Analyzing models with more than two levels is not so straightforward.
4) The power of exponentiation cripples other than 2^k designs when the number of factors increases. To see this, note that four factors require only $2^4 = 16$ runs if each factor has two levels. If each factor has three levels, then four factors require $3^4 = 81$ runs.

So useful is the 2^k factorial design that one expert has commented, "Two level designs are very useful. There are statistical consultants that, except for occasional forays, use nothing else" (Wheeler, 2009, p. 21).

We do not wish to give the impression that the experimental design world begins and ends with the 2^k design. 3^k designs have their place, and mixed-level designs

Business Experiments with R, First Edition. B. D. McCullough.

Companion Website: www.wiley.com/go/mccullough/businessexperimentswithr

where one factor has two levels, another factor has three levels, etc. arise very naturally in applied problems. We will consider such a mixed-level problem in Chapter 9. We do wish to convey the impression that, for all their simplicity and ease of use, 2^k factorial designs are most appropriate for second course in statistics.

In the last chapter we developed all the machinery to analyze full factorial designs. In fact, we analyzed full factorial designs in the last chapter; we just didn't call them that. Remember the credit card example and the direct mail example? Both of those were full factorial designs.

By the end of this chapter, readers should:

- Be able to specify, estimate, and interpret a 2^k factorial model.
- Know what the determinant of a matrix is, and why it's important.
- Know what aliasing is.
- Be able to block a factorial design.
- After an experiment, select the best levels of the factors.
- Appreciate the benefits of including interactions in a model.
- Apply a modeling strategy to a saturated design.
- Apply a modeling strategy to a non-saturated design.

7.1 Case: The Postcard Example

Burnham (2004) wrote an article for a trade journal that described an interesting experiment. A business regularly sent postcards to its customers to inform them of new products and sales. The business wondered whether it could do anything to increase the response rate of its customers. To this end, the company considered three factors, each at two levels, low (L) and high (H):

- Factor A: Number of colors, two (L) and four (H)
- Factor B: Size, 4" × 6" (L) and 5.5" × 8.5" (H)
- Factor C: Type of paper, light (L) and heavy (H)

The number of different versions of the postcard is $2^3 = 8$, and an equal number of each type was sent to randomly selected samples of the customer mailing list. Since we're going to run a regression, let us represent the L with a "−1" and the H with a "+1." The data are presented in Table 7.1.

Notice the pattern of the 1 and −1 indicators in the columns: first, each observation alternates, then each pair of observations, and, finally, each quadruple alternates. Recall that this way of organizing the data is called *standard order*. We do not conduct the eight experiments in this order; we will randomly draw numbers one through eight and assign them to each row, as shown in the first column of the

Table 7.1 Postcard data.

Random order	Std order	*A*	*B*	*C*	Response
3	1	−1	−1	−1	34
4	2	1	−1	−1	18
7	3	−1	1	−1	44
6	4	1	1	−1	26
8	5	−1	−1	1	26
2	6	1	−1	1	17
5	7	−1	1	1	29
1	8	1	1	1	21

table. For example, the first run was with all factors at the hi level, and the second run was with hi, lo, and hi. After all the runs have been executed and the responses observed, we can account for all interactions by specifying the following model:

$$\text{response} = b_0 + b_1A + b_2B + b_3C + b_4AB + b_5AC + b_6BC + b_7ABC \quad (7.1)$$

where AB, AC, BC, and ABC are the interaction variables, e.g. $AB = A \cdot B$, etc. This is a 2^3 *full factorial model.* The astute reader will realize that the direct mail case study from Section 6.4 was also a 2^3 full factorial design. In the present case, there are eight observations and eight parameters to be estimated, so this case is a saturated model, and there are no degrees of freedom left over to estimate the error variance, i.e. we will have zero error degrees of freedom; we will have no standard errors for the coefficients and cannot test the coefficients.

The purpose of running an experiment is to be able to set the factor levels to maximize (or minimize, if that is the goal) the response. Doing so is not always straightforward, especially with many factors and many interactions.

Looking at Equation 7.1 and recognizing that we want to maximize the dependent variable, we would like each term in Equation 7.1 to be positive. For example, if $b_1 > 0$, we want A to be positive, which we would achieve by setting A to its high level, +1. Similarly, if $b_2 < 0$, we would choose $B = -1$ so that their product, b_2B, would be positive. The interaction terms are another matter. Suppose we have already chosen A, B, and C, so the interaction variables are determined. If we have chosen $A = +1$ and $B = -1$, then $AB = -1 < 0$. The corresponding interaction term b_4AB will be positive only if $b_4 < 0$. This type of analysis is useful when trying to determine, after an experiment has been run, whether to set variables at high or low levels in order to maximize (or minimize) the value of the dependent variable.

Table 7.2 Full factorial regression coefficients for postcard experiment.

Int	**26.875**
A	−6.375
B	3.125
C	−3.625
$A : B$	−0.125
$A : C$	2.125
$B : C$	−1.375
$A : B : C$	0.375

We learned in the previous chapter how to estimate and analyze Equation 7.1. The results of this regression are given in Table 7.2. Remember, these regression coefficients are *half* the effect sizes.

A half-normal or normal plot will give us some insight into the possible significance of these coefficients, as shown in Figure 7.1. The level of significance can be adjusted with the `alpha` = option, and here we adjusted it to 0.10. It appears that only factor A is significant. "Significant" factors fall far from the line, and the `halfnormal` command automatically prints the factor name next to significant points. Remember that half-normal plots are only suggestive, not definitive.

Burnham summarized his results: "Factor A (number of colors) stands alone in significance. Factor B (postcard size) and Factor C (card stock) hold no significance in influencing response. Interactions between factors are statistically insignificant." He concluded: "The results are now saving the software company thousands of dollars that would have been wasted with four-color printing."

Burnham did not apply a modeling strategy to his data. No doubt this is because he was writing for a trade journal, not a statistical journal. If he had applied a modeling strategy to his data, his final model would have been much different, and so would have been his conclusions. Let us explore this important idea.

Look at the regression coefficients in Table 7.2. The two smallest coefficients are for AB and ABC, both of which are an order of magnitude smaller than the other coefficients (even if AB and ABC are significant, they won't have much an effect on the predictions). As usual, drop the highest interaction term, the three-way interaction term, and run the regression again. Notice how the standard error of the coefficients has decreased. This means that we get more precise estimates of the coefficients. Now the largest p-value is on AB, so drop it and run the regression again. Notice again that the standard error of the coefficients has decreased. In

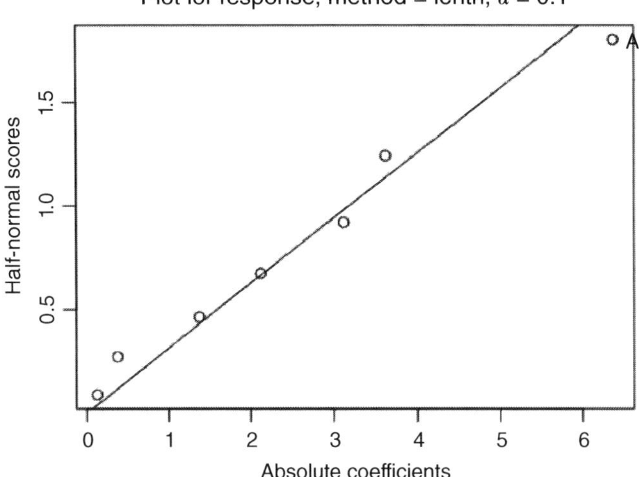

Figure 7.1 Half-normal plot of postcard data.

Table 7.3 Postcard regression, reduced model.

Variable	Estimate	Std. error	*p*-Value
Int	26.87	0.28	0.00
a	−6.37	0.28	0.00
B	3.13	0.28	0.01
C	−3.62	0.28	0.01
AC	2.12	0.28	0.02
BC	−1.38	0.28	0.04

general, dropping terms increases the precision of the remaining coefficients (this means we get better standard errors for the coefficients). Now all the coefficients are significant, as shown in Table 7.3. This is tentatively our reduced model, but of course we must do model checking before we accept this reduced model.

Try it!

Make sure you can apply the modeling strategy to obtain the reduced model in Table 7.3.

To model check factorial designs, we repeat the recommendations mentioned in the previous chapter (Mee, 2009, §2.6):

1) Plot residuals against $\hat{y}$.
2) Plot residuals against run order. (We really can't do this, because there is no run order for most of our data. For pedagogical reasons, we store the data in standard order.)
3) Create a normal quantile plot of the residuals.
4) Identify outliers, if any.

Since none of these points raises any concerns, we use the reduced model shown in Equation 7.2. Since we want to maximize the response, we would like each term in the equation to be positive, as a term that is negative would decrease the response rather than increase it:

$$\text{response} = b_0 + b_1 A + b_2 B + b_3 C + b_4 AC + b_5 BC \tag{7.2}$$

Let's pretend that model checking raised no red flags and interpret the coefficients and try to deduce the best model, e.g. what values we should select for A, B, and C so as to maximize the response? To contribute to a larger predicted value of y, we want the product of A (which is either +1 or −1) and b_1 to be positive. Since $b_1 < 0$, we need A to be negative also, so we'll set A at its low value (−1). Similarly, we'll set B at a high value (+1) and C at a low value (−1). By our choices of levels for A, B, and C, the interaction variables AC and BC will be positive and negative, respectively. Thus, $AC \times b_4$ and $BC \times b_5$ will both be positive, contributing to a higher predicted response. What is this predicted response when we set A, B, and C at their best levels?

Let's take another look at choosing optimal levels of the factors, this time using a table. First, for the main effects, we have the results shown in Table 7.4, from which we deduce the tentatively desired levels for A, B, and C.

Now that we have chosen levels for A, B, and C, next let's see if these levels are consistent with the coefficients we want for the interaction terms. Table 7.5 shows that our choices of A, B, and C are consistent with interaction terms that contribute positively to the dependent variable, i.e. both b_4AC and b_5BC are positive.

We will see in the next section that the requirements for main effects can be at odds with the requirements for interaction effects. In such a situation, rather

Table 7.4 Choosing A, B, and C for main effects.

Since	We will set	So that
$b_1 < 0$	$A = -1$	$b_A \cdot A > 0$
$b_2 > 0$	$B = +1$	$b_B \cdot B > 0$
$b_3 < 0$	$C = -1$	$b_C \cdot C > 0$

Table 7.5 Are interactions consistent with main effects?.

Since	We need	Is this true?
$b_4 > 0$	$A \cdot C > 0$	Yes, because $A = -1$ and $C = -1$
$b_5 < 0$	$B \cdot C < 0$	Yes, because $B = +1$ and $C = -1$

Table 7.6 Design matrix and predictions.

	A	*B*	*C*	Response	$\hat{y}$
1	−1.00	−1.00	−1.00	34	34.50
2	1.00	−1.00	−1.00	18	17.50
3	−1.00	1.00	−1.00	44	43.50
4	1.00	1.00	−1.00	26	26.50
5	−1.00	−1.00	1.00	26	25.75
6	1.00	−1.00	1.00	17	17.25
7	−1.00	1.00	1.00	29	29.25
8	1.00	1.00	1.00	21	20.75

than figure out what the best treatment assignments are, we can just print out the design matrix with their predicted values and then choose the row that has the best predicted value.

Now that we have found the best combination, we can do the math to predict the response rate at this combination:

$$\hat{y} = 26.87 - 6.37(-1) + 3.13(1) - 3.62(-1) + 2.12(-1)(-1)$$
$$-1.38(1)(-1) = 43.49 \tag{7.3}$$

Alternatively, we could have the computer do the math and compute $\hat{y}$ for all combinations of A, B, and C, one of which will be the best combination. Our desired combination is row three of Table 7.6, which comports with the calculation in Equation 7.3. We will see in the next section that there can be advantages to having the computer do this, rather than doing the math ourselves.

After applying our modeling strategy, we won't wind up with a full factorial model if some of the main effects or interaction effects are insignificant.

Try it!

Reproduce the results in Table 7.6.

Table 7.7 Postcard regression results, with replication.

Variable	Estimate	Std. error	p-Value
Int	26.937 5	0.826 8	0.59E−10
A	−4.062 5	0.826 8	0.001 17
B	2.937 5	0.826 8	0.007 48
C	−3.812 5	0.826 8	0.001 73
$A : B$	−0.062 5	0.826 8	0.941 60
$A : C$	2.187 5	0.826 8	0.0294 5
$B : C$	−1.312 5	0.826 8	0.151 07
$A : B : C$	0.187 5	0.826 8	0.826 28

We considered an experiment with replication in Section 5.2.2, and we showed how to arrange the data for importation into a statistical program. With replication, we no longer have a saturated design, so we get standard errors for all the coefficients when we run a full factorial. Moreover, with replication, we have a large error degrees of freedom, so our modeling strategy for an unsaturated design (that has a reasonable number of degrees of freedom to estimate the error) will consist of simply dropping statistically *and* practically insignificant variables all at once. Suppose the postcards experiment had been replicated, with two vectors of responses. The first, from `Postcards.csv`, is $34, 18, 44, 26, 26, 17, 29, and 21$. Suppose the second was $31, 23, 39, 31, 21, 22, 24, and 25$. The results of running the regression for this replicated full factorial design are given in Table 7.7.

Try it!

Make sure you can analyze data that have been replicated. Getting the data into a format that R can use may require data munging. If you need a hint, something similar was done in Section 5.2.2. Reproduce the results in Table 7.7.

Let us apply the modeling strategy to the results in Table 7.7. Remember, we have two modeling strategies: one for saturated designs and another for designs with replication. Here we use the latter. Clearly, *AB* and *ABC* need to be dropped – not only are the coefficients statistically insignificant, but also they are practically insignificant, being an order of magnitude smaller than significant coefficients. Factor *BC* is a judgment call; it's half the size of significant coefficients, so it's in the region where it may be practically significant, and the p-value isn't that tiny. In this case, let us make predictions with and without *BC*

Table 7.8 Replicated postcard predictions with and without *BC*.

	With *BC*	Without *BC*
1	32.75	34.06
2	20.25	21.56
3	41.25	39.94
4	28.75	27.44
5	23.38	22.06
6	19.62	18.31
7	26.62	27.94
8	22.88	24.19
9	32.75	34.06
10	20.25	21.56
11	41.25	39.94
12	28.75	27.44
13	23.38	22.06
14	19.62	18.31
15	26.62	27.94
16	22.88	24.19

and see what happens to the predicted value of the response. Results are given in Table 7.8. Note that the first eight predictions are the same as the second eight predictions, because the design points are the same (this is a replicated design, remember). The predictions seem to be comparable, and what's really important is that adding or dropping BC does not change the best design, which is the third row. So we'll drop BC from the model.

Now we have to discuss *effect heredity*, which is the idea that if a higher-order term is included in the model, then some or all its related lower-order terms must also be included, whether or not they are significant (practically or statistically). There are two forms of effect heredity, *strong effect heredity* and *weak effect heredity*. Strong effect heredity requires that all related lower-order component must be included. For example, if *ABC* is included, then so must *A*, *B*, *C*, *AB*, *AC*, and *BC* be included. Weak effect heredity requires that at least one of each lower-order type must be included and they must share a factor. For example, if *ABC* is included, then it is sufficient to include only *B* and *BC* or *A* and *AB* (notice that the same factor appears in both pairs of first-order and second-order terms). For example, if *BC* is active, including *A* would not satisfy the requirement that a factor be shared. In

Table 7.9 Standard order for 2^3 design.

	A	*B*	*C*
1	−1	−1	−1
2	+1	−1	−1
3	−1	+1	−1
4	+1	+1	−1
5	−1	−1	+1
6	+1	−1	−1
7	−1	+1	+1
8	+1	+1	+1

this book we will only require weak effect heredity. Our final model for the replicated postcard example includes A, B, C, and AC; weak effect heredity is satisfied.

It can be tedious to create the patterns of +1 and −1 that make up the independent variables in standard order. Software packages that accommodate experimental methods have procedures for automatically producing these variables, and R is no exception. An example of such a design, in standard order, is shown in Table 7.9.

Software Details

To create the orthogonal variables for factorial designs...

First, create the design matrix. For three variables this is `dm <- expand.grid(A=c(-1,1),B=c(-1,1),C=c(-1,1))`

Use this design matrix in the `model.matrix` command to create model matrices. To the right of the tilde (~), write a formula.

For only a main effects model `model.matrix(~A+B+C,data=dm)`

For a full factorial model `model.matrix(~A*B*C,data=dm)`

For all main effects and second-order interactions `model.matrix(~.^2,data=dm)`

It is obvious how to add variables D, E, etc.

These model matrices include not just the +1 and −1 values for the variables but also a column of +1 values for estimating the intercept.

Exercises

7.1.1 Burnham's analysis showed that only factor A (two or four colors) mattered and that it should be set at the low level (-1, two colors). What is the predicted response rate? You'll have to make assumptions about the other two factors that have no effect. In particular, you should set them at the levels that minimize cost. For example, if there's no difference between the large and small postcard, the small postcard costs less, so use it rather than the large postcard.

7.1.2 Interactions can have powerful effects on the ability of a regression model to make accurate predictions. Regress `response` on factors A, B, and C. Note the R^2. Compare this with the R^2 of the reduced model. Which model would you rather use for making predictions?

7.1.3 Apply model checking to the model presented in Table 7.3.

7.1.4 Consider the results in Table 7.7. Apply the modeling strategy to the replicated data set. What is the reduced model? Perform model checking.

7.1.5 Write out 2^3 full factorial design matrix, and then write out the model matrix, both in standard order.

7.1.6 Write out 2^4 full factorial design matrix, and then write out the model matrix, both in standard order.

7.1.7 Write the model matrix for a 2^4 factorial with second-order (i.e. two-factor) interactions.

7.2 Case: Email Campaign

We are interested in designing an email that will maximize the response rate. We are very sure, based on experience, that there are four drivers for this process. However, we have no idea how these four factors may interact. These factors are shown in Table 7.10.

A 2^4 factorial design will require 16 different treatment combinations. Each treatment combination was sent to a random sample of several thousand recip-

Table 7.10 Factors for email campaign.

Factor	Type	Hi (+1)	Lo (−1)
A	Subject line	"Message A"	"Message B"
B	Hero image	Picture of product	Picture of person using product
C	Discount	5%	10%
D	When sent	Weekday	Weekend

ients to ensure that the estimates of the response rates are reasonably accurate. The design matrix (in standard order) and the percentage response variable are in `EmailCampaign.csv`.

Estimating the full factorial model, we see that the main effects are nonzero in the first decimal, whereas most two-way interactions and half of the three-way interactions, as well as the four-way interaction, are zero in the first decimal. The half-normal plot indicates that only the factor D is important, but this is merely suggestive. To gain a degree of freedom for estimation of the error, we drop the four-way interaction and then apply our modeling strategy. The resulting model is presented in Table 7.11. Observe that all the coefficients are roughly the same order of magnitude, so we probably won't have to worry about the distinction between statistical significance and practical significance.

Table 7.11 Regression coefficients email campaign.

Variable	Estimate	Std. error	*p*-Value
Int	2.029 53	0.025 06	5.44E−09
A	−0.125 52	0.025 06	0.004 07
B	−0.155 72	0.025 06	0.001 58
C	0.123 64	0.025 06	0.004 34
D	0.387 02	0.025 06	2.06E−05
AB	0.064 83	0.025 06	0.049 01
AC	−0.155 91	0.025 06	0.001 57
BC	0.091 06	0.025 06	0.014 99
ABC	−0.107 69	0.025 06	0.007 73
ABD	0.138 38	0.025 06	0.002 67
ACD	0.072 69	0.025 06	0.033 75

It's going to be tedious to write out "two-way interaction" and "three-way interaction," so in the sequel we'll also refer to them as "2FI" (two-factor interaction) and "3FI."

Try it!

Apply the modeling strategy to the email campaign data and make sure you wind up with the final model in Table 7.11.

Looking just at the main effects, we will want to set $A = -1, B = -1, C = +1$, and $D = +1$. Will these settings be consistent with what we need for the interaction effects?

See that $b_{AB} > 0$. For the term $b_{AB} \cdot AB$ to make a positive contribution, we want the product of A and B to be positive so that the factor AB is positive. We have already set $A = -1$ and $B = -1$, so this AB is positive. This is good.

See that $b_{AC} < 0$. For the term $b_{AC} \cdot AC$ to make a positive contribution, we want the factor AC to be negative. We have already set $A = -1$ and $C = +1$, and this product is negative, so AC is negative. This is good.

See that $b_{BC} > 0$. For the term $b_{BC} \cdot BC$ to make a positive contribution, we want BC to be positive. We already have set $B = -1$ and $C = +1$, so the factor BC is negative! Our results are laid out in Table 7.12. There is a contradiction between what the main effects require of the levels of the factors and what the interaction terms require. We will not be able to reason our way to finding the best combination. Therefore, we will have to look at the design matrix augmented with the responses and the fitted values, given in Table 7.13, and choose the combination that has the highest fitted value.

Treatment combination 13 gives the highest predicted response, so that's the combination we'll use: subject line message B, a hero image of the person using the product, a 5% discount, and sent on a weekday. Observe that treatment combination 9 is practically the same. Since this is a digital campaign, there is no cost difference between combinations 9 and 13. However, were treatment combination 9 much cheaper than combination 13, we'd of course save money by using

Table 7.12 Contradictory interactions in email campaign.

Since	We want	Is this true?	Because
$b_{AB} > 0$	$AB > 0$	Yes	$A = -1$ and $B = -1$
$b_{AC} < 0$	$AC < 0$	Yes	$A = -1$ and $C = +1$
$b_{BC} > 0$	$BC > 0$	NO!	$B = -1$ and $C = +1$

Table 7.13 Email campaign fitted values with design matrix and response.

	A	*B*	*C*	*D*	*y*	**Predictions**
1	−1	−1	−1	−1	1.67	1.70
2	1	−1	−1	−1	1.93	1.83
3	−1	1	−1	−1	1.12	1.13
4	1	1	−1	−1	1.39	1.41
5	−1	−1	1	−1	1.95	2.00
6	1	−1	1	−1	1.60	1.66
7	−1	1	1	−1	2.21	2.24
8	1	1	1	−1	1.26	1.17
9	−1	−1	−1	1	2.92	2.89
10	1	−1	−1	1	2.09	2.19
11	−1	1	−1	1	1.79	1.78
12	1	1	−1	1	2.33	2.31
13	−1	−1	1	1	2.96	2.91
14	1	−1	1	1	2.36	2.30
15	−1	1	1	1	2.61	2.59
16	1	1	1	1	2.27	2.36

treatment combination 9. Notice the large difference between the two best combinations and the remaining combinations. The levels of the factors really do make a difference!

Exercises

7.2.1 Again let us take notice of the power of interactions. Run the regression for the data set `EmailCampaign.csv` with just the four factors and observe the R^2. Compare this with the R^2 of the reduced model. With which will you obtain more accurate predictions?

7.2.2 A small Alaskan fishing company encountered hard times and had to decide how best to compete in the future. The company considered three options. First, the company might purchase a new sonar system for each boat that would enable to the boats to find fish better, but it is an expensive system. Second, the company might move to a new fishing area that is farther away. This would entail an additional expense due to the time and fuel required for travel. Third, the company considered switching to a different type of fish. This would require severing ties with established buyers and acquiring new relationships with new fish buyers. These methods are not mutually exclusive, and, indeed, it is possible to make all three changes at once. Hence, a full factorial design is necessary to account for possible interactions. To measure the effectiveness of each proposed method, the company used weekly earnings before interest and taxes (EBIT) in thousands of dollars, a common accounting measure. The data are in `Fishing.csv`. The experiment was run for four weeks, so this is a full factorial with replication. You will have to rearrange the data before analyzing it. Sonar is hi, no sonar is lo; new fishing area is hi, current fishing area is lo; different type of fish is hi, current type of fish is lo.

Start with a full factorial model and apply the modeling strategy; since the error degrees of freedom is large, just drop all insignificant variables. You will observe that the three-way interaction is significant, as is one two-way interaction, so what are the implications of weak heredity for this?

Perform model checking. Determine the optimal combination of factors and predict the EBIT for this combination.

7.2.3 The point of this exercise is that getting to a reduced model is not always straightforward. Observe how the half-normal plot and the modeling strategy work in tandem to help identify the model. It also shows the hazard of relying on an estimate of the standard error with only one or two degrees of freedom. Without an accurate estimate of the standard error, inference can be compromised.

A direct marketing by mail experiment studied the effect of four factors on the response rate: (A) payment, whether monthly ($-$) or annual ($+$); (B) whether a valuable coupon was included in the letter ($+$) or not ($-$); (C) whether the letter included an entry into a sweepstakes ($+$) or not ($-$); and (D) whether the trial membership card was paper ($-$) or plastic ($+$). The data are in `DirectMail.csv`. Each of the responses was the result of 10 000 mailings. This is a saturated design. At the beginning of the modeling strategy, after dropping $ABCD$, everything is significant with p-values that all are machine zeros (e.g. 3.4E−15).

Look at the R^2. Do you really believe all the variables are significant and you have a perfect fit? ("Perfect fit" means $R^2 = 1$). Examine the half-normal plot. A, B, C, and D are flagged as significant, but other points fall not on the line. To identify these points, change α from 0.05 to 0.10 (use the option `alpha=0.10` in the half-normal command).

Drop variables in the following order: $ABCD, AC, AD, BD, AB, BCD, ABC, CD, ABD$, and ACD. As you drop variables, keep track of the degrees of freedom, the residual standard error, and the R^2. Notice how the variables start off all significant, then many lose significance, then some gain it again, and finally we are left with only A, B, C, D, and BC. How many degrees of freedom do you need before the residual standard error stabilizes? Especially for saturated models with many variables, an automated method of searching for a parsimonious model can be useful. The *stepwise regression* procedures are one such class of automated methods. See the Learning More section for further details.

7.2.4 Numerous small vendors push carts and sell food to tourists in Washington, DC, especially near The Mall. At one end of The Mall is the Lincoln Memorial, at the other end is the Capitol Building. A man owns several of these carts and deploys them at either end of The Mall. He wants to maximize his profits and so conducts an experiment. The levels of the factors are given below.

Treatment	Variable	Low	High
Hot dog	A	Budget	Premium
Drinks	B	Canned only	Canned and bottle
Location	C	Near Capitol	Near Lincoln Memorial
Vendor	D	Students	Retirees

The file `HotDogCarts.csv` contains the design matrix for this 2^4 factorial and seven replications. Remember the data may have to be rearranged in order to read them in properly. Start with a full factorial model and apply the modeling strategy. Describe any interactions. Perform model checking. Determine the optimal combination of factors and the predicted profit for this combination.

7.3 The Determinant of a Matrix

When we estimate parameters of a statistical model, e.g. the slope in a regression of Y on X, we like the range of X to be big. Go back and take another look at Figure 3.1. Note that the range of X corresponds, loosely, to the "length" of the X data. The reason we want the length of X to be big is because it leads to a more precise estimate. Our goal is to get an estimate of the slope of the line, e.g. running a regression of Y on X. If we restrict the range of X to the interval $[X_0, X_1]$, our estimates of the true slope can vary widely, from a steep positive slope (line A) to slightly negative (line B). On the other hand, if X is allowed to vary over a much larger interval $[X_0, X_2]$, all the estimated slopes will be roughly the same and close to the true slope. It should be intuitively obvious that a greater length for X implies a more precise estimate of the slope.

Length is what we measure when we have one X variable. If we have two X variables, for example, running the regression $Y = \beta_0 + \beta_1 X_1 + \beta_2 X_2 + \epsilon$, then we will need both X_1 and X_2 to have a big length, which implies that the factors will vary over a large *area*. Thus, to estimate the slopes β_1 and β_2 precisely, the data for X_1 and X_2 must span a large area. If we have three X variables, we will need each to have a big length, which implies the factors will vary over a large *volume*. If we have more than three X variables, we want them to vary over a big *hyper-volume*.

All you really need to know about the *determinant* of a matrix is that it encompasses this idea of length, area, volume, or hyper-volume depending on how many variables are in the data matrix. You don't need to know how to calculate it or how to use it. Rather than focus on the determinant of the matrix X, for technical reasons that concern us not, we will focus on the matrix $X'X$, which is sort of like X squared in a matrix sense. This is okay because in general, if the matrix X is large, then so is $X'X$, it's just that $X'X$ is mathematically easier to work with than X. Maximizing $(X'X)$ is sometimes referred "maximizing the information" in the design.

If $X'X$ is large, this is the same thing as saying its inverse $(X'X)^{-1}$ is small. Thus, if we want precise estimates, we want the determinant of $(X'X)^{-1}$ to be small. We have seen this matrix before, in the definition of the covariance matrix of the regression coefficients. $\text{cov}(b) = \sigma^2 (X'X)^{-1}$ where X is the model matrix (that includes a column of ones for the intercept). We want the determinant of $(X'X)^{-1}$ to be small, because that means the standard errors are small, so we can better detect any effects.

Consider two parallelepipeds in two dimensions, A and B, as shown in Figure 7.2. The determinant of A is bigger than the determinant of B. We won't concern ourselves with the formula for computing a determinant, but we will need to compute determinants for our design matrices of effect-coded variables – but we'll have the computer do it for us.

The 2^k design is optimal because it maximizes the value of the determinant of $X'X$. It achieves this in part because it is orthogonal and balanced and uses all the points in the design space. For example, with four factors a 2^k design requires 16 runs so that it can test every combination of factors once. But what if runs were expensive, and we could only afford 10 runs? Which 10 would we want? Obviously, we wouldn't want to run the same combination of factor levels 10 times; we would

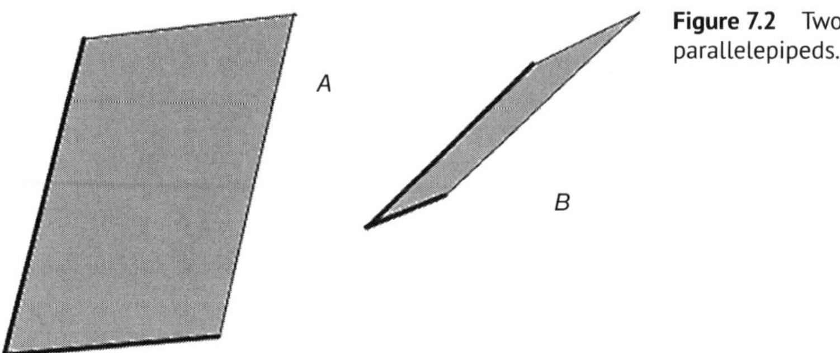

Figure 7.2 Two parallelepipeds.

learn nothing that way. The answer to this question is that we want to choose the 10 combinations to maximize the determinant of a design matrix that has only 10 runs. Since the value of the determinant of $(X'X)$ is a function of the design matrix X, different design matrices will yield different determinants of $(X'X)$. We can change the determinant by changing the points of the design matrix.

We can't always run 2^k designs. If we have four factors, each with two levels, and we can only afford 10 runs, our design will not be optimal. Nonetheless, when we have to use less-than-optimal designs, we still want these less-than-optimal designs to be a close to optimal as possible. Therefore, we want to find a design with 10 runs that has a determinant of $(X'X)$ that is as close as possible to the determinant of $(X'X)$ for the 2^4 design with 16 runs. Such a design is called D-optimal, and it makes the standard errors of the regression coefficients as small as possible (given the fact that we can only use ten runs) so that the effects are estimated as precisely as possible.

There are many other optimality criteria. G-optimal designs minimize the width of the prediction interval for an individual value. V-optimal designs minimize the width of the prediction interval for an average predicted value. There are still other optimality criteria, e.g. A, C, E, and T optimality. Taken together, all these methods are sometimes referred to as "alphabetic optimality." When two D-optimal designs are equally good, appealing to other optimality criteria can help decide which one to choose; how to do this is discussed in Robinson and Anderson-Cook (2011).

So far we have only considered 2^k experiments. To better understand determinants, let us briefly consider a 3^2 experiment that has two factors, each of which has three levels: −1, 0, and 1. A factorial design would use nine experiments with the levels of x_1 and x_2 given in Table 7.14.

Suppose, though, that we could only choose three runs for our experiment. (We choose three here because it keeps the math and geometry easy to visualize.) There are $\binom{9}{3} = 84$ possible choices for three points. Which three should we choose? We

Table 7.14 Design points for a 3^2 experiment.

Obs	x_1	x_2
1	−1	−1
2	−1	0
3	−1	1
4	0	−1
5	0	0
6	0	1
7	1	−1
8	1	0
9	1	1

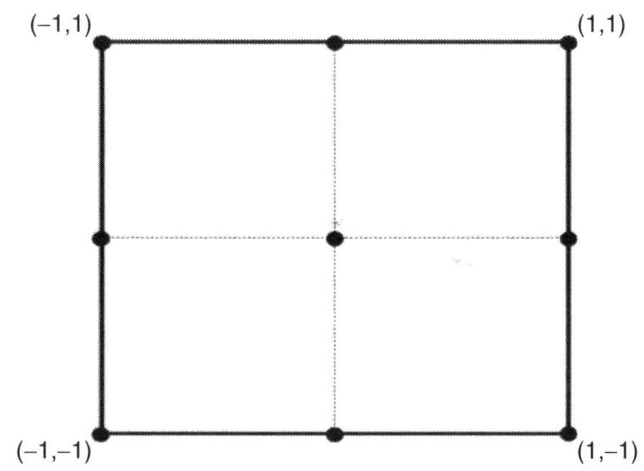

Figure 7.3 All possible design points.

want to choose the three points so that the area enclosed by the three points covers as much of the design space as possible. A graphical representation of the design space is shown in Figure 7.3.

Let's consider five possible choices of three design points:

- Design 1: $(-1,-1), (0,0), (1,1)$
- Design 2: $(-1,-1), (0,0), (-1,0)$
- Design 3: $(0,-1), (-1,0), (0,1)$
- Design 4: $(-1,-1), (0,1), (1,0)$
- Design 5: $(-1,-1), (-1,1), (1,-1)$

Note the three points of Design 1 constitute a straight line, which encloses no area. Running this experiment wouldn't teach us very much.

Let X be the model matrix for using a regression to estimate effects, so the first column is all ones to account for the constant term. The second and third columns represent the points of the design. Thus, the model matrix corresponding to Design 3 is given by

$$\begin{bmatrix} 1 & 0 & -1 \\ 1 & -1 & 0 \\ 1 & 0 & 1 \end{bmatrix} \tag{7.4}$$

Software Details

To create the above matrix, premultiply it by the transpose, and compute the determinant...

```
mat3 <- matrix(c(1,0,-1,
          1,-1,0,
          1,0,1),nrow=3,byrow=TRUE)

det(t(mat3)%*%mat3)
```

We can analyze these designs by forming the model matrix X, computing $(X'X)$, and then calculating the determinant of $(X'X)$. Graphically plotting the three points in the design space enables us to see how much of the design space is covered by the three points. We can compute the area covered by the three points and compare it to the determinant. We will see that there is a 1:1 correspondence between the area and the determinant: as the determinant gets bigger, so does the area. This is illustrated in Figure 7.4 for three of the designs; the analysis of the remaining two three-point designs is left as an exercise. Study this figure carefully. It makes the case that, in two dimensions, a larger determinant corresponds to a larger area. This larger area leads to better estimates, as we argued in Figure 3.1. The same argument holds for determinants of larger matrices.

For a given number of rows, the design that maximizes the determinant is called a *D-optimal design*, where the "D" stands for "determinant." As noted already, there are only 84 possible three-point designs, so it would be a simple matter to compute the determinant for all 84 designs and pick the one that is largest. What we would find is that there is more than one design that attains the maximum. This is a general rule when finding D-optimal designs: a D-optimal design is usually not unique.

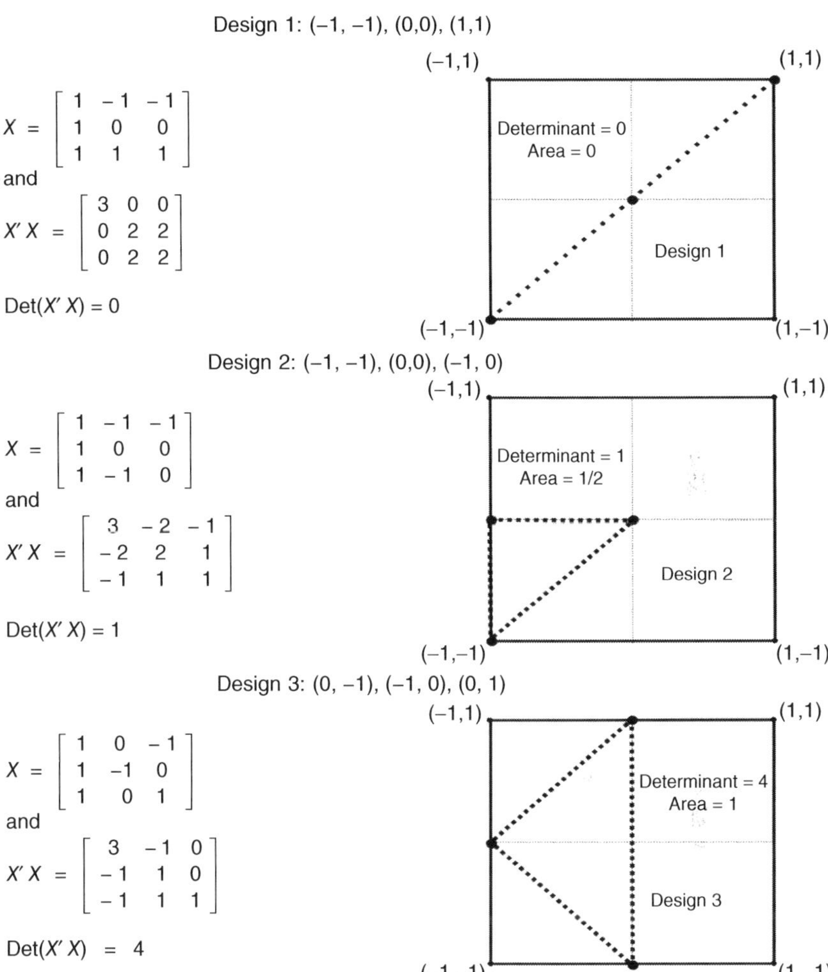

Figure 7.4 Determinants and areas of two three-point designs.

Exercises

7.3.1 We noted in the text that if $(X'X)^{-1}$ is large, we are prone to type II errors. Explain why.

7.3.2 Write the model matrix X for Design 4. Compute the determinant. Draw a picture like those in Figure 7.4 that shows the design points, the determinant, and the area enclosed by the design points. If you need a hint, look at the last page of this chapter.

7.3.3 Write the model matrix X for Design 5. This is a D-optimal design. Draw a picture like those in Figure 7.4 that shows the design points, the determinant, and the area enclosed by the design points. If you need a hint, look at the last page of this chapter.

7.3.4 Another D-optimal design for the "three points" problem is $(1,1)$, $(-1,1)$, and $(1,-1)$. Write the design matrix X, and without computing $(X'X)$ or the determinant of $(X'X)$, show that it is D-optimal. (Hint: Graph it and compare this graph to that of a design known to be D-optimal.)

7.4 Aliasing

Aliasing occurs when two columns of the matrix are identical *or* when one column is equal to a linear combination of other columns. In either case we cannot obtain clean estimates (unbiased estimates) of the regression coefficients corresponding to the aliased variables. We have already encountered the former case in Table 5.5, which you may wish to review. In that case, the city and theme variables were confounded, so their separate effects could not be determined. In particular, we could not tell whether the theme "Story" was better than other themes or whether people in Philadelphia just happen to like DigiPuppets. As an example of the second case, consider the design presented in Table 7.15.

One column is not an exact copy of any other, but it is easy to verify that $D = A + B + C - 2E$. This means that we cannot separate the effect of D from the effect of the other variables. For example, if A, B, C, or E changes, so will D. Further, it is easy to verify that not all pairs of variables are orthogonal. Remember, we like our design matrices to be orthogonal so that all pairs of variables have a dot product of zero. When the variables are not orthogonal, they are correlated. The dot products of A and B, A and C, A and D, B and C, and B and D all are zero; the following dot products are not zero: A and E, B and E, and D and E. This shows that E is the problematic variable: E is aliased with A, B, and D.

Table 7.15 An aliased design.

	A	*B*	*C*	*D*	*E*
1	−1	−1	−1	−1	−1
2	1	−1	−1	1	−1
3	−1	1	−1	1	−1
4	1	1	−1	−1	1

What we will see in a later chapter is that we can control the aliasing to some degree. As a short example of this, suppose that we really wanted clean estimates of A and B, but we didn't really care about D and E. If we changed E to $-1, 1, 1, -1$, then E would be aliased only with D, and we would get clean estimates for the effects of A and B.

Determining the precise nature of all the aliasing relationships in a matrix is beyond the scope of this course. To be precise, we will not concern ourselves with figuring out that $D = A + B + C - 2E$ in Table 7.15. We will use software, when possible, to determine the aliasing structure of a particular design. What we need to know is that when the design variables are aliased, we will not get clean estimates of the parameters, i.e. the estimates will be biased.

Suppose we wanted to run a 2^3 full factorial, the standard order for which is given in Table 7.16, but because each experiment is very expensive, we could only run four experiments. If we used the design in the top half of the table, we would have a severe aliasing problem, which we summarize like this: $A = -AC, B = -BC$, and $AB = -ABC$ (when A is positively aliased with AC, we write this as $A = AC$; when negatively aliased, we write $A = -AC$). If we used the design in the bottom half of the table, we would still have an aliasing problem: $A = AC$, $B = BC$, and $AB = ABC$. Notice that in both cases we can't get clean estimates of either A or B, and obviously C is aliased with the intercept. We wouldn't want to use either the top half or the bottom half. As we'll see, judiciously choosing a few rows from the top and a few rows from the bottom will be a good choice.

Sometimes the aliasing is not so simple that we can detect it by looking at the design matrix; we saw this in Table 7.15. Moreover, there are times when aliasing is not in the design matrix, but is in variables that are not in the design that we estimate! To see how this can happen, take another look at Table 7.16. Suppose we estimated a main effects model that included only the intercept and A, B, and

Table 7.16 2^3 Design with interactions.

	A	*B*	*C*	*AB*	*AC*	*BC*	*ABC*
1	−1	−1	−1	+1	+1	+1	−1
2	+1	−1	−1	−1	−1	+1	+1
3	−1	+1	−1	−1	+1	−1	+1
4	+1	+1	−1	+1	−1	−1	−1
5	−1	−1	+1	+1	−1	−1	+1
6	+1	−1	−1	−1	+1	−1	−1
7	−1	+1	+1	−1	−1	+1	−1
8	+1	+1	+1	+1	+1	+1	+1

C and we used only four runs. A would still be aliased with AC, so the estimated coefficient for A would include the effects of both A and AC. Similarly, the estimated coefficient for B would include the effects of both B and BC. If (and this is a big "if") we knew that the interaction terms AB and BC were negligible, then we would have clean estimates of A and B, but C would still be aliased with the intercept.

Understanding these alias patterns and how to minimize the effect of aliasing is very important, and we shall be preoccupied with these ideas in the next two chapters.

Software will produce an "alias matrix" that tells us the extent to which main effects are aliased with interaction effects. For example, consider the design in Table 7.17.

The model we estimate is

$$y = \beta_0 + \beta_1 A + \beta_2 B + \beta_3 C \tag{7.5}$$

which is valid if there are no higher-order interactions. If second-order interactions are active (if the true coefficients on the interaction variables are nonzero), then the model that actually governs the response is

$$y = \beta_0 + \beta_1 A + \beta_2 B + \beta_3 C + \beta_{12} AB + \beta_{13} AC + \beta_{23} BC \tag{7.6}$$

The effect of these "omitted variables" (variables that should be in the model we estimate but that we have failed to include in the model) is imputed to the variables that are included in the model, and the coefficients of the included variables will be biased by the omitted variables. The alias matrix corresponding to this design is given in Table 7.18. The alias matrix tells us what coefficients we can estimate cleanly and which are confounded. Notice that it has main effects for the rows and interactions on the columns, since we are interested in finding out if any of the interactions main effects is aliased with the interactions. When an entry in the alias matrix equals one, there is perfect confounding (also known as perfect collinearity).

We interpret this alias matrix as follows:

Table 7.17 Four runs to estimate main effects of 2^3 design.

Run	*A*	*B*	*C*
1	−1	−1	+1
2	+1	−1	−1
3	−1	+1	−1
4	+1	+1	+1

- The intercept is aliased with *ABC*, but this doesn't matter because *ABC* is not in the true model, so the true coefficient on *ABC* equals zero.
- *A* is aliased with *BC*.
- *B* is aliased with *AC*.
- *C* is aliased with *AB*.

We can verify this by writing out the full factorial and inspecting the columns to find $x_1 = x_2x_3$, $x_2 = x_1x_3$, and $x_3 = x_1x_2$, which we leave as an exercise for the reader. Perform the requisite multiplication and see that $x_1x_2x_3$ is aliased with the intercept.

The alias matrix shows the potential bias to the main effect estimates as a result of two-factor interactions that are omitted from the model. The estimated regression equation, if biased due to the existence of an interaction effect that is omitted from the model, can be written as

$$\text{biased regression coefficient} = \text{good estimate} + \text{bias due to aliasing} \quad (7.7)$$

where the bias due to aliasing can be decomposed as follows:

$$\text{bias due to aliasing} = \text{interaction term coefficient} * \text{alias matrix value} \quad (7.8)$$

As an example, suppose that we were concerned with factor *A* from Table 7.18 and suppose that the coefficient on *BC* was nonzero so that aliasing is present. Let's give some numbers to these coefficients. Let the estimated coefficient for *A* be 50. Suppose we happened to know that the coefficient on *BC* is 20 (remember, we haven't estimated this, so we just have to know it). Because the aliasing coefficient is 1, the full effect of *BC* is imputed to *A*, i.e. the bias due to aliasing $= 1 * 20 = 20$. Therefore, $50 = \text{good estimate} + 20$, from which we can deduce that a good estimate of the coefficient for factor *A* is really 30. Of course, we never know the coefficients for the omitted interaction terms, but we can take guesses for them and deduce plausible ranges for good estimates of aliased coefficients. We shall do this in Chapter 9.

Table 7.18 Alias matrix for 2^3 design.

	AB	*AC*	*BC*	*ABC*
Int	0	0	0	1
A	0	0	1	0
B	0	1	0	0
C	1	0	0	0

We could simply write the full factorial design corresponding to Table 7.17 and see, by inspection, the pattern in the above alias matrix. We leave this as an exercise.

An easier way is to just have a function that computes the alias matrix for a given design; here we give one. To make life easy, the factors should be labeled "A," "B," "C," etc. If factors have long names, then representing interaction terms in tables and graphs becomes problematic.

Software Details

To get the alias matrix…

We had to write our own function to create the alias matrix. See the appendix to this chapter for details. The function is read into R the same way that you read in a data file. The file `aliasMatrix.R` should be in the same directory as the data. Then just use the `source` command to read it in.

```
source("aliasMatrix.R")
```

You can use it like any other R command. The argument is a design matrix, but the factor names have to be "A," "B," "C," etc.

`aliasMatrix.R` will check for second-order interactions. If you want to check for third-order interactions, use `aliasMatrix3.R`.

Get the alias matrix for the design in Table 7.9 by issuing the following commands:

```
dm <- expand.grid(A=c(-1,1),B=c(-1,1),C=c(-1,1))
aliasMatrix(dm)
```

The alias matrix for the design in Table 7.9 is a very boring alias matrix shown in Table 7.19. No main effect is aliased with any of the two-way interactions. We will discuss exciting alias matrices in the next chapter, though in this business, boring is better than exciting.

Try it!

Use the command `aliasMatrix3` to check for aliasing between second-order effects and third-order effects for the design in Table 7.9.

Table 7.19 A boring alias matrix.

	$A:B$	$A:C$	$B:C$
Int	0	0	0
A	0	0	0
B	0	0	0
C	0	0	0

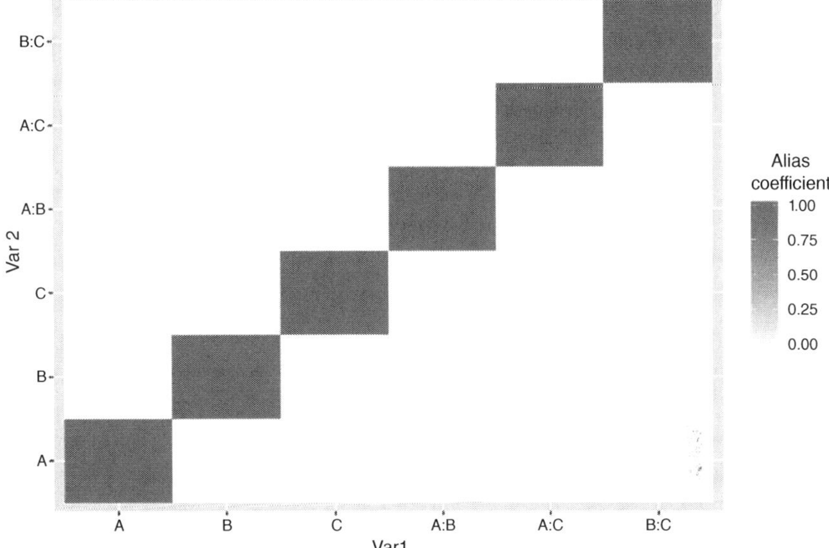

Figure 7.5 A boring color map of the correlations.

Another useful way to analyze a design is via the color map (more precisely, a heat map) of the correlations of the estimated coefficients. We know that the full factorial design is optimal and has a diagonal covariance matrix so that the estimated coefficients are uncorrelated. We can expect that the color map of correlations for a full factorial design will be very boring, and, indeed, it is, as shown in Figure 7.5.

Each cell of the color map represents a level of correlation between the coefficients in the model, between the absolute value of the correlation, since our primary concern is whether the correlation exists, not whether it is positive or negative. White is zero correlation, and dark gray is a correlation of unity. The main diagonal is all dark gray, which is to be expected since, e.g. the coefficient on A is perfectly correlated with itself. Because this is a full factorial design and all the coefficients are uncorrelated (we can add and drop variables without changing the estimated coefficients), the off-diagonal cells all are white.

Try it!

Create the color map to check for correlations between second-order effects and third-order effects for the design in Table 7.9.

Software Details

To create a color map of the correlations...

Our color map function depends on two packages, `ggplot2` and `reshape2`, and these have to be loaded using the `library` command.

```
library(ggplot2)
library(reshape2)
```

If you get an error message, it's probably because at least one of these packages is not installed on your computer. If so, to install a package that has not been installed, issue one (or both, if necessary) of the below commands:

```
install.packages("ggplot2")install.packages("reshape2")
```

Then you can just load the function and execute it on the design in Table 7.9.

```
source("colorMap.R'')
colorMap(dm)
```

where "dm" is the design matrix and this should reproduce the boring color map in Figure 7.5.

If you want the actual correlation matrix that underlies the color map,

```
source("corrOfCoeff.R")
corrOfCoeff(dm)
```

Correlation matrices are straightforward to interpret. If you want some help interpreting color maps, see the appendix for a couple of examples.

Exercises

7.4.1 Below are four variables. Verify that $z = x + y$, i.e. z is a linear combination of x and y. Regress w on x, y, and z. What happens?

w	x	y	z
8	6	2	8
1	3	9	12
6	5	5	10
1	8	6	14
9	4	8	12
2	6	7	13

In the old days, the regression routine would simply produce an error message to the effect that the right-hand side variables had a perfect linear correlation. Nowadays, most statistical package will automatically drop one of the perfectly correlated variables so that the regression can be run.

7.4.2 Below is an alias matrix. Interpret it and determine which factors are aliased with which interactions?

	A : B	A : C	A : D	A : E	B : C	B : D	B : E	C : D	C : E	D : E
Int	0	0	0	0	0	0	0	0	0	0
A	0	0	0	0	0	1	0	0	1	0
B	0	0	1	0	0	0	0	0	0	0
C	0	0	0	1	0	0	0	0	0	0
D	1	0	0	0	0	0	0	0	0	0
E	0	1	0	0	0	0	0	0	0	0

7.5 Blocking (Again)

Blocking for reducing variation in the error variance was employed in 5.1.1. You may wish to review that section before proceeding.

A pizza parlor that is open from noon to 2 a.m. seven days a week wants to increase sales and to this end wants to conduct an experiment. Three promotional offers are considered, the high levels being presented in Table 7.20 and the low levels all are "nothing."

Since we have three factors at two levels, a 2^3 experiment is called for. Based on experience, the proprietors are relatively sure that there are no interactions but wish to confirm their intuition, so a full factorial design is in order. We'll need eight runs; how do we conduct the runs? Eight days in a row is a possibility, but some days are better for sales than others. In fact, there is a regular pattern to daily sales, as shown in Figure 7.6. If the promotion was run for eight consecutive days, this would introduce a great deal of variation into the response; it would make for a lot of noise that would make it hard for us to see the signal.

Table 7.20 Pizza parlor promotions.

Factor	Promotion
A	Free order of fried mozzarella sticks
B	15% off
C	Free liter of soda with order

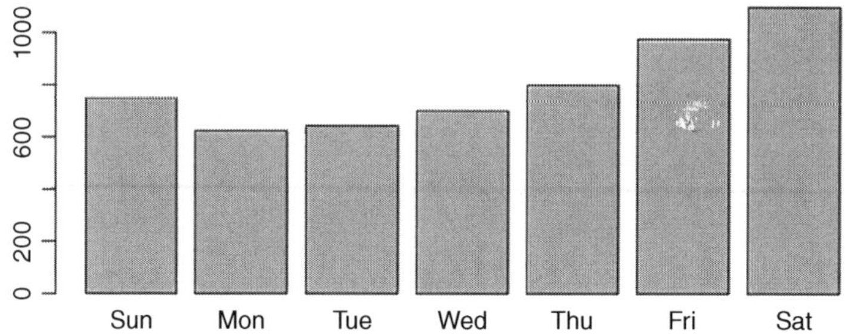

Figure 7.6 Average pizza sales by day of week.

We could do this on the same night every week, but which night? We'd like to do this on the busiest night (more sales equal more transactions equal more opportunities for the promotions to be used), which is Saturday. However, eight Saturdays is two months, and that will take too long; the proprietors want answers faster than that. The proprietors would like to have the results in a month so they can start making more money right away. If we run the experiment on Fridays and Saturdays, we'll be done in a month, but there's a problem: Saturday sales are typically higher than Friday sales. The day of the week is a nuisance variable, and we ought to control for it. We'll have two levels for this nuisance factor: -1 is Friday, and $+1$ is Saturday. What sequence of -1 and $+1$ shall we use for the nuisance factor? There are only 2^3 ways to arrange the -1 and $+1$, all of which are represented in a full factorial. We'll have to confound the nuisance factor with one of the interaction factors. In accordance with the *effect hierarchy principle*, we'll confound the nuisance factor with the ABC interaction term. Remember, the effect hierarchy principle says that lower-order effects are typically stronger than higher-order effects. So if we want a variable that is likely to have a small coefficient (so we can ignore that variable), we should choose *ABC* rather than *AB* or *C*.

The design matrix for this experiment, shown in Table 7.21, has a special feature, and it is the run order. The runs have to be randomized, but what happens if a Saturday night is run order 1 and Friday is run order 2? Then we have to skip the first Friday and conduct the first experiment on a Saturday night and wait till next Friday to run the second experiment. But what if a Friday is run order 3? Then we skip over that Saturday and run the third experiment the following Friday. In the first five weekend nights, we have run only three experiments. If, instead, we randomize according to blocks, then we don't have to skip any nights. So we randomize the four Fridays and get (2, 1, 4, 3); randomizing the four Saturdays, we get (4, 3, 1, 2). These run numbers are appended with an "f" or an "s" to indicate the block. For example, "1f" signifies Friday of the first weekend, while "3s" signifies Saturday of the third weekend. The result is shown in 7.21.

We rearrange the rows of Table 7.21 into the run order shown in Table 7.22 and execute the experiment on four successive weekends.

Table 7.21 Design for the blocked pizza experiment: standard order.

Run	Std	*A*	*B*	*C*	bv
2f	1	−1	−1	−1	−1
4s	2	1	−1	−1	1
3s	3	−1	1	−1	1
1f	4	1	1	−1	−1
1s	5	−1	−1	1	1
4f	6	1	−1	1	−1
3f	7	−1	1	1	−1
2s	8	1	1	1	1

The experimental data are in `PizzaPromotion.csv`; to protect the financial information of the establishment, the sales data have been rescaled. When applying the modeling strategy (when estimating the full factorial model), remember that the blocking variable *bv* is confounded with the *ABC* interaction, so you cannot include the *ABC* interaction in your model.

The results of applying our modeling strategy to the data are presented in Table 7.23. The *AB*, *AC*, and *BC* interaction terms are insignificant, so we're pretty sure the ABC term is also. This confirms our decision to choose ABC as the confounding factor. We apply the modeling strategy to the data – we leave this as an exercise for the reader – and wind up with a reduced form model presented in Table 7.24. Promotion A is clearly the best: customers prefer free mozzarella sticks to either 15% off or a free liter of soda. Notice that the coefficient for the blocking

Table 7.22 Design for the blocked pizza experiment: experimental order.

Run	Std	*A*	*B*	*C*	bv
1f	4	1	1	−1	−1
1s	5	−1	−1	1	1
2f	1	−1	−1	−1	−1
2s	8	1	1	1	1
3f	7	−1	1	1	−1
3s	3	−1	1	−1	1
4f	6	1	−1	1	−1
4s	2	1	−1	−1	1

Table 7.23 Reduced form of pizza experiment.

	Estimate	Std. error	*t* Value	Pr(> |*t*|)
(Intercept)	1176.6250	21.7298	54.15	0.0000
A	158.8750	21.7298	7.31	0.0053
B	61.1250	21.7298	2.81	0.0671
C	79.1250	21.7298	3.64	0.0357
bv	88.3750	21.7298	4.07	0.0268

Residual standard error: 61.46 on 3 degrees of freedom

Multiple R-squared: 0.9681, adjusted R-squared: 0.9257

Table 7.24 Pizza results without the nuisance variable.

	Estimate	Std. error	*t* Value	Pr(> |*t*|)
(Intercept)	1176.6250	48.0278	24.50	0.0000
A	158.8750	48.0278	3.31	0.0297
B	61.1250	48.0278	1.27	0.2721
C	79.1250	48.0278	1.65	0.1748

Residual standard error:135.8 on 4 degrees of freedom

Multiple R-squared: 0.7925, adjusted R-squared: 0.6369

variable equals the difference between average Saturday sales and average Friday sales.

What would have happened had we not included the blocking variable? The results of applying our modeling strategy *without* the blocking variable would be identical to those in Table 7.23, except "bv" would be replaced with "*ABC*," and we would mistakenly conclude that there exists a huge three-way interaction! This would be curious, since if there is a strong three-way interaction, we would expect to see *some* significant two-way interaction.

Suppose that we simply estimated a main effects model. We would obtain the results shows in Table 7.24. We would find only a single significant main effect and mistakenly conclude that the *B* and *C* promotions had no effect. Look at the huge reduction in the standard error and the corresponding increase in the R^2 that we get from including the blocking variable. This nuisance variable (the blocking variable) soaked up variation in the response and enabled us to see significant effects that we otherwise would have missed.

Exercises

7.5.1 Reproduce Table 7.23. Perform model checking.

7.5.2 Reproduce Table 7.24. Perform model checking. Are the residuals in this model better or worse than the residuals in the model for Table 7.23? (Table 7.23 is a better model, so it should have better residuals. How are the residuals better?)

7.6 Mee's Blunders

Mee (2009) wrote the bible on 2^k factorial designs, appropriately entitled *A Comprehensive Guide to Factorial Two-Level Experimentation*, which weighs in at 545 pages. In Section 14.7 thereof he describes "Four Blunders to Avoid," and they all concern computation of the standard errors of the estimates. Here we briefly describe these blunders. The reader is referred to the source (Mee, 2009, p. 466) for elaboration.

1) **Make sure that replication captures all the sources of run-to-run variation.** Specifically, if you are replicating, make sure that you have true replication, not pseudo-replication. Pseudo-replication occurs when individual observations are not independent of each other. The simplest example is when multiple measurements are taken on each unit. If the effect of diet on the weight gain of mice is of interest, let the weight of each mouse be measured before and after a week of the diet. If each mouse is weighed several times before and several times after, there are many more observations, but they are not independent.
2) **Avoid basing your estimation of the error variance on only 1 or 2 degrees of freedom.** This effect is clearly demonstrated in Exercise 7.2.3, where with one degree of freedom the residual standard error is machine zero (e.g. 3E−17) and everything is significant and with two degrees of freedom it jumps orders of magnitude to 0.007 and nearly half the factors are insignificant, but with three degrees of freedom or more it jumps another order of magnitude to roughly 0.01.
3) **Remember that the mean square error for a reduced model can seriously underestimate the error variance.** Especially if there is no replication so that the previous point applies, use the half-normal plot to inform your judgment as to the significance of effects.
4) **Recognize randomization restrictions when conducting tests for effects.** In this book our only randomization restriction has been blocking. Recall that the blocking variable uses up a degree of freedom. With

sophisticated blocking schemes, several degrees of freedom can be used up. This point reminds us not to forget to count the degrees of freedom due to blocking.

Mee reminds us that "plotting the data and the effect estimates is always recommended." It would be a blunder not to make these plots.

7.7 Chapter Exercises

7.1 Producers of flavored sodas (e.g. Coca-Cola, Pepsi, etc.) are concerned with how much their sodas foam. Some sodas are known to foam a lot; some foam only a little. An experiment was conducted to determine the factors that affect the volume of foam produced by a soda. The factors and levels are given below.

Letter	Factor	Lo	Hi
A	Soda	Pepsi	7Up
B	Temperature	Room temperature	Refrigerated
C	Quantity	40 ml	60 ml
D	Diameter of glass	3.6 cm	4.8 cm

The data for this replicated experiment (the experiment was run twice for each combination) are in file `SoftDrinks.csv`. Develop a model to predict the volume of foam. Be sure to look at the interaction plots! Describe any large interactions you find. What levels of the factors produce the most foam? The least?

7.2 A website owner wants to optimize his website to maximize clickthroughs. He has identified four factors that he thinks are relevant.

Label	Factor	Hi	Lo
A	Background	Light yellow	Light blue
B	Font size	16pt	22pt
C	Location of button	Left side	Right side
D	Size of button	Small	Large

Each combination was run on the website several thousand times in order to get a good estimate of the proportion of clickthroughs. This is a busy website, so it was not difficult to replicate; the experiment is replicated. The data are in file `WebDesign.csv`. Analyze the data and find the best combination. Does your final model satisfy weak heredity? (It should!)

7.8 Learning More

Previously we have talked about orthogonality and balance in a design. There are balanced non-orthogonal designs and unbalanced orthogonal designs. For a non-technical discussion of these topics that features rafts and an alligator, see Kuhfeld (2010, pp. 63-67).

- The soda foam example is based on Maqsood and Shafi (2017).
- We covered the basics of design in Chapter 3. More details can be found in Freeman et al. (2013) and Simpson et al. (2013).

Section 7.1 "The Postcard Example"

- There are three primary reasons for choosing full factorial designs over other types of designs.

1) Full factorial designs can account for interactions.
2) There is no confounding (or aliasing).
3) Full factorial designs are optimal – the estimates are as precise as they can be, given the amount of data (we can always make estimates more precise by increasing the number of observations). All the regression coefficients have the same standard error, and it is as small as it can be.

- When replicating an experiment, it is very important to distinguish between replicates and duplicate measurements. We already mentioned weighing mice multiple times before and after an experiment. As another example, imagine a process in which a small rectangular metal plate is produced and the factors are the speed at which the rectangles are cut, the pressure to which they are subjected when pressed, and the curing time after having been heated. Four plates are produced on each run. The experiment is repeated 10 times. You do *not* have a 10-run experiment that has been replicated four times, because the exact same experimental conditions were employed for each group of four. What you have is *duplicate measurements*, and the correct thing to do is take the average of the four and use that as the response variable.

Section 7.2 "Email Campaign"

- The direct marketing exercise 7.2.3 is from Tamhane (2009, p 295).

• The fishing exercise 7.2.2 is based on a case study in Frigon and Mathews (1997, p. 273).

• The hotdog exercise 7.2.4 is taken from Frigon and Mathews (1997, p. 161).

• Of the email campaign we remarked that each treatment combination was sent to "several thousand" recipients to ensure that the estimates of the proportions were accurate. To do this, one typically creates a CI for the proportion. Question: How large does the sample have to be so that the margin of error for estimating the proportion is, say, ± 0.001? If we think the population proportion is about $1\% = 0.01$, we want to be sure that our estimate is in the range $(0.009, 0.011)$. To achieve this, we employ the usual sample size formula:

$$n = \frac{Z^2_{\alpha/2}\pi(1-\pi)}{e^2}$$

where $Z_{\alpha/2} = 1.96$, say, for a 95% interval, π is the hypothesized population proportion (in this case 0.01), and e is the margin of error (in this case 0.001). So we would calculate $n = 38\ 032$. If this was too big, we might have to use a 90% CI instead of 0.05, or a large margin of error, e.g. 0.002 instead of 0.001. And yes, this formula is based on the Wald interval, but this is only approximate anyhow, and until someone programs a better method into software, we'll use this one. Persons who want to use the more accurate methods can consult Goncalves et al. (2012).

• When examining the coefficients in Table 7.11, we remarked on the distinction between statistical significance and practical significance. When modeling, we may sometimes have tiny coefficients that are statistically significant that we nonetheless drop from the model because they have no practical effect.

• Sometimes we want to apply ANOVA to the variances to see what is causing variation. In such a case, don't use the variances as the response, but, instead, use the natural log of the variances, perhaps multiplied by some factor of 10 to avoid decimals and make interpretation easier. There are two reasons for taking the log transform: variances are always positive, and the natural log transform will map variances to both positive and negative values; additionally, variances are usually non-normal, and the natural log transform tends to induce normality. A technical justification for this approach is given in Section 4.11 of Wu and Hamada (2009). An example of checking variances can be found in Box (2006, p. 151).

Section 7.3 "The Determinant of a Matrix"

• The "Five Designs" example is adapted from Eriksson et al. (2000).

Section 7.4 "Aliasing"

Aliasing is also called confounding, sometimes with the adjective "perfect" in front to distinguish it from "partial confounding" that can arise in more complicated models that are not covered in this textbook.

Section 7.5 "Blocking (Again)"

• Consider a 2^k experiment with replication. Suppose one replicate was done the first day and second replicate done the second day. We could treat each day as a block. Perhaps in an industrial process, we use two machines to conduct an experiment; then the output from each machine might constitute a block.

• We said, "(more sales equals more transactions equals more opportunities for the coupons to be used)." Having more sales rather than fewer will reduce variation, making it easier to see the signal.

• The covariance matrix of the regression coefficients for a regression that has an intercept and two independent variables on the left, the special case when the regression is for a 2^2 factorial design is on the right.

$$cov(b) = \begin{bmatrix} var(b_0) & cov(b_0, b_1) & cov(b_0, b_2) \\ cov(b_1, b_0) & var(b_1) & cov(b_1, b_2) \\ cov(b_2, b_0) & cov(b_2, b_1) & var(b_2) \end{bmatrix} = \begin{bmatrix} \sigma^2 & 0 & 0 \\ 0 & \sigma^2 & 0 \\ 0 & 0 & \sigma^2 \end{bmatrix}$$

On the right, notice that the variance of the intercept equals the variance of b_1 equals the variance of b_2; they all have the same variance or, what is the same thing, they all have the same standard error. And isn't that what we see in the regression output from a full factorial? What we don't see in the standard regression output, because standard regression output doesn't include covariances of the coefficients, is that all the covariances are zero. This is because the regressors (the independent variables) are effect-coded and therefore uncorrelated (which is the same thing as have zero covariance).

• We know that the full factorial designed is not aliased. If we don't have enough runs to execute a full factorial, if we have fewer than 2^k runs, then there will be aliasing.

7.9 Appendix on aliasMatrix and colorMap

If you know some linear algebra, then the below derivation of the alias matrix will be of interest. If you don't know linear algebra, then skip this paragraph. Let X be the design matrix for the included factors and interactions. Let Z be the even higher-order interactions that are not included in the design. The model we estimate is

$$Y = X\beta + \epsilon$$

while the true model is

$$Y = X\beta + Z\beta_Z + \epsilon$$

Some matrix algebra shows

$$E[\wedge\beta] = \beta + (X'X)^{-1}X'Z\beta_Z$$

If you took an econometrics course, this is a traditional "omitted variables" argument. If we define

$$A = (X'X)^{-1}X'Z,$$

then A is called the *alias matrix*. If none of the higher-order interactions has any effect on the model, then $\beta_Z = 0$ and $E[\wedge\beta] = \beta$. If any of the higher-order interactions really has an effect on the response, then the alias matrix tells us which higher-order interactions are confounded with which included main effects and interactions.

```
aliasMatrix <- function(design){
  X <- as.matrix(design)
  X <- cbind(rep(1,nrow(X)),X)
  colnames(X)[1] <- "int"
  Z <- model.matrix(~.^2,data.frame(design))
  A <- solve(t(X)%*%X)%*%t(X)%*%Z[,-seq(1,ncol(X))]
  return(A)
}

aliasMatrix3 <- function(design){
  X <- model.matrix(~.^2,data.frame(design))
  Z <- model.matrix(~.^3,data.frame(design))
  A <- solve(t(X)%*%X)%*%t(X)%*%Z
  return(A)
}

colorMap <- function(dm){
  dm <- data.frame(dm)
  mm <- model.matrix(~.^2,dm)
  cormat <- cor(mm[,-1],mm[,-1])
  cormat <- abs(cormat)
  melt_cormat <- melt(cormat)
  ggplot(data=melt_cormat,aes(x=Var1,y=Var2,fill=value)) +
    geom_tile(color = "white")+ labs(x="",y="")+
    scale_fill_gradient(low = "white", high = "red",
space = "Lab",
                                   limit = c(0,1),
                                   name="Correlation
\nCoefficient")
}
```

```
colorMap3 <- function(dm){
  mm <- model.matrix(~.^3,dm)
  cormat <- cor(mm[,-1],mm[,-1])
  cormat <- abs(cormat)
  melt_cormat <- melt(cormat)
  ggplot(data=melt_cormat,aes(x=Var1,y=Var2,fill=value)) +
    geom_tile(color = "white")+ labs(x="",y="")+
#     theme(axis.text.x=element_text(angle = -45,
hjust = 0))+
    scale_fill_gradient(low = "white", high = "red",
space = "Lab",
                        limit = c(0,1),
                        name="Correlation
\nCoefficient")
}


corrOfCoeff <- function(dm){
  df <- as.data.frame(dm)
  mm <- model.matrix(~.^2,df)
  cormat <- cor(mm[,-1],mm[,-1])
  cormat <- abs(cormat)
  round(cormat,3)
}
corrOfCoeff(df)
```

All of these functions have been checked against results from the software package JMP.

The article by Jones and Nachtsheim (2011) compares three alternative five-factor designs using color maps to aid their decision making. Here we present two of the three color maps and discuss them briefly. The interested reader is referred to the article for a complete analysis of the designs.

A color map is a symmetric about the main diagonal, so the upper left is a mirror image of the lower right. A color map that includes two-way interactions (the most common type) can be divided into three regions:

1) The lower left 5×5 corner, which shows the correlation pattern between the main effects.
2) The upper right 10×10 corner, which shows the correlation pattern between the 2FIs.
3) The off-diagonal rectangle (top left or lower right), which shows the correlation between the main effects and the 2FIs.

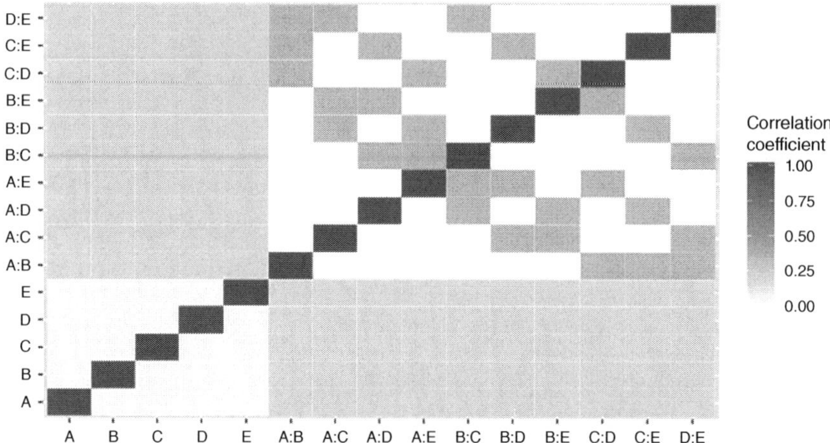

Figure 7.7 Color maps for Bayesian D-optimal design.

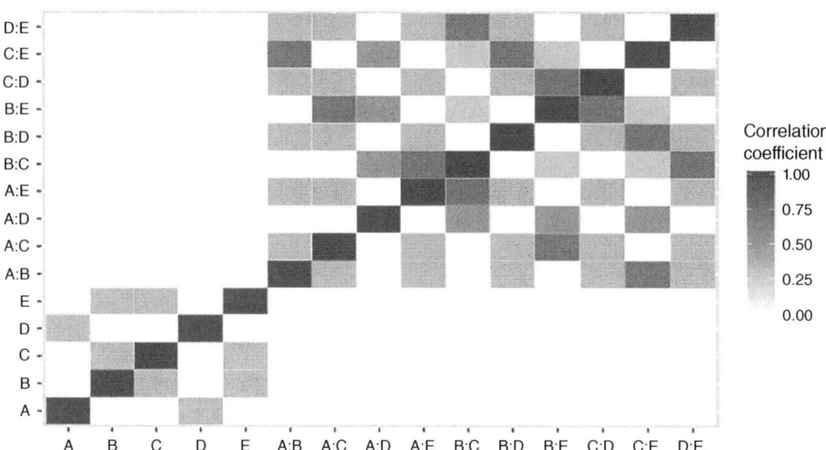

Figure 7.8 Color maps for minimal aliasing design.

It is best if these are in color, to better show the degree of correlation, but on these pages we must work with black and white. The white cells indicate a correlation of zero, the darkest cells indicate a correlation of unity, while various shadings represent correlations between zero and unity.

Looking at Figure 7.7 we can make some observations about the correlations of the Bayesian D-optimal design:

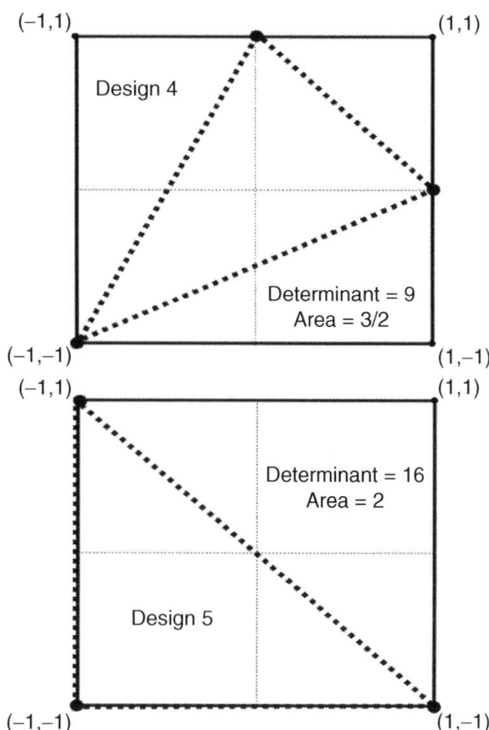

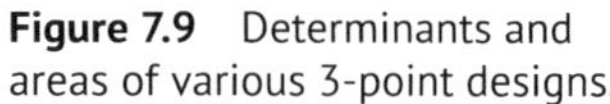
Figure 7.9 Determinants and areas of various 3-point designs.

1) The main effects are uncorrelated with each other.
2) Some of the 2FIs are correlated with each other.
3) Each main effect is correlated with *all* of the 2FIs.

Similarly, we can draw some conclusions about the minimal aliasing design shown in Figure 7.8:

1) Some of the main effects are correlated with each other.
2) *None* of the main effects is correlated with any 2FI.
3) *Most* of the 2FIs are correlated with each other.

Printing the correlation matrices can make it easier to interpret the correlation maps, because sometimes it is necessary to know the precise value of the correlation when looking at a correlation map.

Figure 7.9 shows a couple more examples of determinants for the three-point example.

8

Two-Level Screening Designs

When building a model about a well-understood phenomenon, we expect that most (if not all) of the main effects will be significant, and the real question concerns the extent of interactions: are any 2FIs or 3FIs active? On the other hand, when we are trying to model a process about which we know very little, we will have to guess at the important factors and then use experiments to let us know which factors really are important. In such a case, we expect that most of the factors will *not* be significant. This scenario gives rise to what are called *screening experiments*, in which many factors are screened to find the important ones. We'll guess at several possible factors and develop an experiment to test them all. Most of these guesses will turn out to be wrong. This is the Pareto principle in action, and the statistician Juran has described this as "seeking the vital few among the trivial many." As we will see, fractional factorials are well suited for conducting screening experiments. There is yet another use for fractional factorials.

The second use of fractional factorials is to conduct experiments when the number of runs is limited. Sometimes experiments are complicated, and you are logistically unable to run a large number of experiments. Sometimes experiments are expensive, and you cannot afford to run a full factorial. To make these ideas concrete, consider the following story from the aftermath of World War II. A famous Japanese statistician, Genichi Taguchi, had developed statistical methods to improve production processes. A local manufacturer of bathroom tile, the INA company, purchased an expensive kiln that heated unevenly, resulting in almost all the tiles either breaking or crumbling. The company could not afford to purchase a new kiln and was in dire straits. Taguchi was called to consult. Rather than focus on fixing the kiln, Taguchi focused on changing the composition of the slurry used to make tiles. He identified eight ingredients used in making the slurry and two levels of concentration (hi and lo) for each of the ingredients. The INA management complained that they couldn't afford to run $2^8 = 256$ batches of slurry to test all combinations. Instead, Taguchi tested a fractional factorial design of only 16 runs (with a few extra runs thrown in), determined a workable

Business Experiments with R, First Edition. B. D. McCullough.

Companion Website: www.wiley.com/go/mccullough/businessexperimentswithr

slurry, and saved the company from bankruptcy. We are not going to address this second use in this chapter, but we will address it the next chapter.

By the end of this chapter reader should:

- Know the assumptions that underlie screening designs and the purpose of screening experiments.
- Be able to analyze the data from a screening experiment and determine the important factors.
- Be able to create a screening design.

8.1 Preliminaries

The purpose of a screening experiment is to answer one question: What factors influence the response variable? The success of a screening experiment depends on three assumptions:

1. Effect sparsity – Most of the variation in the response is explained by a small number of effects (this is just a restatement of the Pareto principle).
2. Effect hierarchy – Main effects explain more variation than 2FIs, 2FIs explain more variation than 3FIs, etc.
3. Effect heredity – Two-factor interactions are more likely if both main effects are present. Corollary to this, if neither factor A nor factor B is significant, then the interaction AB probably isn't significant.

A meta-analysis of scores of published studies using factorial experiments by Li et al. (2006) documented several empirical regularities that support the ideas of effect sparsity, hierarchy, and heredity:

- Sparsity – 41% of main effects were active, 11% of 2FIs, 7% of 3FIs, and 3% of 4FIs.
- Hierarchy – Boxplots of the standardized effect sizes from the many studies showed main effects are generally larger than 2FIs, which, in turn, are larger than 3FIs.
- Heredity – Is demonstrated by the sequence of conditional probabilities shown in Table 8.1 where, as usual, "|" means "given."

Suppose we have to investigate a new phenomenon, and we really don't understand it; we really don't know what the important factors are. Think about this for a minute. Suppose we have ten possible factors, only four of which might matter. We design an experiment to test 10 factors. Now, a full factorial will require $2^{10} = 1024$ runs. It doesn't make much sense to build a full factorial model when most main effects – and therefore higher order effects – are null.

Maybe instead we'll do a fraction of that, say, $2^6 = 64$ runs. We'll have lots of aliasing, but suppose we have a good aliasing pattern so that we can get

Table 8.1 Conditional probabilities for effect heredity.

Condition			Probability
AB is active	|	Both A and B are active	0.33
AB is active	|	Either A or B (not both) is active	0.045
AB is active	|	Neither A nor B is active	0.005
ABC is active	|	A, B, and C all are active	0.15
ABC is active	|	Two of A, B, and C are active	0.07
ABC is active	|	One of A, B, and C is active	0.035
ABC is active	|	none of A, B, and C is active	0.01

good estimates of the main effects – for screening purposes we don't care about interactions unless they are aliased with main effects. As we apply the modeling strategy and drop the factors that don't have important main effects, we will be left with 4 of the 10 factors, and 64 runs of data are more than enough to estimate a full factorial on 4 factors!

Screening experiments are rarely replicated and are often saturated, so the degrees of freedom will be small or nonexistent. When we create screening experiments, we like to have a few extra degrees of freedom to estimate the error. There is a large difference between applying the modeling strategy to a well-defined model (when most of the factors are known to be relevant) and to a screening design (when most of the factors are probably irrelevant). There is a risk of fitting the model to the noise, and, since there are many irrelevant variables, there will be a lot of noise. See Exercise 8.3.2 for an extreme example of this.

Another source of noise is the factor levels. To minimize this source of noise, it is customary to be aggressive in setting the factor levels for a screening experiment; choose the levels to be wide, rather than narrow, as discussed in Chapter 3. Of course, if nobody really knows much about the process under investigation, it can be difficult to know what is wide and what is narrow. This quandary is part of what makes experimental design fun and exciting. Screening experimentation is usually followed up by refining experimentation with helpful factors that were significant in the screening design.

Let us now define a fractional factorial. Suppose we had three factors, we knew there were no interactions, and we could only use $2^2 = 4$ runs; we couldn't use the $2^3 = 8$ runs that a full factorial would require. We might decide to use, for example, the top or bottom half of the 2^3 design shown in Table 8.2 (we have included the intercept, even though, technically speaking, it's not part of the design matrix). That is, we would use a *fraction* of a factorial design.

Table 8.2 Two halves of a 2^3 factorial.

Std. order	Int	*A*	*B*	*C*
1	1	−1	−1	−1
2	1	1	−1	−1
3	1	−1	1	−1
4	1	1	1	−1
5	1	−1	−1	1
6	1	1	−1	1
7	1	−1	1	1
8	1	1	1	1

Table 8.3 Two halves of a 2^3 full factorial.

Std. order	Int	*A*	*B*	*C*	*AB*	*AC*	*BC*	*ABC*
1	1	−1	−1	−1	1	1	1	−1
2	1	1	−1	−1	−1	−1	1	1
3	1	−1	1	−1	−1	1	−1	1
4	1	1	1	−1	1	−1	−1	−1
5	1	−1	−1	1	1	−1	−1	1
6	1	1	−1	1	−1	1	−1	−1
7	1	−1	1	1	−1	−1	1	−1
8	1	1	1	1	1	1	1	1

Immediately we see a problem: C is aliased with the intercept. If we use the top half, $-C =$ the intercept; if we use the bottom half, then $C =$ the intercept.

Let us take another example to drive home the point that fractional factorials are confounded. Consider the model matrix for the 2^3 full factorial given in Table 8.3. We know that all the columns are orthogonal and that there is no aliasing. Suppose we could use only the top half or the bottom half. What difficulties would we encounter?

By inspection, we see that there would be substantial aliasing:

- The intercept is confounded with C.
- A is confounded with AC.
- B is confounded with BC.
- AB is confounded with ABC.

When we create a fractional factorial, we don't simply take the top half or the bottom half of a design because to do so creates a big aliasing problem; the main effects are aliased with the interactions so we can't get clean estimates of the main effects. What works better is to alias the interaction terms with each other rather than with the main effects so that the main effects are not aliased. To do this, we strategically choose some of the runs so as to minimize the aliasing problem while simultaneously maximizing the determinant of $(X'X)$. To create a half-fraction design, we take the highest interaction term, in this case ABC, and make one half of the runs correspond to $ABC = +1$ and other half of the runs correspond to $ABC = -1$, as shown in Table 8.4. The information in each half is the same, just expressed differently. In this particular case, the half-fraction design is indicated by 2^{3-1}. We're not going to do this by hand – we'll use software to generate fractional factorial designs.

In general, the 2^{k-p} is also called a $(1/2)^p$ fraction of the 2^k design where k is the number of factors. If $p = 1$ then it is a "half-fraction" design, if $p = 2$ it is called a "quarter-fraction" design, etc. Thus, in the present case, we would say that 2^{3-1} represents a half-fraction of a 2^3 design. The 2^3 full factorial design has eight design points, while the 2^{3-1} has four design points, but the four are chosen to be as far away from each other as possible and so as to maximize the determinant of $(X'X)$. It is also possible to make one-quarter, one-eighth, and even one-sixteenth fractional designs. We won't have to worry about whether the screening designs we create are one-quarter or one-eighth, because we'll be using software to create them. More specifically, don't worry about knowing precisely what a 2^{7-4} or 2^{3-1} looks like or how to compute one, because the software will do it for us.

Suppose we wanted to study seven factors in eight runs; we could do so with a one-sixteenth 2^{7-4} fractional factorial design. The complete seven-factor full

Table 8.4 Two halves of a 2^3 full factorial.

Std. order	*A*	*B*	*C*	*ABC*	Std. order	*A*	*B*	*C*	*ABC*
1	−1	−1	−1	−1	2	1	−1	−1	1
2	1	−1	−1	1	3	−1	1	−1	1
3	−1	1	−1	1	5	−1	−1	1	1
4	1	1	−1	−1	8	1	1	1	1
5	−1	−1	1	1	1	−1	−1	−1	−1
6	1	−1	1	−1	4	1	1	−1	−1
7	−1	1	1	−1	6	1	−1	1	−1
8	1	1	1	1	7	−1	1	1	−1

Table 8.5 The 2^{7-4} fractional factorial design.

	A	*B*	*C*	*D*	*E*	*F*	*G*
1	−1	−1	−1	1	1	1	−1
2	1	−1	−1	−1	−1	1	1
3	−1	1	−1	−1	1	−1	1
4	1	1	−1	1	−1	−1	−1
5	−1	−1	1	1	−1	−1	1
6	1	−1	1	−1	1	−1	−1
7	−1	1	1	−1	−1	1	−1
8	1	1	1	1	1	1	1

factorial design has 128 runs, and one-sixteenth of 128 equals eight. The 2^{7-4} design has eight runs and is shown in Table 8.5. For the interested reader, the associated data file is `Two7Minus4.csv`.

This design is quite aliased, and this aliasing is manifested in the color map, which is given in Figure 8.1. By looking in the lower left-hand corner, we can see that the main effects are uncorrelated with each other. Further, each main effect is aliased with more than one two-way interaction. As long as the main effects aren't aliased with each other, we aren't so concerned about main effects being aliased with interactions. The hierarchical ordering principle says that higher-order effects are smaller than main effects. If the higher-order effect is small, even if it is aliased with a main effect, we'll still be able to identify the main effects. Of course, we don't place blind faith in this principle.

The savings from running a fractional factorial as opposed to a full factorial can be dramatic, as seen in Table 8.6. A problem with the factorial designs, and also

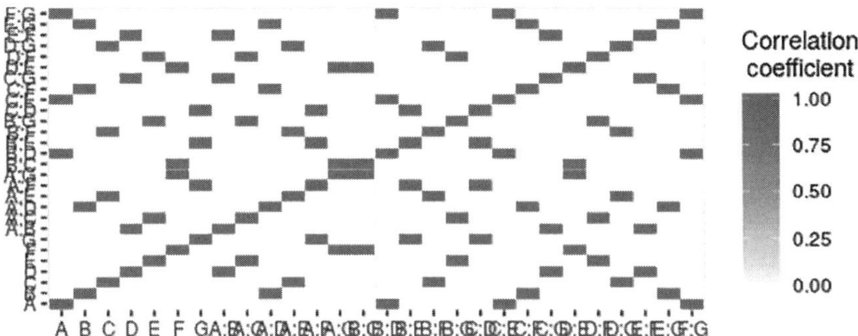

Figure 8.1 Color map of the correlations for 2^{7-4} fractional factorial.

Table 8.6 Number of runs for full factorial and screening designs.

	No. of runs in a DOE	
No. of factors	**Full factorial**	**Smallest screening design**
4	16	8
5	32	8
6	64	8
7	128	8
11	2 048	12
15	32 768	16
19	524 288	20
23	8 388 608	24

with the fractional factorial designs, is that the number of runs required is always a power of two. This can present difficulties.

Suppose we wanted to execute a screening design with 17 variables. A 2^4 design allows only 16 runs, which is too few, while a 2^5 design requires 32 runs, which is much more than 17. Hence, the smallest fractional factorial that can accommodate this is a 2^{17-12} with 32 runs. That's a lot of extra runs. To deal with this situation, Plackett and Burman invented the Plackett–Burman design, which requires that the number of runs be a factor of four. So, if we had 17 factors, we would need only 20 runs, which is a lot better than 32. Still, the Plackett–Burman designs have a lot of aliasing, though the main effects are uncorrelated with each other, so they can be useful for screening designs.

All of these screening designs are aliased to some degree, and they are categorized according to the degree of aliasing by reference to the "resolution" of the design:

- Resolution II designs: Some main effects are aliased with other main effects. In order for this to happen, we need to have more parameters than runs. In this book we never have fewer runs than factors, so we don't have to worry about this.
- Resolution III designs: No main effect is aliased with any other main effect, but main effects are aliased with two-factor interactions, and some two-factor interaction may be aliased with other two-factor interactions. These are to be avoided if possible, because we at least want clean estimates of main effects. Sometimes, however, experiments are expensive, and we can't avoid having aliased main effects. The classic case of this is the Plackett–Burman design.

- Resolution IV designs: No main effect is aliased with any other main effect or with any two-factor interaction, but two-factor interactions are aliased with other two-factor interactions. We choose this when we really want clean estimates of main effects. The cost is that the required sample size can be large.
- Resolution V designs: No main effect is aliased with any other main effect or two-factor interaction, and no two-factor interaction is aliased with any other two-factor interaction, but two-factor interactions are aliased with three-factor interactions. We choose this when we want clean estimates of first- and second-order effects.

There are entire books of screening designs (e.g. Wheeler (1990) and Connor and Young (1957)). In the traditional method of experimental design, what you do is try to look up a design that meets your experimental needs. We won't need a catalog because we have software. All we'll need to know is how many factors we want to screen and how many runs we want to use (as long as the number of runs is a power of two or a multiple of four).

Exercises

8.1.1 Aliasing is not always obvious and cannot always be deduced simply by looking at the design matrix. Suppose we can only afford eight runs, yet we have four factors. Any possible choice we make for the variable *D* will be aliased. What should we do?

Int	*A*	*B*	*C*	*AB*	*AC*	*BC*	*ABC*	*D*
1	−1	−1	−1	+1	+1	+1	−1	−1
1	+1	−1	−1	−1	−1	+1	+1	−1
1	−1	+1	−1	−1	−1	−1	+1	−1
1	+1	+1	−1	+1	+1	−1	−1	+1
1	−1	−1	+1	+1	+1	−1	+1	−1
1	+1	−1	+1	−1	−1	−1	−1	−1
1	−1	+1	+1	−1	−1	+1	−1	+1
1	+1	+1	+1	+1	+1	+1	+1	−1

What combinations of + and − shall we give to *D*? It turns out that any possible combination we can give to *D* is equal either to one of the existing columns or some linear function of the existing columns. Hence, linear regression will not be able to separate the effects of *D* and the existing column(s). Our choice for *D* certainly doesn't look like one of the existing columns, but it is equal to $-0.5 * \text{int} + 0.5 * B - 0.5 * AC - 0.5 * ABC$. Confirm this. The data are in file `aliasedD.csv`. (If you have taken linear algebra, the first eight columns are linearly independent, but adding any ninth column induces a linear dependence of some form.)

8.1.2 Using the data in file `aliasedD.csv`, regress D on the full factorial model matrix for A, B, and C. What happens? Why? (Hint: `lm(D~A*B*C,data=df)`)

8.2 Case: Puncture Resistance (Small Screening Experiment)

In response to customer complaints, a paper products manufacturer wanted to increase the puncture resistance of a particular type of cardboard. The amount of force required to puncture the cardboard was the response variable. Brainstorming activities produced a list of seven possible factors, and the product engineering team was brought in to help determine the high and low levels for the experiment.

To make cardboard, pine chips have to be pulped in a high-pressure tank for several hours. The pulped chips have to be roll-pressed into sheets, and this is done with some pressure. Before pressing, an additive is mixed into the pulp to make the final product moisture resistant. The paste used to glue the sheets together (what is cardboard but several sheets of paper pressed together?) can be applied at various temperatures, and the paste can be made with clay or without. The finished paperboard must be cured for several days. The machine can be operated as various speeds. A summary of the factors and their levels is given in Table 8.7.

The design and response variable for the experiment are presented in Table 8.8; the data are in `PunctureResistance.csv`. The aliases for each factor are included at the bottom of each column. For example, factor A is aliased with the

Table 8.7 Factors for puncture resistance experiment.

Factor	Description	Low	High
A	Paste temperature	130 F	160 F
B	Moisture inhibitor	0.2%	0.5%
C	Press roll pressure	40 psi	80 psi
D	Pulping time	6 hours	8 hours
E	Paste type	No clay	Clay
F	Cure time	5 days	10 days
G	Machine speed	120 fpm	200 fpm

Table 8.8 Data for puncture resistance experiment.

A	*B*	*C*	*D*	*E*	*F*	*G*	*y*
+	−	−	+	−	+	+	14.67
+	+	−	−	+	−	+	40.87
+	+	+	−	−	+	−	51.33
−	+	+	+	−	−	+	48.53
+	−	+	+	+	−	−	54.12
−	+	−	+	+	+	−	38.77
−	−	+	−	+	+	+	54.53
−	−	−	−	−	−	−	14.63
−*BD*	−*AD*	−*AG*	−*AB*	−*AF*	−*AE*	−*AC*	
−*CG*	−*CE*	−*BE*	−*CF*	−*BC*	−*BG*	−*BF*	
−*EF*	−*FG*	−*DF*	−*EG*	−*DG*	−*CD*	−*DE*	

negative of the interaction term *BD*. Simply looking at the alias matrix for this experimental design will verify these aliases; see Exercise 8.2.3. There are ways to figure this out by hand, for example, a method that employs *alias generators* and *defining relations* is taught in many experimental design textbooks, but we think it much easier simply to employ the alias matrix.

The main effects plots for this experiment (see Exercise 8.2.1) show that only three of the factors have an effect of appreciable size. Application of the modeling strategy produces the results shown in Table 8.9.

It never hurts to plot the effect sizes, especially for a reduced model, and such a plot (see Exercise 8.2.2) shows that, while all five factors are statistically significant, only three of them really matter. For the two that don't matter, what shall we do? Set them at the lower-cost option.

If we were to continue with this example, we could now set up a full factorial experiment with replication to really investigate how these three factors relate to the response.

Exercises

8.2.1 Create the main effects plots for the puncture resistance example. Be sure to keep the y-axis the same for all the plots.

8.2.2 Make a bar plot of the effect sizes for each factor in the puncture resistance example (remember to double the regression coefficient!).

Table 8.9 Reduced model of puncture experiment.

	Estimate	Std. error	t Value	Pr(> |t|)
Int	39.680	0.1047	378.98	0.0000
A	0.565	0.1047	5.40	0.0327
B	5.195	0.1047	49.62	0.0004
C	12.445	0.1047	111.86	0.0000
D	−0.660	0.1047	−6.30	0.0243
E	7.390	0.1047	70.58	0.0002

Residual standard error: 0.2961 on 2 degrees of freedom

Multiple R-squared: 0.9999, adjusted R-squared: 0.9997

F-statistic: 4328 on 5 and 2 DF, p-value: 0.000 231

8.2.3 Compute the alias matrix for the puncture resistance design in Table 8.8 and verify the aliases at the bottom of the table. Recognize that, though there is much aliasing, the main effects are not aliased with each other, which is what we want for a screening experiment.

8.2.4 Researchers were interested in determining which factors affect the amount of lactic acid produced from wheat bran. Fifteen possible factors were identified: 3 physical factors (A–C), 1 buffer (D), and 11 nutrients (E–O). A Placket–Burman design in 16 runs was used. The design and response are in file `LacticAcid.csv`. Analyze the data using a half-normal plot and the estimated coefficients, and determine which factors should be used in a follow-up experiment. (The precise number is a judgment call, so don't worry if you're off by one or two, but most of the factors should be obvious.)

8.3 Case: College Giving (Big Screening Experiment)

A small college wanted to increase its alumni giving. The consulting firm QualPro, which specializes in statistical experiments to improve business performance, was engaged. The key measures were established:

- Increase total gift dollars.
- Increase the total number of gifts.

QualPro led the university staff in several sessions and generated 91 ideas to increase giving. These were then culled to 7 ideas to increase large gifts and 23 ideas to increase small gifts. All the ideas were practical (easy to implement), fast (could be implemented quickly), and essentially cost-free. The database of givers (and potential givers) numbered approximately 20 000. Givers were not randomly assigned to a design point, but instead were stratified in an attempt to avoid covariate imbalance. The key characteristics that were distributed evenly across the experimental groups were:

1. large donor
2. receivers of alumni email letter
3. alumnus of a top-25-ranked reunion class for the previous year
4. nursing, business, or education graduate
5. recent undergrad
6. current university faculty/staff

For the small gift ideas, these 23 factors, labeled A–W, are described in Table 8.10.

The experimental design and the responses are in the file `SmallCollegeA.csv`. The two responses correspond to the key measures and are called GiftCount and GiftDollars. There are 24 parameters to estimate, yet the file has 30 observations –30 experiments were run. So there are six runs to estimate the error; this is definitely *not* a saturated design. We expect that we have many variables in the model that really don't matter; this will cause an increase in the standard error, making it harder to detect significant effects. Rather than use a p-value cutoff of 0.05, we'll increase it to 0.15. Remember, this is just a screening experiment, and any variables we identify now will be confirmed in a later experiment. Let's analyze the data for GiftCount.

Software Details

To analyze the data for GiftCount ...

```
df <- read.csv("SmallCollegeA.csv")
lm1 <- lm(GiftCount~A+B+C+D+E+F+G+H+I+J+K+L+M+N+O+P+Q+R+
          S+T+U+V+W,data=df)
summary(lm1)
```

It can be hard to look for small p-values when there are so many coefficients. A nice way to do it is to make a bar plot of the p-values. To do this, issue the command

```
barplot(summary(lm1)$coefficients[,4])
```

How do the significant factors increase or decrease the number of gifts? Hint: Do *not* just look at the sign of the coefficient. Be sure to take into consideration how the high and low levels of the factors are defined.

Table 8.10 Factors for college giving experiment.

A	Solicitation envelope type	Standard	Vibrant colored
B	Envelope message sticker	None	Yes
C	Letter font	Current font size	Large font size
D	Letter sent on	December 2	November 20
E	Letter signer	University president	Program dean
F	Letter format	Card	Letter
G	Online giving link	Included in letter	Not included
H	Survey insert	No	Yes
I	Mention tax benefit of giving	No	Yes
J	President's list of desired gifts	Big gifts	Little gifts
K	President's vision insert	No	Yes
L	President's vision ask	$20	$200
M	President's ask special gift	No	Yes
N	Special donor's letter	1 page	2 page
O	Museum wish list	No	Yes
P	Special letter ask	No	Yes
Q	Promote alumni community	No	Yes
R	Lifetime alumni offer	Status quo	New version
S	Video of famous alumnus	No	Yes
T	End of year mailing	Regular letter	With video link
U	Homepage link	Don't promote link	Promote link
V	Special gift recognition by level	No	Special insert
W	Special gift recognition by class	No	Special insert

Look at the R^2. It's rather high, but the adjusted R^2 is negative; taken together these suggest that most of the variables are insignificant. Indeed, look at the *p*-value on the *F*-statistic for testing the null hypothesis that all the slopes are equal to zero – it's large, which suggests again that most of the variables really don't matter.

Try it!

Analyze the data for the response GiftDollars. How many of the factors are important? Do these factors increase or decrease the number of dollars?

This was just one of many preliminary screening experiments. After a final model was tentatively developed, QualPro ran more experiments to verify conclusions. One interesting experiment involved *reversing* some conditions to see whether the opposite results would be produced (which should happen if the model is correct).

In the end, only two ideas for increasing large donations worked: telling a prospective donor that "We need your support" and bringing along the school yearbook from the year that the prospective donor had graduated. Factors for increasing small gifts were *D*, sending the letter before Thanksgiving; *N*, sending a shorter letter rather than a longer letter; and *O*, including a wish list for the school's museum in the mailing. Implementing these ideas the next year increased the gift count by 13% and the dollar amount by 13%. Particularly noteworthy is that these ideas were implemented in the 2009–2010 at the beginning of the great recession, when charitable giving was down across the entire country.

Exercises

8.3.1 A large consumer organization had eight million customers and processed four million bills per month. However, only 60% of the bills went out on time. In addition to creating problems with the revenue stream, large numbers of customers were irritated by late fees for not paying bills they never received. Two key measures were identified: rejects per 1000 (a reject is a bill that cannot be processed by the computerized system and has to be processed by hand) and cycle time (the amount of time from the proposed billing date to the date the invoice was actually mailed out). Brainstorming produced 80 ideas that were culled to 23 ideas (*A*–*W*) that were practical, fast, and cost-free. Some of these ideas were as follows:

Factor	Description	Lo	Hi
A	Pre-bill online	No	Yes
D	Plan supplies	No	Yes
F	Verify errors	No	Yes
G	Create two smaller billing cycle	No	Yes
H	Email between departments to	Everyone	Specific persons
M	Old vs. new paper	Old	New
Q	Designate a rework person	No	Yes

Analyze the data in file `LateBillingsScreening.csv`. This is a saturated design so we don't have an estimate of the error. Our modeling strategy won't work here, as shown in the next exercise. Therefore, QualPro used an advanced method based on control limits. The control limit for cycle time is 0.82 and for rejects is 9.86. These control limits are used as cutoffs – if any effect exceeds the control limit, then that variable is deemed significant. Identify important factors for both cycle time and rejects. (Remember to double the regression coefficients! A good way to do this is to issue the command `2*lm1$coefficients`.)

8.3.2 This exercise shows why our modeling strategy won't necessarily work when screening many variables, most of which don't matter, and why QualPro had to use an advanced method. Apply the modeling strategy for a saturated design and look at the regression output each time you drop a variable. (In order, drop the variables H, W, R, Q, and D.) Look at the R^2. Do you really believe all these variables are significant? What has happened here is that including so many irrelevant variables has soaked up much of the variance, making the standard error artificially small, creating spuriously significant coefficients. This suggests that, when creating screening designs, we always include extra runs to have enough degrees of freedom to get a good estimate of the error. Remember Mee's Blunders!

8.3.3 Based on the results of the late billings screening experiment, a refining experiment was carried out. Though "old vs. new" paper was important, it turned out there was not enough old paper left for a second experiment, so this factor could not be included in the refining experiment. As a practical matter, the designated rework person was reducing the billing backlog, even if the effect wasn't statistically significant; so factor Q was included in the refining experiment. Based on expert knowledge, the combination of F high and G high was known to be good for reducing cycle time. The factor F was significant in the screening design, but the factor G was not. However, screening designs aren't designed to pick up interactions. To check for interactions, a model that includes interactions is necessary. Therefore, the factor G was included in the refining experiment.

The refining experiment is a full factorial experiment with all interactions and four replicated points. The replicated points are included so that there are enough degrees of freedom to estimate the error. The design and responses are in `LateBillingsRefining.csv`. Analyze the data. The control limit for cycle time is 1.37 and for rejects is 15.40.

For each key measure, which factors should be implemented and at what levels?

After implementing recommended changes, cycle time was reduced from 5.88 to 4.48 days. Rejects per 1000 were reduced from 10.5 to 1.4. Overall, percent on-time for standard billing runs increased from 51 to 100%.

8.3.4 Suppose the average bill is \$50 – some four million are sent out every month. If the interest rate on short-term money is 5%, how much money does the company save each year by reducing the cycle time by 1.4 days? (Assume that getting the bill sent a day and half earlier means the company will be paid a day and half earlier.)

8.3.5 If each reject costs \$5 in time and materials to reprocess, how much money does the company save each year by reducing the rejects per 1000 from 10.5 to 1.4?

8.4 How to Set Up a Screening Experiment

Now that we know what fractional factorials are, and we've taken two examples of screening experiments, it's natural to wonder, "How does one go setting up a screening experiment?"

The steps are:

1. determine the experimental variables (more about this later)
2. rank the variables which
   (a) are known to affect the response
   (b) are believed to affect the response
   (c) may affect the response
3. define the variables precisely, including how to measure them
4. set the high and low levels for the variables
5. identify possible interactions
6. choose an experimental design – make sure to include enough degrees of freedom to estimate the error!
7. run the experiment

With respect to points (1) and (3), if you don't really know very much about the process, how can you select variables? You can't. At this stage, you're guessing; perhaps some of the guesses are educated, but some will be just guesses. Methods such as five whys, brainstorming, etc. can be invaluable in this situation. See the discussion of "Knowledge-Based Tools" in Hoerl and Snee (2012, pp. 172-205), which tools include:

1. flowcharting
2. brainstorming
3. affinity diagram

4. interrelationship digraph
5. multivoting
6. cause-and-effect diagram
7. five whys

See Holland and Cochran (2005) for more examples of how to generate ideas. Use all available sources of knowledge, and never forget the adage, "A week of experimentation can save an hour of research."

Of course, after the screening experiment has been set up and then run, the story does not end. Proper follow-up is a necessary part of any experiment. Screening experiments can suffer from a small number of runs, aliasing, and many other difficulties. Therefore they have to be interpreted cautiously. Tentative conclusions from the screening design ought to be verified with confirmatory experiments that account for interactions. Other types of confirmatory experiments can be minimal, for example, using the screening model to make a prediction about a design point that is not part of the experiment and then running a follow-up experiment at that point to see if the prediction and the actually observed value are reasonable close. Adding and dropping factors as well as rescaling variables (changing the high and low levels) may be in order.

Exercises

8.4.1 A favorite design for analyzing 11 factors in 12 runs is the Plackett–Burman design, which accommodates a number of runs that is a multiple of four. It is a non-regular fractional factorial, and it has a complicated alias structure. The file `PB11-12.csv` contains such a design. What is its alias structure? In particular, how are the main effects aliased?

8.4.2 Examine the alias structure for the late billings screening experiment. After loading the data frame, extract the columns that constitute the design matrix as follows: `design1 <- df[,2:24]`. Look at both the alias matrix and the color map. What do you see?

8.5 Creating a Screening Design

We will use the package "FrF2" to create fractional factorial designs, so the first thing to do is install the package; remember, this only has to be done once on your computer. Next, load the package into R using the `library` command.

The algorithm depends on the random number generator, so for reproducibility purposes it is important to set the seed before using the command. While there are

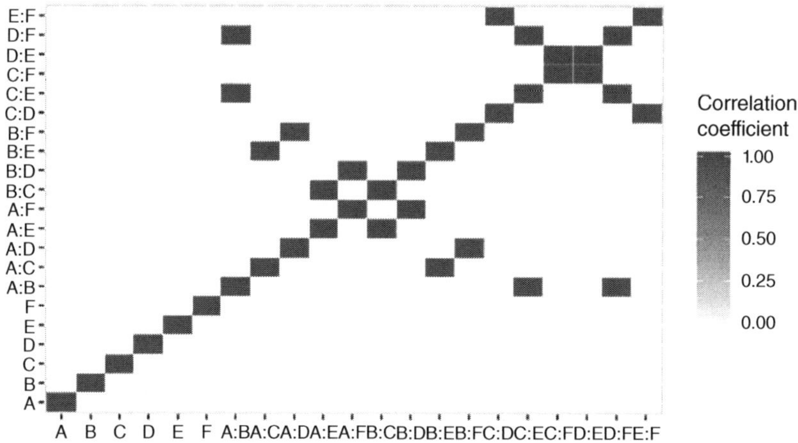

Figure 8.2 Color map of the correlations for six factors in 16 runs, fractional factorial.

many options for this command (consult the reference manual for further details), we will only concern ourselves with two of them: the number of runs and the number of factors. Hence, the form of the command will be `FrF2(nruns = 16, nfactors = 8)`, which will create a design to screen eight factors in 16 runs.

We have previously discussed the 2^{6-2} fractional factorial for screening six factors in $2^6/2^2 = 64/4 = 16$ runs. Let's use FrF2 to create a fractional factorial for screening six factors in 16 runs and look at its color map, which is shown in Figure 8.2. This is an excellent screening design: the main effects are uncorrelated with the main effects and the 2FIs. The 2FIs are all aliased correlated with each other, which is fine because for a screening design we aren't interested in estimating the coefficients for 2FIs. We will get nice, clean estimates of the main effects.

Software Details

To create a fractional factorial design...

The objects created by `FrF2` do not allow us to access them by use of the $ sign, the way we access the coefficients of an `lm` object after running a regression. So we'll have to stick an extra line of code in to extract the design from the object we create.

```
set.seed(123)
ff1 <- FrF2(nruns=16, nfactors=6) # create the object
ff1design <- attr(ff1,'desnum') # extract the design
colorMap(ff1design)
aliasMatrix(ff1design)
```

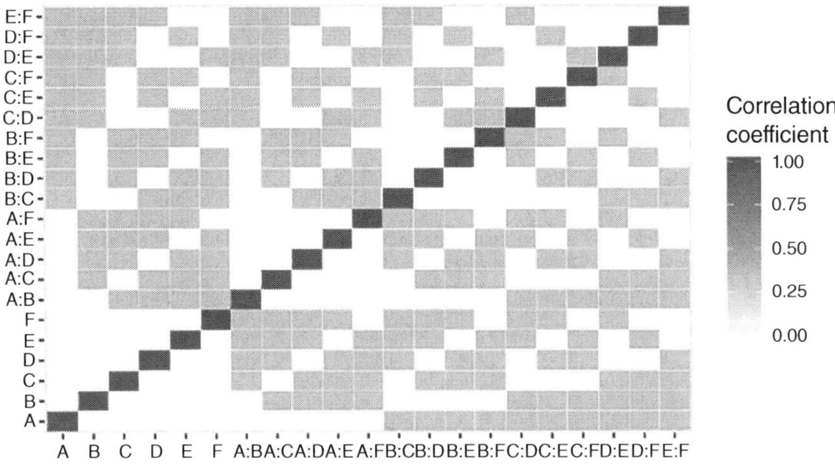

Figure 8.3 Color map of the correlations for six factors in 12 runs, Plackett–Burman.

Suppose we wanted to screen six factors in 12 runs. We couldn't use a fractional factorial design, because 12 is not a power of two. It is a multiple of four, however, so we could use a Plackett–Burman design. The color map is shown in Figure 8.3, and it's not bad. The main effects are uncorrelated with each other, and the degree of correlation with the 2FIs is not too severe; consulting the alias matrix, one can see that the aliasing coefficients off the diagonal are either zero or 0.333. Notice, however, how much cleaner the estimates of the main effects are using the above fractional factorial: that's what four extra runs (the difference between 12 and 16) can buy you.

Software Details

To create a Plackett–Burman design...

```
set.seed(123)
ff2 <- pb(nruns=12, nfactors=6) # create the object
ff2design <- attr(ff2,'desnum') # extract the design
colorMap(ff2design)
aliasMatrix(ff2design)
```

Finally, we may need to block a screening design. Suppose that half the runs would be done one on day and the other half would be done the next day. Or maybe half the runs would be done in one factory and the other half in another factory.

If either the day or the factory was a source of variation, we'd have to block on that factor. Creating blocked screening designs is very easy for fractional factorials, since the `FrF2` command has a "blocks" option and all you have to do is specify the number of blocks, which must also be a power of two.

Software Details

To create a blocked fractional factorial...

```
set.seed(123)
ff3 <- FrF2(nruns=8, nfactors=3,blocks=2) # create the object
ff3design <- attr(ff3,'desnum') # extract the design
ff3design # print the design

  Blocks1  A  B  C
1      -1 -1  1  1
2      -1  1 -1  1
3      -1 -1 -1 -1
4      -1  1  1 -1
5       1  1  1  1
6       1 -1 -1  1
7       1 -1  1 -1
8       1  1 -1 -1
```

Observe that the blocks are labeled in the first column.

Sometimes there is no way to block a certain number of runs, in which case the software will give you an informative error message.

There is no option to block a Plackett–Burman design.

Now we know how to create screening designs as long as the number of runs is a power of two or a multiple of four. Don't forget, after creating a design, the experiments must be run in a random order. If you have created a blocked design, you probably should randomize within each block.

Exercises

8.5.1 Create a fractional factorial to screen six factors in eight runs. Set the seed to 123. If you were going to use a full factorial, how many runs would you need?

8.5.2 Create a Plackett–Burman design to screen six factors in 12 runs. Set the seed to 123. If you were going to use a full factorial, how many runs would you need?

8.5.3 Compare the above two designs.

8.5.4 Create a fractional factorial to screen six factors in 32 runs with four blocks. Set the seed to 123. The four blocks will be represented by three dummy variables (Blocks1, Blocks2, and Blocks3). Note: Converting these to a single categorical variable for use in a regression is left as an advanced exercise and is beyond the scope of this class.

8.6 Chapter Exercises

8.1 A screening experiment to determine the factors that affect a measure of the yield of peanut oil per batch was conducted. The potential factors were as follows:

Label	Factor	Lo	Hi
A	Carbon dioxide measured in bars	415	550
B	Temperature in degrees Celsius	25	95
C	Moisture as a percentage of weight	5	15
D	Flow in liters per minute	40	60
E	Particle size in millimeters	1.28	4.05

A 2^{5-1} design was used. The data are in `peanuts.csv`. Determine the important factors.

8.2 Create a screening design of your own choosing, setting the seed to 123. For whatever number of factors you choose, let the number of runs be the number of factors plus three or four. Get a color map for the design. Recreate the design three more times, each time changing the seed. Create a color map each time. Compare the four color maps. What do you see?

8.3 The making of paper helicopters is a very popular experiment for learning experimental design. There is a standard form for cutting a helicopter from a standard sheet of paper and then folding it. However, the length of the cuts and the folds can be varied. The object of the exercise is to create a helicopter that takes the longest to drop. Below are some factors relating to the construction of such a helicopter; the factors are part of a screening experiment.

A	Rotor length	5.5	11.5
B	Rotor width	3.0	5.0
C	Body length	1.5	5.5
D	Foot length	0.0	2.5
E	Fold length	5.0	11.0
F	Fold width	1.5	2.5
G	Paper weight	Light	Heavy
H	Fold direction	Against rotation	With rotation

The data are in `HelicopterScreening.csv`. Based on advice from a subject matter expert, some interactions may well be important in this problem, so in your screening design, include all main effects and the following interactions: AB, AC, AD, BC, BD, CD, and $ABCD$. This design is fully saturated, so you may wish to make use of a half-normal plot to assist you. Don't forget, in the `half-normal` command, you can induce the program to label points by increasing alpha. Which of the eight factors appear to be important?

8.7 Learning More

Designs with few runs and no replications are often very low powered in environments with any noise. The engineers love these designs because (a) their runs are often expensive and (b) the measurement noise is very, very small. You don't see many fractional factorials in marketing, medicine, social science, or economics, where there is a lot more noise.

The Italian economist Vilfredo Pareto (1848–1923) observed that 80% of the land in Italy was owned by 20% of the people. This observation gave rise to the "80/20 rule," also known as the Pareto principle.

- The helicopter screening problem is based on Erhardt (2007).
- For the complete story of Taguchi and INA, read Taguchi (1986).

Section 8.1 Preliminaries

- We have focused only on screening for main effects. It is possible to create screening designs that estimate both first- and second-order effects, but that is beyond the scope of this course. Consult the `FrF2` documentation for details. In particular, look at the options `resolution`, which allows you to specify the resolution of the design you want. The various resolutions were defined at the end of Section 8.1.

Table 8.11 Number of parameters for 2^7 full factorial design.

Type of parameter	Number
Intercept	1
Main effects	7
2FIs	21
3FIs	35
4FIs	35
5FIs	21
6FIs	7
7FIs	1
Total	128

- Fractional factorials are very useful when many factors are being analyzed. Consider a full factorial with seven factors, which has 128 parameters as shown in Table 8.11. Three-way interactions are rare; four-way interactions are rarer still; no one has ever seen five-, six-, or seven-way interactions in a business experiment, so we are estimating a lot of unnecessary parameters. If we assume that all four-way and higher interactions are zero, we could estimate all the relevant parameters with just 56 runs instead of 128. If we wanted a good estimate of the error, we could throw in another eight runs, and 64 (which is a power of two, so we can use a fractional factorial) is still a lot less than 128.

Section 8.2 "Small Screening"

- The puncture resistance experiment was adapted from Example 4 in chapter 4 of Barrentine (1999).
- The lactic acid exercise is taken from Dean et al. (2017, p. 531).
- Figure 8.4 is more or less what the color map of an ideal screening design looks like, which is a Resolution IV design. The problem is that it requires many observations to achieve this degree of resolution. Great progress is being made in producing good screening designs that require few observations, but the technical demands are severe – right now, such designs are only for experts. For example, in Figure 8.5 is a color map for screening five factors with only 14 observations that is based on the design given in table 4 of Errore et al. (2017).

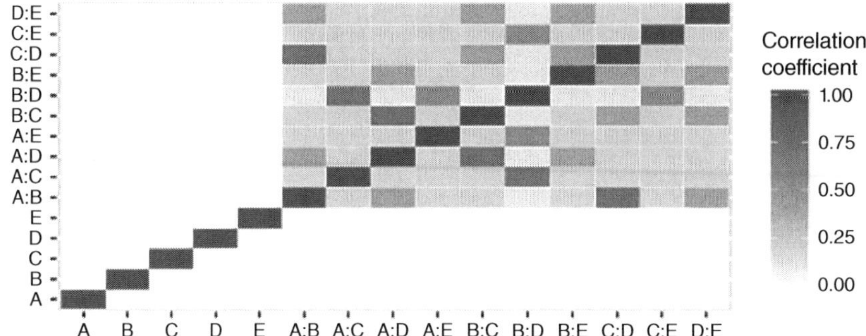

Figure 8.4 Color map of ideal screening design.

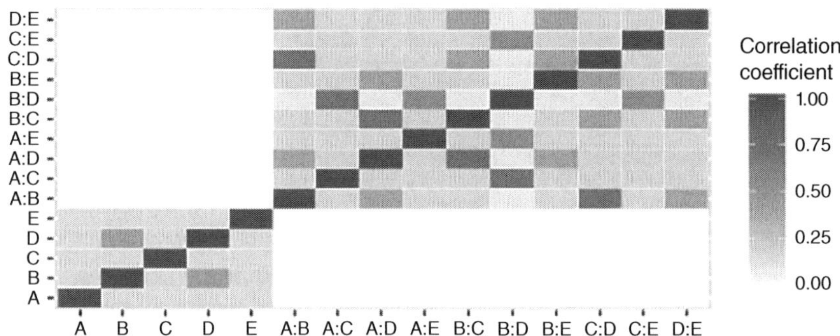

Figure 8.5 Color map of a really good screening design.

Section 8.3 "The College Giving Experiment"

• We are indebted to QualPro for providing us the college giving experiment and the late billings experiment.

Section 8.4 "How to Set Up a Screening Experiment"

• The book by Holland and Cochran (2005) has a wealth of detail on the practicalities of setting up and executing experiments, and Chapter 13 is devoted to screening experiments.

More details on setting up a screening experiment can be found in chapter 4 of Carlson (1992).

Two decent articles on screening experiments are Wass (2010) and Wass (2011).

See Montgomery and Jennings (2006) for even more details on screening experiments; §3.4 addresses the issue of follow-up experiments.

Section 8.5 "Creating a Screening Design"

- You can actually tell FrF2 that you want a Resolution IV design, but you'll have to read the reference manual to figure out how to do this.
- The peanuts exercise is based on Example 8.5 from Tamhane (2009).

9

Custom Design of Experiments

Full factorial designs can be very expensive in the sense that they may require more runs than are necessary. Fractional factorial designs, too, can be similarly expensive.

We know that factorial designs are "optimal," as mentioned back in Section 6.8. We didn't really define optimal at that time; we'll do so in this chapter. Frequently, the custom designs we create will not be optimal, so we'd like to know how close to optimal a custom design is. If it's close to optimal, we'll be comfortable using it. If it's far from optimal, we might not want to use it.

Orthogonal designs, by definition, do not suffer from aliasing, and in that sense are optimal. Nonoptimal designs suffer from aliasing. To the extent that a custom design is not orthogonal/optimal, it will be aliased, so it would behoove us to study the aliasing pattern of a custom design to determine whether the aliasing is so severe as to create a problem.

We also know that for an orthogonal design, the coefficients of the model are not correlated so that when a variable is dropped, the other coefficients do not change. Again, this will concern us with our custom designs; it would behoove us also to study the correlation of the coefficients to see whether this is a problem. We will do so by examining the color map of the correlations of the regression coefficients.

By the end of this chapter, reader should:

- Create a custom 2^k design.
- Create a blocked custom 2^k design.
- Interpret the alias matrix for a custom design.
- Analyze the color map for a custom design.
- Create a custom design for designs with more than two levels, e.g. a $4^4 3^3 2^2$ design that has four factors each with four levels, three factors each with three levels, and two factors each with two levels.

Business Experiments with R, First Edition. B. D. McCullough.

Companion Website: www.wiley.com/go/mccullough/businessexperimentswithr

9.1 Case: Selling Used Cars at Auction I (Small Custom Screening)

There is a large "dealer-only" auction market for used cars; sometimes thousands of cars will be sold at a single auction. The Acme Company specializes in selling used cars at these auctions. A new car dealer might take a used car as a trade-in. The new car dealer, not in the business of selling used cars, will sell the used car to Acme. A used car dealer that specializes in cars that are only a couple years old might take a 10-year-old car as a trade-in. This used car dealer, not wanting to tarnish its reputation by selling older cars, will sell the 10-year-old car to Acme. Acme will sell both cars at the auction. Used car dealers will purchase them, take them to their lots, and sell them to the public.

Used cars often are not in pristine condition, and they can all use some reconditioning to enhance the car's appeal to prospective buyers. The employees of Acme brainstormed some ways to recondition the used cars and came up with the following list (which is summarized in Table 9.1):

- Touch-up paint – A complete paint job is expensive and often unnecessary. However, used cars can have dings and scratches that can be eliminated by touch-up painting.
- Clean tires and hubcaps – Dirty tires and hubcaps cannot simply be cleaned with soap and water; it requires special treatment to make them look nearly new.
- Steam cleaning the engine – Car engines can get quite dirty, and a steam cleaning can remove much of the accumulated grime and grease.
- Detailing – To detail a car means to give it an extremely thorough cleaning, including spot remover on fabrics, and shampooing of all fabrics, conditioning leather seats, and all plastic and vinyl being properly cleaned and dressed. It can make the interior of the car look practically brand new.
- Change the oil.

Table 9.1 Factors for used car experiment.

Factor	Treatment	Hi	Lo
A	Detail	Yes	No
B	Oil change	Yes	No
C	Steam clean engine	Yes	No
D	Paint touch-up	Yes	No
E	Tires and hubcaps	Yes	No

Now, this is not the type of brainstorming we did in Chapter 8. In Chapter 8, we really had no idea what the drivers of the process were, and so we generated a list of possible drivers, most of which would turn out to be ineffectual. In the present case, we have substantial experience in the industry, and we're pretty sure most of these possible drivers are relevant, but before we build a full-fledged experiment, we want to make sure we haven't included any irrelevant factors.

The company wonders which of these methods, if any, should be applied to its cars in order to increase the price fetched by the cars at auction. Right now the company is not worried about the costs of these methods. Applying these methods on an ad hoc basis is not the cheapest way to use these methods. If any of the methods is found to be useful, the company will then investigate how low it can drive the cost of a method by employing it on a large scale and then determine whether the increase in auction price will more than cover the per-unit cost of the method. Since this is a screening experiment, we don't worry about interactions – we're only trying to find out which factors matter, we'll design another experiment to model the possible interactions.

A large source of variation in car price is the type of car. To eliminate this, Acme procured several of the same type of car, in this case a Ford Taurus. Since the age of the car and the mileage are also large sources of variation, to minimize this source of variation, the company procured cars all with the same model year and all having mileage within a few thousand miles of each other. All the cars are 2010 models with between 100 000 and 105 000 miles. There are nine of these Ford Tauruses. With five factors, each having two levels, what can we do with only nine runs? We can create a custom design! First we will create the design using an algorithm. Then we will evaluate the design to determine whether or not it is good enough to use. Then, if it is good enough, we will actually use it.

9.1.1 Create the Design

Given that we have only nine runs, we won't be able to estimate all the interactions. A 2^5 full factorial has 32 parameters to estimate: an intercept, $\binom{5}{1} = 5$ main effects, $\binom{5}{2} = 10$ second-order interactions, $\binom{5}{3} = 10$ third-order interactions, $\binom{5}{4} = 5$ fourth-order interactions, and $\binom{5}{5} = 1$ fifth-order interaction. With nine runs, we can estimate at most nine parameters if the design is saturated. However, this is a screening design, and we really aren't interested in interactions. We just want good estimates of the main effects so we can know which factors are important and which ones aren't. With nine runs, after estimating the intercept and five main effects for the parameters, we'll have three degrees of freedom left over to estimate the error term.

We could use a 2^{5-2} fractional factorial design that uses eight runs, but to do so we'd have to throw away one-ninth of our data (we could only use eight cars, not all nine) and have only two degrees of freedom to estimate the error instead of three. The custom design will let us use all our data and give us a better estimator of the error.

We can conduct an experiment with at most nine runs. The 2^5 factorial design matrix is given in Table 9.2; the corresponding model matrix, X, for a screening design that estimates only main effects will add a column of ones to the design matrix. Since we're not actually going to use this X to do modeling, it's not really a model matrix; instead let's call it a *candidate set*. We're going to select nine rows from the candidate set and use those as our model matrix, X. Based on our knowledge of determinants, what we need to do is pick the nine rows of the candidate set that will produce the largest determinant of $X'X$. Such a design is called a *D-optimal* design; it is optimal in the sense that it maximizes the determinant in the class of designs that has that many rows.

One way to find the optimal design matrix, and we have computers at our disposal, is to create every possible nine row matrix and corresponding $X'X$ matrix, compute the determinant of each $X'X$ matrix, and pick the one with the largest

Table 9.2 Design matrix in standard order for 2^5.

	A	*B*	*C*	*D*	*E*		*A*	*B*	*C*	*D*	*E*
1	−1	−1	−1	−1	−1	17	−1	−1	−1	−1	1
2	1	−1	−1	−1	−1	18	1	−1	−1	−1	1
3	−1	1	−1	−1	−1	19	−1	1	−1	−1	1
4	1	1	−1	−1	−1	20	1	1	−1	−1	1
5	−1	−1	1	−1	−1	21	−1	−1	1	−1	1
6	1	−1	1	−1	−1	22	1	−1	1	−1	1
7	−1	1	1	−1	−1	23	−1	1	1	−1	1
8	1	1	1	−1	−1	24	1	1	1	−1	1
9	−1	−1	−1	1	−1	25	−1	−1	−1	1	1
10	1	−1	−1	1	−1	26	1	−1	−1	1	1
11	−1	1	−1	1	−1	27	−1	1	−1	1	1
12	1	1	−1	1	−1	28	1	1	−1	1	1
13	−1	−1	1	1	−1	29	−1	−1	1	1	1
14	1	−1	1	1	−1	30	1	−1	1	1	1
15	−1	1	1	1	−1	31	−1	1	1	1	1
16	1	1	1	1	−1	32	1	1	1	1	1

determinant. However, this is an inefficient method, and for bigger or more complicated design problems, it takes far too long; see Exercise 9.4.1 on this point. The space of possible matrices is too large to enumerate all possible matrices and choose the best one, so algorithms have been invented to search the space in an intelligent way. These search methods will not search the entire space, and so may miss the absolute best design, but usually they will come close to the best design. Some of the more popular methods are:

- sequential method
- simple exchange method
- DETMAX algorithm
- modified Fedorov algorithm

Maximizing the determinant of $X'X$ has three direct consequences, all of which are very desirable:

1) Small variances for the coefficients – Coefficients are precisely estimated.
2) Small correlations between the coefficients – When dropping one variable, the coefficients on remaining variables don't change as much.
3) Small variances for the predicted responses – Accurate predictions.

R offers several packages that create optimal designs, none of which is the equal of a commercial package (e.g. JMP or Design-Expert). Each of the R packages has its own advantages and disadvantages. We use `AlgDesign` because it is one of the simplest to use. It does not offer two features we need, the *alias matrix* and the *color map of the correlations*, so we have written R code for each of these. Instructions for using `AlgDesign` are given below.

Software Details

To create a custom design...

```
# make sure you have already installed the
AlgDesign package
# if it's been installed, then load the package
library(AlgDesign)
# create the design matrix/candidate set with 32 rows
dm <- expand.grid(A=c(-1,1),B=c(-1,1),C=c(-1,1),
     D=c(-1,1),E=c(-1,1))
# select 9 rows to maximize "D" the determinant of (X'X)
fedorov1 <- optFederov(~.,data=dm,nTrials=9,criterion="D")
fedorov1 # display the output
design1 <- fedorov1$design # extract the custom design
```

(Continued)

(Continued)

Note the "~." notation in the `optFedorov` command. From running linear regressions, we know that the "~" symbol separates the left-hand side of a formula (the dependent variable) from the right-hand side (the independent variables).

To create a design, we don't need to know the values of the dependent variable, so there is no dependent variable in this formula. We could write "~A+B+C+D+E," but instead we write "~," which is just R shorthand for "all the variables in the data frame," which, given the data frame we're using in this case, means "all the main effects."

We don't save much time or room by using this shorthand, but when we want to include interaction terms, shorthand notation will save us a *lot* of typing, as we'll see shortly.

Be aware, though, that if there are other variables in the data frame besides the factors, then "~." will include those other variables, too, and that would be something you probably don't want.

The result of the last command is given in Figure 9.1, which shows nine rows of Table 9.2 that maximize the determinant of $(X'X)$. Your results will be different because we did not set the seed for the random number generator before running the program. (Of course, the experiments should not be executed in the order shown; the order should be randomized!)

In general, the custom design is not unique; there will be other designs that have about the same value of "D" but different rows. We ran the above command five times, each time getting the same "D" but different rows. The results are presented in Table 9.3. We are indifferent between using any of these five designs.

What we have just produced is a computer-generated design, but the methodology is called *computer-aided design*; note the difference. In general, it is not safe to

Table 9.3 Five equivalent custom designs.

$D	$rows
0.975 784 3	3 6 12 13 17 24 25 26 31
0.975 784 3	2 7 11 14 17 24 28 29 31
0.975 784 3	4 5 10 15 19 22 25 28 32
0.975 784 3	1 4 14 15 21 24 26 27 30
0.975 784 3	4 7 9 14 17 22 28 31 32

Figure 9.1 Output of optFedorov command.

```
> design1
$D      0.9757843
$A      1.044643
$Ge     0.889
$Dea    0.882

$design
    A  B  C  D  E
6   1 -1  1 -1 -1
19 -1  1 -1 -1  1
15 -1  1  1  1 -1
21 -1 -1  1 -1  1
32  1  1  1  1  1
4   1  1 -1 -1 -1
11 -1  1 -1  1 -1
26  1 -1 -1  1  1
9  -1 -1 -1  1 -1

$rows     6 19 15 21 32 4 11 26 9
```

accept the computer-generated design as "good"; it should be used to inform the analyst as she constructs a design. See the Learning More section for more about this distinction between "computer-aided" design and "computer-generated" design. We refer to them as "custom designs." Indeed, when we consider the use of custom design to create fractional factorial designs, we will encounter this problem: custom design will give us a design that we really shouldn't use, and we'll be able to discern this by looking at the color map.

Let us briefly discuss the output shown in Figure 9.1. The value "D" multiplied by 100 is the D-efficiency of the design, where the letter "D" signifies "Determinant." The D-efficiency roughly tells us that the $X'X$ matrix of our design is 97.578 43% as large as the corresponding matrix from a factorial design. The other design efficiencies (A, Ge, Dea) can be used to break a tie when two different custom designs have the same D-efficiency. We refer the interested reader to the `AlgDesign` documentation for further details. The custom design is given by "design," and "rows" tells us which rows of the candidate set (original design matrix) (e.g. Table 9.2) were selected by the algorithm.

We have found a nine-run design, which is in `TaurusDesign1.csv` along with the response variable as the first column. Suppose you have read this file in as `df`. To extract the design itself, drop the first column (which is the response variable) using the command `design1 <- df[,-1]`. How good is this design? Will it do what we want? What do we want it to do?

9.1.2 Evaluate the Design

There are three primary considerations in evaluating a custom design: (1) efficiency (how close is it to the optimal factorial design?), (2) aliasing, and (3) the correlation of the regression coefficients.

We answer the first point by appealing to the concept of D-efficiency. The D-efficiency is given by

$$DE = 100\frac{|X'X|^{1/p}}{N}\% \tag{9.1}$$

where X is the custom model matrix, $|X'X|$ is the determinant of $X'X, p$ is the number of parameters in the model, and N is the number of rows in the custom design matrix.

As an example of how formula 9.1 is applied, consider the simple 2^2 full factorial so that the model matrix (not the design matrix) X is

Int	*A*	*B*	*AB*
1	−1	−1	1
1	1	−1	−1
1	−1	1	−1
1	1	1	1

Since this is a factorial design, we know it is optimal, and therefore its D-efficiency must be 100%. Now, the determinant of $X'X$ is 256. We estimate four parameters for this model: an intercept and coefficients for A, B, and AB; hence $p = 4$. The fourth root of 256 is 4. There are $N = 4$ rows in the model matrix, and $4/4 = 1.0$. Thus, we have shown that the D-efficiency of the 2^2 full factorial equals 100%.

On a related note, the D-efficiency equals 100% for both fractional factorial and Plackett–Burman designs for main effects models (that have no interactions). This fact will be of importance in Section 9.4 when we discuss screening because, as we know from Chapter 8, screening models are just main effects models.

To obtain the D-efficiency of the design, issue the command `design1$D`, and your result should be 0.975 748 3 or something close to this value; multiply this by 100 to turn the proportion into a percentage.

In the present case, when we are creating a screening design for used cars, p is six (an intercept with five main effects), and N is nine. What this formula does is (roughly) compare the hyper-volume of the custom design matrix to the hyper-volume of a factorial design that is orthogonal and balanced (and so is optimal). Since $D = 0.975\ 784\ 3$, this says that the non-orthogonal custom design with nine observations is 97.6% as good as an orthogonal factorial design, which has a D-efficiency of 100%.

What does this 97.6% mean in practical terms? We know that for an orthogonal design when D-efficiency = 100%, dropping variables does not alter the remaining coefficients. Our design is not orthogonal, so the remaining coefficients will change when we drop variables. However, they will not change by much because 97.56% is so close to 100%.

We also have to know the *aliasing pattern* of the design, which requires that we answer two questions:

1) Which main effects and interactions are confounded with which main effects and interactions? Recall that if we had a full factorial design, there would be no aliasing.
2) How strong is the confounding? Is it perfect confounding, with a value of 1.0? Or is it something much less that might be safely ignored, like 0.10?

D-optimal designs are usually not factorial designs because the number of runs is usually less than the number of runs required for a factorial design (but see Exercise 9.4.2), so we can expect that there will be aliasing with this design. In general, we don't have to worry about main effects being confounded with each other, since that would require fewer runs than factors, and in this book we always have at least as many runs as factors. The alias matrix will show the confounding pattern of the factors included in the model with other factors (especially interaction terms) that are not included in the model.

Software Details

To compute the alias matrix...

Remember, the design we are considering is in the file `TaurusDesign1.csv`. You will have to drop the first column to get the design.

```
df <- read.csv("TaurusDesign1.csv")
design1 <- df[,2:5]
source("aliasMatrix.R") # load the command
aliasMatrix(design1) # execute the command
```

Depending on your computer, you may see a "−0" in the output. Where does that come from? The first time you printed the alias matrix, you may have seen some very tiny numbers in scientific notation, like "−2.775 58E−17." This is what is known as a *machine zero*; it is as close to zero as the computer could get. When this number gets rounded to zero, the minus sign comes along for the ride.

(Continued)

(Continued)

There are many unnecessary decimals that we can get rid of by issuing the command `round(aliasMatrix(design1),2)` and make the matrix easier to read.

In addition to `aliasMatrix.R`, which compares included main effects to unincluded second-order effects, there is also the program `aliasMatrix3.R`, which compares included first- and second-order effects to second- and third-order effects. Both these programs came with the data for this book.

The alias matrix for our custom design is given in Table 9.4, and big values have been underlined.

Reading across rows, we see that each factor is confounded with some two-way interactions, usually just a small amount (±0.07 or ±0.14), but sometimes a large amount (0.86 or 1). By the hierarchical ordering principle, we expect the coefficients for the interactions to be appreciably smaller than the coefficients for the main factors, and multiplying them by 0.07 or 0.14 makes their effects even smaller.

The 1 on the far right of the second row of Table 9.4 means that A is perfectly correlated with DE. If the DE interaction is active (i.e. it is non-zero), it will bias our estimate of the coefficient for A by the full amount of the coefficient for DE. As an example, suppose the true coefficient on DE equals 100 (we didn't estimate it; we're just assuming that we happen to know the truth), and we estimated the coefficient on A to be 150. What we estimated is "true value of A + 100," so we should really estimate the value of A as 50, because our estimate is contaminated by the aliasing. It may be useful to refer to Equations 7.7 and 7.8 and the surrounding text; this is an important idea that we'll use again to assess the effect of aliasing.

Table 9.4 Alias matrix for TaurusDesign1.csv, large values underlined.

	AB	*AC*	*AD*	*AE*	*BC*	*BD*	*BE*	*CD*	*CE*	*DE*
Int	−0.07	0.07	0	0	−0.14	0.14	−0.07	−0.14	0.07	0
A	0.07	−0.07	0	−0	0.14	−0.14	0.07	0.14	−0.07	<u>1</u>
B	−0.07	0.07	−0	0	−0.14	0.14	−0.07	<u>0.86</u>	0.07	−0
C	0.07	−0.07	0	−0	0.14	<u>0.86</u>	0.07	0.14	−0.07	0
D	−0.07	0.07	−0	<u>1</u>	<u>0.86</u>	0.14	−0.07	−0.14	0.07	−0
E	0.07	−0.07	<u>1</u>	−0	0.14	−0.14	0.07	0.14	−0.07	0

Software Details

To create a color map of the correlations...

We presume that `ggplot2` and `reshape2` have been installed. Load them.

```
library(ggplot2)
library(reshape2)
```

Then you can just load the function and execute it on the design.

```
source("colorMap.R")
colorMap(design1)
```

The corresponding color map of the correlations is shown in Figure 9.2, which depicts the absolute values of the correlations between the estimated regression coefficients. Issuing the command `colorMap(design1)` creates the color map. Remember that if the design were orthogonal, then the coefficients would be uncorrelated (i.e. the covariance matrix would be all zeroes off the main diagonal). The correlation shows the bias associated with the parameter estimates. We see, for example, there is a strong correlation between *D* and *AE*, and *D* and *BC*, and also between *C* and *BD*. If we wanted to see the actual numbers, we could print out the correlation matrix with the command `corrOfCoeff(design1)`

Try it!

Reproduce the color map in Figure 9.2.

Notice how closely the color map corresponds to the alias matrix. For example, look at row *D* of Table 9.4, which has large values for the *AE* and *BC* interactions. Now look at the row for *D* in Figure 9.2; it, too, shows large values for *AE* and *BC*.

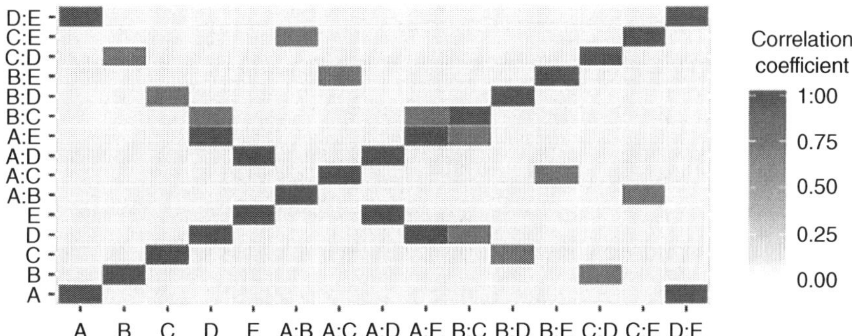

Figure 9.2 Color map for Taurus design.

We can look at either the alias matrix or the color map. In general, the color map is much easier for getting a first impression of the quality of a design.

We can hope that our screening design will correctly identify main effects for two reasons. First, we can expect that most of the interactions won't be active. Second, we can expect that the coefficients on the interactions will be smaller than the coefficients on the main effects, so even if an interaction is active, it probably won't change the estimate of the main effect enough to matter. What do we mean by "enough to matter"? We mean "to change an insignificant coefficient to significant" or "change a significant coefficient to an insignificant coefficient." Of course, to sustain our hope, we'll have to look at the experimental results. Thus, even if the coefficients are biased, often we can determine what plausible good estimates might be and whether they are significant or insignificant. We will take examples of this in Section 9.1.3.

Having evaluated the design, we can now implement the design and run the experiment.

9.1.3 Use the Design

The order of experiments provided in `TaurusDesign1.csv` was randomized, the experiment was conducted, and the design and resulting auction prices are in file `TaurusDesign2.csv`. Regression results are presented in Table 9.5.

Let's analyze Table 9.5 in view of the alias matrix given in Table 9.4. We are trying to assess the effect of any possible aliasing on our estimated model. Remember, we don't know what the coefficients are for the factors that we did not include in our model. We are only asking, "How big would those coefficients have to be before we couldn't trust our model?"

To assess the effect of aliasing on our model, we proceed as follows. Suppose the estimated coefficient on A is 200. We know that A is perfectly aliased with DE. We ask ourselves, "How large does the coefficient on DE have to be to change our

Table 9.5 Results of Taurus screening experiment.

	Estimate	Std. error	*t* Value	Pr(> |*t*|)
Int	8316.59	146.08	56.93	0.00
A	398.66	146.08	2.73	0.07
B	108.59	146.08	0.74	0.51
C	334.66	146.08	2.29	0.11
D	524.34	146.08	3.59	0.04
E	435.91	146.08	2.98	0.06

Multiple R-squared: 0.9188, adjusted R-squared: 0.7833

conclusions?" Usually the answer is "pretty large," and due to the hierarchical principle, interaction effects might not be that large. Of course, subject matter expertise would be necessary to make such a determination.

As a more concrete example of assessing the effect of aliasing, we know that A is aliased with DE with aliasing coefficient 1.0. What this means is that the value we estimate for A, 398.66, is really composed of the true value of A plus the coefficient on DE. Suppose the coefficient on DE was +200. Then we could surmise that the true coefficient on A is closer to 200 than to 400, since $200 * 1.00 = 200$ and $398.66 - 200 = 198.66$. If the coefficient on DE was +200, would the "true" value of the coefficient on A (198.66) be significant? We know the coefficient on C is 334.66 has a p-value of 0.11, we can deduce that the p-value on 198.66 would be much larger, so *if* the true value of DE were +200, *then* the factor A would not be significant. If the true value of DE is as large as +200, then aliasing will affect our modeling.

Similarly, D is aliased with BC with aliasing coefficient +0.86. If the regression coefficient on BC was −100, we would expect that the true coefficient on D is probably about $524.34 + 86 = 610.34$, which would be even more significant. Remember:

$$\begin{aligned}\text{biased estimate from regression} &= \text{good estimate} + \text{effect of aliasing}\\ 524.34 &= \text{good estimate} + (-100) * (0.86)\\ 524.34 + 86 &= \text{good estimate}\\ 610.34 &= \text{good estimate}\end{aligned}$$

What we see here is that *if* the coefficient on BC is negative, it will only make factor D *more* significant, so negative values on the coefficient for BC will not change the factors that we include in our model.

What we are trying to do in the above is figure out how large the coefficients on the omitted factors have to be in order to change our results. Then we will have to use expertise to determine whether the coefficients on the omitted factors are likely to be that large. In this way we can assess the effect of the aliasing.

The reduced form of the model is given in Table 9.6. Observe that the values of the remaining coefficients changed when we dropped factor B. This is because the custom design is not orthogonal. Because the design is close to orthogonal (D-efficiency = 96.86%), the values of the coefficients did not change by much. We have a good R^2, all the coefficients are of the expected sign, and of course we keep factor C, even though the p-value exceeds 0.05 for two reasons. First, we are not slaves to p-values. Second, we realize that our sample is small, and we have expert reason to believe that factor C matters.

Let's consider the effect of possible aliasing on the coefficient for factor D. Suppose that AE is active and the coefficient on AE is +130 (we pulled this number almost out of thin air, and it happens to be about 25% of the coefficient on D. We could have chosen 100 or 150, but we didn't). Since AE is perfectly correlated with

Table 9.6 Reduced form of Taurus screening experiment.

	Estimate	Std. error	*t* Value	Pr(> |*t*|)
Int	8324.94	137.26	60.65	0.00
A	390.31	137.26	2.84	0.05
C	326.31	137.26	2.38	0.08
D	532.69	137.26	3.88	0.02
E	427.56	137.26	3.11	0.04

Multiple R-squared: 0.9038, adjusted R-squared: 0.8076

D, we know that our biased estimate of the coefficient on D is too high, and a good estimate is probably about 400. But a coefficient of 400 would still be significant (we know this because factor E is significant), so we would still correctly identify D as an active main effect even if AE is active with a coefficient of +130.

Now let's consider the effect of possible aliasing on the coefficient for factor A. If DE is active with a coefficient of, say, 100, we might have a problem. DE is perfectly correlated with A, and so our best guess is that the true coefficient on A is closer to 300 than 400 – and a coefficient of 300 will have a p-value of about 0.10 – we can deduce this by looking at factor C. Remember, we don't know whether DE is active or, if it is active, what its coefficient is. We are just exploring the limits of the design we have estimated.

In getting to the reduced model, we dropped factor B, which has an estimated coefficient of about \$100 (108.59 to be precise). Could this have been a mistake due to aliasing? Should we really have kept factor B in the model? Now, B is aliased with CD with a value of 0.86. For us to decide that B might be an active main effect worth including in the follow-up experiment, it should have a coefficient of at least 300 (we included C with a coefficient of 334 and a p-value of 0.11). Could aliasing have caused a true coefficient of 300 to fall to 100? Remember:

$$\text{good estimate} = \text{biased estimate from regression} - \text{effect of aliasing}$$
$$300 = 100 - \text{effect of aliasing}$$
$$300 = 100 - 0.86(\text{coefficient on CD})$$

For the above to be true, the effect of aliasing would have to be about negative \$200. For this to happen, the coefficient on CD would have to be negative, and 0.86 times the coefficient would have to equal negative 200, so the coefficient on CD would have to be at least $-200/0.86 = -233$. Is it likely that any two-way interaction has an effect size of \$233 when the active main effect sizes range from \$300 to \$500? We'd need more experience with this model to make such a determination with certainty, but the principle of effect hierarchy is reassuring.

All in all, we have determined that four of the five factors have an effect on price. We are pretty sure that the factor we dropped, B, needed to be dropped. We are reasonably sure that the factors we decided to keep needed to be kept. If we are wrong about keeping factors, we'll find out in the follow-up experiment that assesses the interactions.

Exercises

9.1.1 Consider a 2^5 factorial design. You can only afford 21 runs. Create a custom design to estimate main effects and 2FIs. Analyze the design using D-efficiency, the alias matrix, and the color map of the correlations. Set the seed of the rng to 1236.

9.1.2 In Section 7.3 we posed the "three-point problem" where three points from a nine-point grid needed to be chosen. Use software to determine an optimal design. Do this five times and compare the designs, perhaps by drawing the points of design on the square given in Figure 7.4. Compare these to designs we know to be optimal (see Exercise 7.3.4). Hint: `dm <- expand.grid(x1=c(-1,0,1),x2=c(-1,0,1)) fedorov1 <- optFederov(~.,data=dm,nTrials=3,criterion="D")`

9.2 Case: Selling Used Cars at Auction II (Custom Experiment)

Returning now to our car selling problem, we have four factors, which we have relabeled A = detail, B = steam clean engine, C = paint touch-up, and D = tires and hubcaps, and we are concerned that there might be interactions. Therefore we want to conduct an experiment that will let us estimate the intercept, four main effects, and the $\binom{4}{2} = 6$ 2FIs. That's 11 parameters. What would such an experiment look like? We want to estimate all the interactions and test them, so we'll need a couple extra runs to estimate the error – how about three more runs. Together with the 11 parameters, we'll need 14 runs. We will create the design with 14 runs and then analyze it.

When invoking the alias matrix, we have to think about what effects outside the model we want to include. If our model includes 2FIs but no 3FIs, then we want the alias matrix to include three-way interactions so that we can determine whether the 3FIs will bias the coefficients of our estimated model. Therefore, we'll have to be sure to load and use `aliasMatrix3` when issuing the relevant commands, which are given below:

Software Details

To create this custom design...

```
library(AlgDesign)
set.seed(123)
source("aliasMatrix3.R") # load the function
dm<-expand.grid(A=c(-1,1),B=c(-1,1),C=c(-1,1),D=c(-1,1))
design1 <- optFederov(~.^2,data=dm,nTrials=14,
criterion="D")
design1$D
design2 <- design1$design
round(aliasMatrix3(design2),2)
```

Note that we used another notation in the `optFedorov` command, this time it was "~. ^ 2," which is R shorthand for "all main effects and two-way interactions." Written out its "~A+B+C+D+A:B+A:C+A:D+B:C+B:D+C:D," but using the shorthand is much easier.

The D-efficiency of the design is 92.2%, and the rows that constitute the design points are 1, 3, 4, 5, 6, 7, 9, 10, 11, 12, 13, 14, 15, 16, so we only drop rows 2 and 8. We are estimating two-way interactions, so we'll want to know whether the 2FIs are aliased with the 3FIs. Thus we use `aliasMatrix3`. The alias matrix is presented in Table 9.7.

This is a beautiful alias matrix. Looking across the rows for the main effects, we see that the main effects are not aliased with the 2FIs. The aliasing with the 3FIs is not greater in magnitude than 0.50, and this is nice, since 3FIs are usually small. If there is a bias from 3FIs, it probably won't be much to worry about. Looking across the rows for the 2FIs that we want to estimate, we see that they are aliased only with 3FIs, and as already noted, these effects are likely to be small. So we think we can get reasonable estimates of the main effects and the 2FIs from our design.

The corresponding color map is shown in Figure 9.3. Again, the color map is very nice. In the lower left corner, we see that there is very little correlation between the main effects (doing this on your computer in color will make it easier to see), and there is very little correlation between the main effects and the 2FIs, or between the 2FIs and the 3FIs. Looking at the actual numbers may make this easier to see and is explored in Exercise 9.4.1.

When this design has been implemented, we'll have a good idea of which methods work and how much they add to the value of the car, so we can decide whether to do them. For example, if having an unblemished paint job is worth $300 and car comes in requires $500 worth of paint work, we know not to bother with painting it.

Table 9.7 Alias matrix for 14 runs.

	Int	*A*	*B*	*C*	*D*	*AB*	*AC*	*AD*	*BC*	*BD*	*CD*	*ABC*	*ABD*	*ACD*	*BCD*
Int	1	0	0	−0	−0	−0	−0	0	0	0	−0	0.33	0	−0.33	−0.33
A	0	1	0	0	−0	−0	−0	−0	−0	0	0	−0	0.50	0	0
B	0	−0	1	0	0	−0	−0	−0	−0	−0	0	−0	0.50	−0	0
C	0	0	0	1	−0	0	0	−0	0	0	0	−0.33	0	0.33	0.33
D	−0	0	0	−0	1	0	0	0	0	0	−0	0	−0.50	0	0
$A : B$	−0	0	−0	0	−0	1	0	0	−0	0	−0	0.33	0	−0.33	−0.33
$A : C$	0	0	0	0	−0	−0	1	0	0	0	−0	0	−0.50	0	0
$A : D$	0	0	−0	−0	0	−0	−0	1	−0	0	0	−0.33	−0	0.33	0.33
$B : C$	−0	0	0	−0	−0	−0	−0	−0	1	−0	−0	0	−0.50	−0	0
$B : D$	0	0	0	0	0	0	−0	0	0	1	0	−0.33	0	0.33	0.33
$C : D$	−0	0	0	−0	0	−0	0	0	0	0	1	0	0.50	0	−0

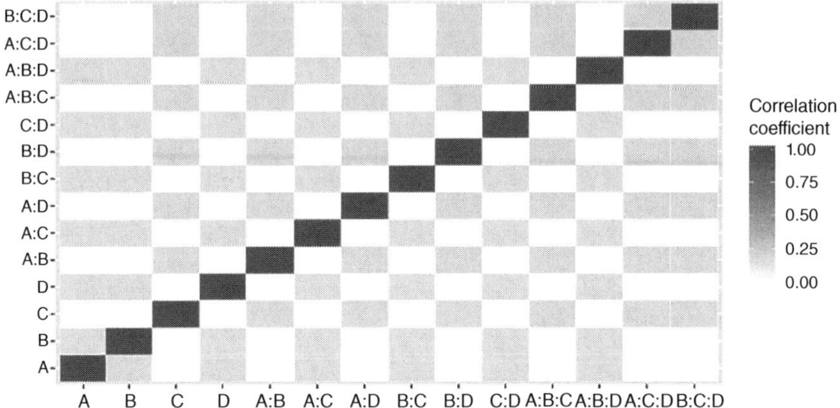

Figure 9.3 Color map for 14 runs.

After going to all this trouble, the manager tells us that we were lucky to get 9 cars for the screening experiment, and we'll never get 14 cars for this experiment. What can we do? We will answer this question in the next section.

Exercises

9.2.1 Get the matrix for the correlation of the coefficients corresponding to the design mentioned in Figure 9.3.

9.2.2 Create three more color maps for 14 runs similar to Figure 9.3 but change the seed each time. Observe how the various color maps are both similar and different.

9.2.3 Use custom design to create a full factorial with four factors (in other words, set nTrials=16). Is the custom design optimal? Set the seed of the rng to 1237.

9.3 Custom Experiment with Blocking

We have the same four factors that we want to test for interactions: A = detail, B = steam clean engine, C = paint touch-up, and D = tires and hubcaps. A 2^4 full factorial would be fully saturated and wouldn't let us test the interactions. Even if we were willing to rely on Lenth's method, a 2^4 would require 16 runs, and we'll never get 16 Ford Tauruses of a similar vintage and similar mileage. A full factorial with replication with require 32 runs. If we'd never get 16 Ford Tauruses, we'll never get 32. What can we do?

Well, we want to estimate a model with four factors. That will be an intercept, four main effects, and $\binom{4}{2} = 6$ two-way interactions, which adds up to 11 parameters. If we wanted to consider three-way interactions, that would be an additional $\binom{4}{3} = 4$ parameters, for a total of 15 parameters.

Maybe we could get close to ten Ford Tauruses again, and if we could get close to ten cars of a similar type, we could then block on car type and conduct our experiment. Indeed, a very similar car type is the Buick LaCrosse. For our next experiment, we were able to collect six Ford Tauruses and seven Buick LaCrosses, all manufactured in 2009 and all with between 108 and 115 thousand miles. We have only 13 cars, so we won't be able to estimate three-way interactions. We'll have to settle for two-way interactions.

Taurus and LaCrosse cars, while similar, are not identical. This constitutes a source of variation in the response that we should model. So let us block on this variable. There is really no good way to create a blocking design for this problem using fractional factorial or Plackett–Burman methods. It is fairly easy using custom design. To do this we use the `optBlock` command in the AlgDesign package, as shown below:

Software Details

To create a custom design with blocking...

```
set.seed(123)
dm <- expand.grid(A=c(-1,1),B=c(-1,1),C=c(-1,1),D=c(-1,1))
block1 <- optBlock(~{}.^{}2,dm,blocksizes=c(6,7))
block1
design1 <- block1$design
```

The output (not shown) presents two separate blocks of 6 and 7 rows each, together with a 13-row design that incorporates both blocks where the first 6 rows are one block and the next 7 rows are the other block, but does *not* have a blocking variable. We can create the blocking variable very easily by `bv <- c(rep("F",6),rep("B",7))` where "F" indicates Ford and "B" indicates Buick.

One of the assumptions of blocking is that the blocking variable doesn't interact with the model, so when we analyze this design by alias matrix or color map, we'll exclude the blocking variable. Of course, when we run the regression to analyze the data, we'll be sure to include the blocking variable! It goes without saying, but we have to say it: we would not run the experiments in the order given by the

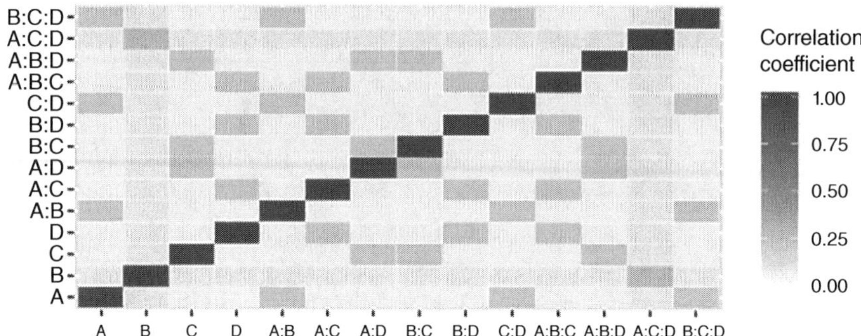

Figure 9.4 Color map for blocked Taurus design.

computer, but would randomize the order in which the cars are put on the auction block.

This alias matrix, shown in Table 9.8, is wonderful, as is the color map in Figure 9.4. The main effects and 2FIs are not aliased with each other, and the aliasing with the 3FIs has many small and only a few large values.

The alias matrix was very good, so it's no surprise that the color map is good, too. The only strong correlations are on the diagonal, meaning that A is correlated with A, B with B, etc. It doesn't get much better than this.

We like the design, so let's append the blocking variable to the data frame: `design1$bv <- c(rep("F",6),rep("B",7))`. The completed design is shown in Table 9.9.

The design has been created, the experiment has been conducted, and the results are in `customblocking.csv`. The statistical analysis is now a mere formality, which we leave for the exercises.

Exercises

9.3.1 Analyze the data in `customblocking.csv`. Run the initial regression, and apply the modeling strategy. What are your conclusions?

9.3.2 Suppose that the blocking variable was omitted. What would you conclude? (Only run a regression for main effects to answer this question.)

9.3.3 Suppose we had six Crown Vics, five Tauruses, and eight Camrys, all of a similar age and mileage. Create a design to test the four factors with 2FIs. Set the seed to 123.

Table 9.8 Alias matrix for the blocked Ford/Taurus design.

	Int	*A*	*B*	*C*	*D*	*A* : *B*	*A* : *C*	*A* : *D*	*B* : *C*	*B* : *D*	*C* : *D*	*A* : *B* : *C*	*A* : *B* : *D*	*A* : *C* : *D*	*B* : *C* : *D*
Int	1	0	0	0	0	0	0	0	0	0	0	−0.14	0.14	0.43	−0.14
A	0	1	0	0	0	0	0	0	0	0	0	−0.29	0.29	−0.14	0.71
B	0	0	1	0	0	0	0	0	0	0	0	−0.14	0.14	0.43	−0.14
C	0	0	0	1	0	0	0	0	0	0	0	0.29	0.71	0.14	0.29
D	0	0	0	0	1	0	0	0	0	0	0	0.71	0.29	−0.14	−0.29
A : *B*	0	0	0	0	0	1	0	0	0	0	0	−0.29	0.29	−0.14	0.71
A : *C*	0	0	0	0	0	0	1	0	0	0	0	−0.71	−0.29	0.14	0.29
A : *D*	0	0	0	0	0	0	0	1	0	0	0	−0.29	−0.71	−0.14	−0.29
B : *C*	0	0	0	0	0	0	0	0	1	0	0	0.29	0.71	0.14	0.29
B : *D*	0	0	0	0	0	0	0	0	0	1	0	0.71	0.29	−0.14	−0.29
C : *D*	0	0	0	0	0	0	0	0	0	0	1	0.29	−0.29	0.14	−0.71

Table 9.9 Blocked design for Taurus/LaCrosse experiment.

Row	*A*	*B*	*C*	*D*	bv
1	−1	−1	−1	−1	*F*
6	1	−1	1	−1	*F*
10	1	−1	−1	1	*F*
11	−1	1	−1	1	*F*
12	1	1	−1	1	*F*
13	−1	−1	1	1	*F*
2	1	−1	−1	−1	*B*
4	1	1	−1	−1	*B*
5	−1	−1	1	−1	*B*
7	−1	1	1	−1	*B*
9	−1	−1	−1	1	*B*
14	1	−1	1	1	*B*
16	1	1	1	1	*B*

9.4 Custom Screening Experiments

We know how to create screening designs when the number of runs is a power of two (use a fractional factorial design) or a multiple of four (use a Plackett–Burman design). Sometimes, though, we need a different number of runs. Perhaps experiments are expensive or time consuming, and we just can't increase the number of runs until we hit a power of two or multiple of four. In such a case we can use custom design to create a screening experiment.

Suppose we are interested in screening eight factors – remember that we are primarily concerned with getting clean estimates of the main effects, so we are not interested in estimating coefficients for interaction terms. We might use a standard fractional factorial design known as a half-fraction 2^{8-1} experiment, which requires $2^8/2 = 256/2 = 2^7 = 128$ runs. If we're only interested in main effects, then we are estimating a constant term and eight coefficients for a total of nine parameters, which leaves $128 - 9 = 119$ degrees of freedom for estimating the error term. Even if we were also estimating the $\binom{8}{2} = 28$ two-way interactions, we'd still have 91 degrees of freedom to estimate the error; that's still many more degrees of freedom than we need. So the fractional factorials can be very profligate with respect to estimating the error term. If we could only afford eight degrees of freedom for the errors and we wanted to estimate all the main effects and two-way interactions, we'd need 38 runs. However, there is no fractional factorial that does this. A fractional factorial has to have a number of runs that is a power of 2, so the closest fractional factorial designs will have 32 runs (not enough) or 64 (way too many). We could use the Plackett–Burman design, which requires that the number of runs is a multiple of four but not a power of two, e.g. 12, 20, 24, 28, etc. The method of custom design, however, can give us a design that requires exactly 38 runs and is nearly as good as an orthogonal design. See Exercise 9.4.3.

We show in the text below and in some of the exercises that, when a fractional factorial is appropriate, the custom approach *can* create such a design. Similarly, when a Plackett–Burman is justified, the custom approach *can* create a Plackett–Burman design (see also Exercise 9.4.6). Notice the emphasis on the word "can" in the previous sentence. As we shall see, it is not guaranteed that custom design *will* create these designs, so we must be careful and always check the color map and alias matrix of any custom design.

What we know is that when we are screening for main-effects and the number of runs is a power of two, fractional factorial designs are D-optimal. When we are screening for main-effects and the number of runs is a power of four, Plackett–Burman is D-optimal. What we want to do first is show that when the number of runs is a power of two, the custom design can produce a fractional factorial. We also want to show that when the number of runs is a multiple of four, the custom design can select a Plackett–Burman.

We presented a 2^{7-4} fractional factorial in Table 8.5 to analyze seven factors (A–G) in eight runs and showed its color map in Figure 8.1. Let's use custom design to create a design to screen seven factors in eight runs.

Software Details

To create custom screening design for seven factors in eight runs...

```
set.seed(123)
dm <- expand.grid(A=c(-1,1),B=c(-1,1),C=c(-1,1),
      D=c(-1,1),E=c(-1,1),F=c(-1,1),G=c(-1,1))
fedorov74 <- optFederov(~.,data=dm,nTrials=8,
criterion="D")
design74 <- fedorov74$design
fedorov74$D
```

The fact that the D-efficiency equals 85.4% tells us that this is *not* a fractional factorial design because the D-efficiency does not equal 100%. If you look at the color map, it's horrible. Rerun the commands except this time add "nRepeats=10" to the `optFedorov` command (see the next "Software" section below for an example). Now the D-efficiency is 100%, and the design is fractional factorial.

Is it the same one we got in Table 8.5? Probably not, because in general, the D-optimal design is not unique. But the one we got by custom design is statistically equivalent in terms of D-efficiency to the one in Table 8.5 (though it could differ in terms of other optimality measures, e.g. A-optimality or C-optimality). It *could* have a worse alias matrix and/or color map, but it doesn't. We can show this easily by comparing the alias matrices or color maps. The off-diagonal ones might be in different places, but there's still the same number of them, so the designs are equivalent.

Try it!

Create the color map for the above custom design, "design74." Compare it with the color map in Figure 8.1. What do you conclude?

Can custom design return a Plackett–Burman design (or something that is statistically equivalent in terms of D-efficiency)? Yes. In Exercise 8.4.1 you obtained the alias structure of the Plackett–Burman design for analyzing 11 factors in 12 runs. Let's do the same thing with custom design.

Software Details

To create a custom design to screen 11 factors in 12 runs...

```
set.seed(123)
dm <- expand.grid(A=c(-1,1),B=c(-1,1),C=c(-1,1),
      D=c(-1,1),E=c(-1,1),  F=c(-1,1), G=c(-1,1),
      H=c(-1,1),I=c(-1,1),J=c(-1,1),K=c(-1,1))
fedorov1112 <- optFederov(~.,data=dm,nTrials=12,
criterion="D")
design1112 <- fedorov1112$design
fedorov1112$D
```

You can see that the alias matrices are substantially the same. Maybe a −0.33 is a +0.33, but the total number of +0.33, −0.33, and 0.00 entries are the same in corresponding rows. The color maps are substantially the same. The designs are statistically equivalent.

Our advice for creating a screening design (e.g. fractional factorial or Plackett–Burman): keep increasing "nRepeats" until AlgDesign returns a D-efficiency of 100%. If it can't do that, get help. We have shown that AlgDesign can return factorial and Plackett–Burman designs (or their equivalents) when appropriate. However, it may be easier just to use the `FrF2` package.

It is not uncommon to use the Plackett–Burman design for 12 runs to screen 11 factors; there is precedent for allowing a single degree of freedom to estimate the error. We prefer more than one degree of freedom to estimate the error – cost permitting – so we generally select the number of runs to equal the number of factors to be screened plus one (for the intercept), plus a few more to estimate the error.

In this section, we have shown how to use the computer to generate screening designs. The time required to find a design increases exponentially with the number of factors, as shown in Table 9.10. These timings were executed on an Intel I-7 chip with 32 gigs of memory. With custom design, we are not limited to having screening designs that are powers of two or multiples of four!

However, and this is a big caveat, custom design will not always return a good screening design, as we now show. Therefore, it is imperative to remember that this is "computer-aided design" and not "computer-generated design." The analyst should never accept a custom design blindly.

Suppose you want to screen four factors in eight runs. A full factorial for four factors implies $2^4 = 16$ runs. Half of a 16 is eight runs, so we want a half-fraction of a 2^4, commonly denoted 2^{4-1}. Create a custom design for four factors with eight trials, estimating only the main effects. Set the seed to 123. Is the result a factorial design? (It is if the D-efficiency = 1.0.) Look at the color map of this design. The

Table 9.10 Timings for optimal screening designs.

Number of factors	Dimensions of factorial design	Dimensions of screening design	D-efficiency	CPU time
8	256× 8	14× 8	0.9704	0.01 s
12	4 096× 12	18× 12	0.9740	0.36 s
14	16 384× 14	20× 14	0.9693	4.24 s
15	32 768× 15	21× 15	0.9668	16.20 s
16	65 536× 16	22× 16	0.9619	48.43 s
17	131 072× 17	23× 17	0.9633	4.4 min
18	262 144× 18	24× 18	0.9606	16.6 min
19	524 288× 19	25× 19	0.9581	1.14 h
20	1 048 576× 20	26× 20	0.9567	4.67 h

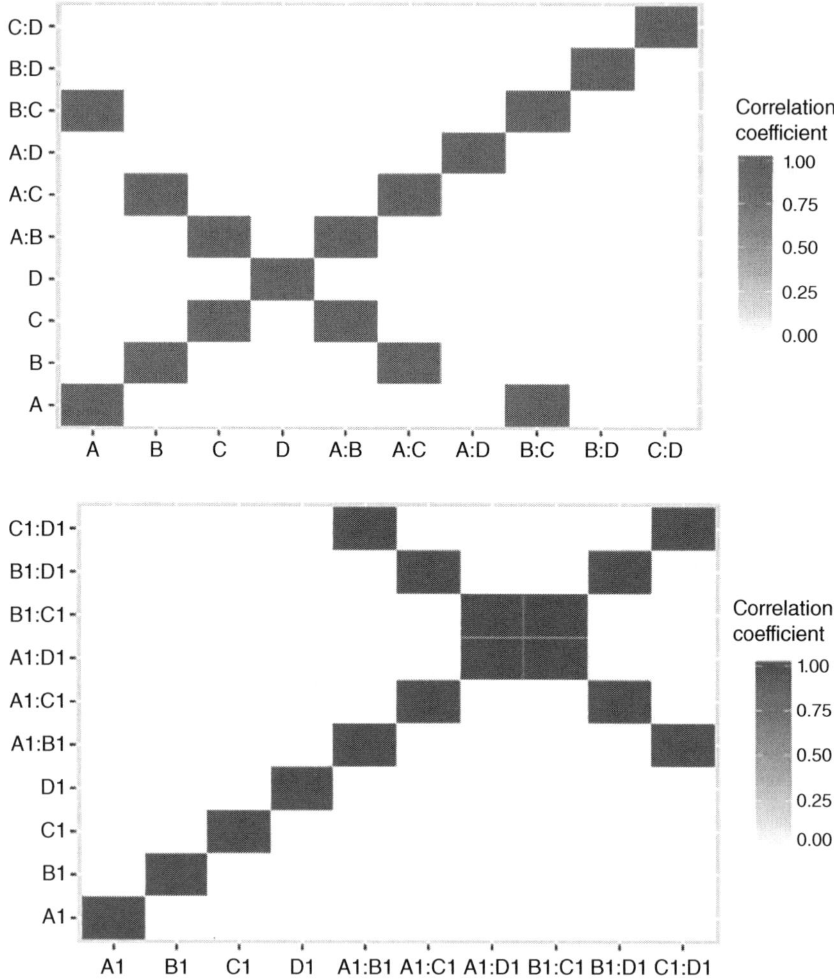

Figure 9.5 Color maps for four factors in eight runs (left is from `AlgDesign`, and right is standard from a catalog).

standard 2^{4-1} fractional factorial is in the file `ff4m1.csv`; get the color map of this design. Both color maps are shown in Figure 9.5.

Notice that for the custom design, the main effects are aliased with 2FIs, while for the standard design, the main effects are *not* aliased with 2FIs. The standard design is clearly preferable. The package FrF2 will produce a design with a color map identical to the one on the right. See Exercise 9.4.4. Thus, we should not trust computer-aided designs blindly.

Our advice for screening designs is as follows. If the number of runs is a power of two or a multiple of four, use the `FrF2` or `pb` command from the "FrF2" package; these are designed to create good color maps. Otherwise, use the "AlgDesign" package, but be sure to analyze the design carefully, always looking at its alias matrix and color map, and perhaps comparing these to alias matrices and color maps from a fractional factorial or a Plackett–Burman design with a nearby number of runs.

Exercises

9.4.1 Enumerating all possible matrices and choosing the best one is not possible, even for simple cases. That's why good search algorithms are necessary and why it's impossible to guarantee that the "best" custom design is chosen. Consider line 4 of Table 9.10. Suppose you wanted to create a screening design for 15 factors but could only afford 21 experiments and your computer could analyze 1 trillion matrices per second. How many hours would it take to analyze all possible matrices with 21 rows from the candidate set of 32 768 rows?

9.4.2 We have seen that AlgDesign can return a Plackett–Burman design when n is a multiple of four that is not a power of two. Can AlgDesign return a full factorial in a situation when a full factorial is called for? Create a custom design for four factors and let the number of runs be 16. We know that the full factorial is optimal, and we know what a 2^4 full factorial looks like. Does the custom design look like a 2^4 full factorial design? (Does the first column alternate in sign, does the second column alternate in pairs, etc.?)

9.4.3 You need to screen eight factors with 38 runs, but you also want to estimate the two-way interactions. Create such a design, and analyze it. Comment on your findings.

9.4.4 Use `FrF2` to create a screening design for four factors in eight runs. Get its color map and compare it to the right side of Figure 9.5.

9.4.5 We saw that, with seed equal to 123, `optFedorov` could not create a good design to screen seven factors in eight runs; at default the D-efficiency was 85.4%. To obtain a D-efficiency of 100%, we set the option `nRepeats=10`. Was 10 really necessary, or would a smaller number have sufficed? Walk it back from 10 to 5 to figure out where it switches from producing a design with D-efficiency of 100% to one with a D-efficiency of less than 100%.

9.4.6 The purpose of this exercise is to see if custom design can produce something equivalent to a Plackett–Burman design. Set the seed to be 123. Create a custom design for 11 factors and 12 runs, and call it `design1112`. What is the D-efficiency of the design you created? What should the D-efficiency be if it is a Plackett–Burman design? Increase nRepeats and get D again. How does it compare? The file `PB11-12.csv` contains a design that we know to be a Plackett–Burman. Get the alias matrix of that and compare it to the alias matrix of the custom design. Another way to check is to verify that the columns are mutually orthogonal: `t(X)%*%X` is a diagonal matrix. Hint: The matrix math in this command only works on matrices, and the designs are not stored as matrices. To see this, issue the command `is.matrix(design1112)`. The design is actually stored as a data frame. To see this, issue the command `is.data.frame(design1112)`. To temporarily convert the data frame to a matrix, issue the command `as.matrix(design1112)` and use this in place of X in the matrix math command.

9.4.7 You have five factors. How many runs does the factorial design require? Create custom designs for main effects that use 9, 15, 21, and 32 runs. Compare the color maps. Set the seed to be 1289 before you create each design.

9.5 More Than Two Levels

As we have noted, the mechanics of creating designs for experiments with more than two levels gets very complicated. However, just as custom design greatly simplified screening designs, so custom design also greatly simplifies creating designs when factors have more than two levels. Here we will take a simplified example of a real problem provided by Stat-Ease, Inc. (Minneapolis), a statistical consultancy, trainer, and publisher of Design-Expert® software specializing in experimental design.

We do not recommend that you attempt creating and using designs with more than two levels based only on reading this section. We present this example just to show you that there is a world beyond 2^k designs. For an introduction to this world, the book by Goos and Jones (2011) is excellent. It contains very little mathematics and shows a wide variety of real problems that are amenable to custom design.

You work for a company that makes wood furniture. A customer has requested a large order of furniture that is as strong as possible. By strong, it is meant that the pieces of wood are glued together as tightly as possible. The adhesion strength of glue is measured by gluing two pieces together and then pulling them apart until the two pieces separate. The response is the percentage of wood failure in the glued area. For example, if the two pieces separate cleanly so that the pieces of wood are intact, then the response is 0%. If the glue holds and one of the pieces of wood ruptures, then the response is 100%, which is perfect. Of course, there can be responses between 0 and 100%.

Based on experience, we believe that there are five factors affecting the adhesion strength:

- *A*: five wood species (chestnut, red oak, poplar, maple, pine)
- *B*: five adhesives (PRF-ET, PRF-RT, RF-RT, EPI-RT, LV-EPI-RT)
- *C*: two adhesive applicators (brush, spray)
- *D*: four types of clamps (pneumatic, manual, spring, mechanical)
- *E*: two pressures on the clamps (firm, tight)

We shall not get into the details of the calculations, but suffice it to say that the coefficients for a full-factorial design include:

- 1 intercept
- 13 main effects
- 63 2FIs
- 139 3FIs
- 136 4FIs
- 48 5FIs

If we were to run a full factorial, it would require $5 \times 5 \times 2 \times 4 \times 2 = 400$ runs. However, we are only interested in the main effects (13 df) plus intercept (1 df) and the 2FIs (63 df). Add 5 df for estimation of the error (although maybe 10 would be better), and we need $13 + 1 + 63 + 5 = 82$ runs. Since the number of degrees of freedom for estimating the error is quite small relative to the number of parameters, we'll treat this sort of like a saturated design and use a half-normal plot to help us select active main effects and interactions. The design and response are in the file `Hardwoods.csv`.

We need to be careful about the contrasts! We haven't had to worry about this before because we always used effects coding with two levels. Recall that there are several types of contrasts (e.g. simple, Helmert, forward difference, etc. See the "Learning More" for Section 6.2 if you want a refresher.). Per page 23 of Wheeler (2009), we'll issue the command `options(contrasts=c("contr.sum","contr.poly"))`. This command need only be issued once per session, and it will apply to the entire session.

Software Details

To analyze a design when a factor has more than two levels....

```
options(contrasts=c("contr.sum","contr.poly"))
library(FrF2) # load this package for the halfnor-
mal command
df <- read.csv("Hardwoods.csv",header=TRUE)
lm0 <- lm(failure~.^2,data=df)
summary(lm0)
halfnormal(lm0,ME.partial = TRUE)
```

We included the "ME.partial=TRUE" option for the `halfnormal` command for technical reasons that arise due to the complexity of this model. Without this option, the names of the observations would not be displayed.

The regression output shows that not many of the coefficients are even marginally significant, but we would not be surprised if dropping some terms from the model would free up enough degrees of freedom to get a better estimate of the standard error that would make many of the coefficients significant.

Examining the half-normal plot, we can identify some possibly significant variables and interactions. We're not going to worry about the numbers, only the letters (if you want to worry about the numbers, consult the documentation for the `FrF2` package). For example, "*B*4" and "*B*2" are identified, so the factor *B* (type of glue) merits inclusion in the model (if only "*B*4" was identified, we'd still include factor *B*). Similarly, factors *A* (type of wood) and *D* (type of clamp) are identified. Interactions *AB* and *BC* are identified, so we'll include them. By the heredity principle, we'll include factor *C* (type of applicator). Factor *E* does not appear to be active.

Thus, the model that we'll estimate includes main effects for *A*, *B*, *C*, and *D* and interactions for *AB* and *BC* – remember that we already issued the command to make sure that orthogonal contrasts are used, so we don't have to issue that command again.

Software Details

To analyze the data...

```
lm1 <- lm(failure~A+B+C+D+A:B+B:C,data=df)
summary(lm1)
```

Of course, model checking is in order; assume performed model checking and that we found nothing of concern. We have a good R^2 and many significant effects.

We would like to find the best combination. To do this, we simply predict all the responses for our model, find the row that has the highest response, and then look at the experimental conditions for that row.

Software Details

To create the predictions and find the row with the largest value....

```
lm1Predictions <- predict(lm1)
maxRow <- which.max(lm1Predictions)
maxRow
print(c(maxRow,lm1Predictions[maxRow])
df[maxRow,]
```

In row 43, for wood = maple, glue = PRF-ET, applicator = brush, and clamp = mechanical, we get a predicted response of 92.91. This is slightly lower than the actual response of 96, but it is still higher than rows 9 or 82, with predicted responses of 88.1 and 89.1, respectively.

However, our 82 runs might not have considered all possible combinations; there might be an untried combination that has an even higher predicted response. To investigate this, we need to make predictions for the entire candidate set, which we do as follows.

Software Details

To make predictions for the entire candidate set...

We have to create a data frame for the candidate set and then use that data frame in the `predict` command.

```
dm <- expand.grid(A=c("chestnut","red oak","poplar"
      ,"maple","pine"),B=c("PRF-ET","PRF-RT","RF-RT",
      "EPI-RT","LV-EPI-RT"),C=c("brush","spray"),
      D=c("pneumatic","manual","spring","mechanical"),
      E=c("firm","tight"))
pred400 <- predict(lm1,dm)
which.max(pred400)
pred400[which.max(pred400)]
dm[which.max(pred400),]
```

The output of the above commands tells us that the maximum value of the prediction occurs on line 29, with a predicted value of 112.4614, and the combination that produces this value is A = maple, B = PRF-ET, C = spray, D = pneumatic, and E = firm. Indeed, this is *not* one of the combinations that the experiment tested. Note that this is over 100, which is the maximum percentage. So we should just interpret this value as 100. However, since we didn't actually test this combination, we are unsure that this is a valid prediction.

Recall from regression a regression of Y on X, you don't want to make predictions outside the range of X; this is called *extrapolation*, and it's dangerous. We may be in a similar situation here. It's entirely possible that this is a combination that is much stronger than the combination that produced the prediction of 92.91. However, before using this combination, we should probably do another experiment to make sure.

If we were going to generate a design for this problem, we would use the following commands:

Software Details

To create a design for factors with more than two levels...

```
 df1 <- gen.factorial(levels=c(5,5,2,4,2),varNames =
        c("wood","adhesive","applicator",
        "clamp","pressure"))
des1 <- optFederov(~.^2,df1,nTrials=82,nRepeats=100)
des1$design
```

Of course, we would check the alias matrix and the color map for the design!

Two caveats are in order.

First, when we don't have a balanced and orthogonal design, the maximum D-efficiency is no longer 100% (a balanced design is one that has the same number of observations for each combination). The maximum efficiency can be larger or smaller than 100%, so the D-efficiency no longer tells us how close we are to an optimal design. However, we can use it to compare two designs. In this situation, we will use it to make sure that we have enough iterations. If we increase the number of iterations and the D-efficiency increases markedly, it's a sign that the first design was not a good one.

Second, drawing color maps and computing alias matrices for complex designs such as these is beyond the scope of the AlgDesign software.

We do not recommend that you undertake a mixed-level design yourself. This section is included just to show the power of custom design. If you really need to execute a mixed-level design, you should hire an expert.

Exercises

9.5.1 We wish to learn about how consumers react to different components of credit card offers by direct mail. These components are "intro" (introductory interest rate), "duration" (how long the introductory rate lasts), "goto" (what the interest rate increases to after the introductory period), "fee" (the annual fee), and "color" (the color of the envelope in which the offer is sent). The levels are as follows:

Intro	Duration	Goto	Fee	Color
0%	6 months	Prime + 4%	$0	Red
1.99%	9 months	Prime +5%	$15	White
2.99%	12 months	Prime + 6%	$45	—

The number of combinations is $3 \times 3 \times 3 \times 3 \times 3 \times 2 = 162$. This is far too many. We are concerned with main effects and two-factor interactions. How many parameters to estimate is tricky. Each three-level factor requires two dummy variables. We need ten runs for the main effects plus the intercept. We need 32 for the 2FIs. There is a total of 42 parameters to estimates. Add five design points to estimate the error. Create a custom design and find its D-efficiency. Be sure to set the seed to 1235 for purposes of reproducibility, and don't forget the command to ensure that effects are orthogonal!

Increase `nTrials` to 162. What is the D-efficiency of this design? What is the relative D-efficiency of the first design to the second? (Just divide the former by the latter.)

9.5.2 A chain of multiplex movie theaters (one location will show several movies at the same time) wants to change its design to increase revenues. Three factors are considered:

a) Price (three levels) – Tickets at $12, $14, and $16.
b) Timing (two levels) – All movies will start within the same 15 minute period, or movies will be staggered, all starting within the same 45 minute period.
c) Food (three levels) – Low quality, hot dogs, and popcorn; high quality, gourmet food; no food.

The factorial with no replication requires 18 runs, but only 12 runs can be accommodated.

a) For purposes of reproducibility, set the seed to 1236. Create a design with 12 runs that estimates only main effects. Be sure to note the D-efficiency of the design.
b) Varying the number of runs from 10 to 18, what is the D-efficiency for each number of runs?

9.6 Chapter Exercises

9.1 In a "Software Details" vignette, we ran a regression labeled "lm1" and obtained the regression coefficients. This was done having issued a `contrasts` command. Take note of the coefficients and the R^2. From a fresh start of R Studio, run the "lm1" regression without invoking the options for orthogonal polynomials. Compare the coefficients and R^2 with those previously obtained. What can you conclude?

9.7 Learning More

What we call "custom design" many instead call "optimal design." There are several R packages that do custom design, but none of them is fully developed. Each offers some useful functionality but lacks other useful functionality. No matter which package was chosen, we'd be missing something that we'd have to do ourselves. We have chosen to use the package "AlgDesign," and the two things it is missing are the alias matrix and the color map of the correlations, so code was written for these two functions. The R package "AlgDesign" was written by Robert E. Wheeler in 2004, and it really hasn't changed since 2010. Wheeler designed and marketed one of the first commercial experimental design packages in the early 1980s, "Echip" (www.echip.com), which is still available today. Wheeler died in 2012. Because none of the R packages for experimental design is fully-developed, we have not been able to give a full treatment of optimal design, but we hope that you have some sense of what it can do for you if you acquire a suitable commercial package like JMP or Design-Expert. An excellent book for becoming acquainted with optimal design is Goos and Jones (2011), which is written in a very conversational style with comparatively little mathematics.

• For details on how to use a spreadsheet for optimal design, see Goos and Leemans (2004).

Section 9.1 "Case: Selling Used Cars at Auction I"

• Commenting on Table 9.3, we remarked that we are indifferent between using any of these five designs. A more advanced student would not be indifferent, see Robinson and Anderson-Cook (2011).

Section 9.2 "Selling Used Cars at Auction II"

• Our expectation that a custom design will return the usual factorial design when the number of trials is a power of two or a multiple of four is on solid ground: "Orthogonal designs for two-level factors are also optimal designs. As a result, a computerized-search algorithm for generating optimal designs can generate standard orthogonal designs" (Goos and Jones, 2011, p. 9). These "standard orthogonal designs" include, of course, fractional factorial designs and Plackett–Burman designs. The important word in the above quote is "can" – custom design can generate standard orthogonal designs, but this is not guaranteed. See Exercise 9.4.5 where, with the number of iterations at default, the generated custom design was not factorial, but when the number of iterations was increased, the algorithm did find the factorial design. Note carefully that the above quote says "can generate" not "will generate." It is not guaranteed that the algorithm will find the optimal design.

• On the topic of computer-aided vs. computer-generated designs, see Snee (1985, pp. 230-231).

Section 9.3 "Custom Experiment with Blocking"

• Usually, randomization in a blocked design occurs within blocks. For example, if we were making production runs over three days and used day as a blocking variable, we randomize the runs on each day.

Section 9.4 "Custom Large Screening Experiments"

• Wheeler (2009, p. 4) makes the same point as the paragraph with which we open this section.

• Suppose that factor A and factor B could not both be "too large," so the upper right corner of the design region is not feasible, as shown in the left panel of Figure 9.6. Neither full factorial nor partial factorial will produce an orthogonal

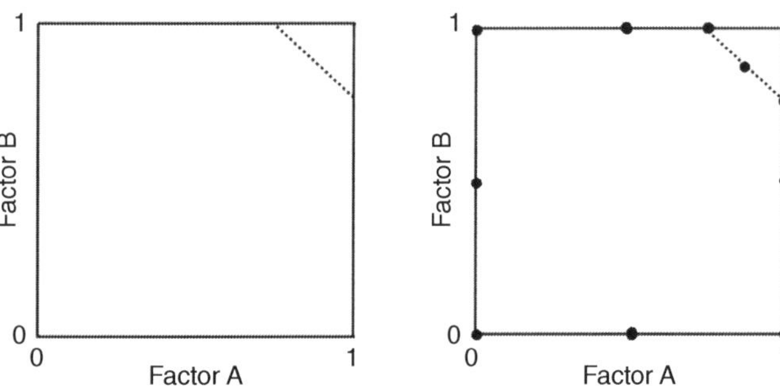

Figure 9.6 Irregular design region.

design, what to do? We might choose some candidate points as in the right panel and then let optimal design choose the best points for us.

Section 9.5 "More Than Two Levels"

• The credit card offer with multilevel exercise is adapted from Jonas Bilenas' paper at NESUG 18. http://www2.sas.com/proceedings/sugi31/196-31.pdf.

• The example in this chapter is a simplified version of a more complicated problem. The real problem was to find a combination that will work well with *all* types of wood. To see how this problem was solved, visit www.statease.com/bdm.

10

Epilogue

This book can serve as a bridge to the more advanced study of experimental design. A person who has had one or two business statistics courses might find it difficult to go beyond this book without serious study of more advanced statistical methods. A person who has a master's in business analytics or a similar discipline can probably dive right into the literature, especially many of the references in the bibliography.

One reason to dive into the literature is because there are better algorithms than those used in this book. Jones (2016) gives a discussion of "definitive screening designs," which represent a great advance over the methods of Chapter 8. Olsen et al. (2016) gives a comparison of definitive screening vs. traditional screening. There are better custom design methods than we have shown, in particular the optimal alias minimizing designs of Jones and Nachtsheim (2011). These methods probably aren't available in R, but some commercial software packages offer them.

Another good reason to open a more advanced text is to learn how to add center points to a model. Center points have two different uses. The first is to add degrees of freedom to a 2^k design, perhaps to get a better estimate of the error. Additionally, the method of center points can also be used to check for nonlinearity. This can be done with R, but it's just a bit beyond the level of this textbook. Consult any standard experimental design textbook.

Be aware that much, if not most, of the literature on experimental design comes from disciplines like engineering where there are well-defined models with small errors. Hence, the literature does not pay much attention to replication. In business applications we really need to emphasize replication for two reasons: (1) we don't have small errors, we have large errors, and (2) usually we don't have well-defined models.

Business Experiments with R, First Edition. B. D. McCullough.

Companion Website: www.wiley.com/go/mccullough/businessexperimentswithr

10.1 The Sequential Nature of Experimentation

One thing that textbook examples and exercises fail utterly to convey is the dynamic nature of experimentation; one-shot experiments are the exception, not the rule. There are preliminary experiments, refining experiments, and follow-up experiments, each one building on its predecessors. If you did not do it as part of your class, you should on your own time conduct Box's helicopter experiment. The original article is Box (1992a), but a web search will turn up many hits for websites devoted to this idea. Cutting all the different helicopter designs is tedious, time consuming, and prone to error. You can use the provided program `Helicopter.R` to specify the parameters and print out a helicopter pattern. There are six factors, so a $2^6 = 64$ run full factorial would be required.

Not all factors matter, however. So first execute a screening design and then run a full factorial to be followed by other experiments. Keep track of the runs on the provided spreadsheet `HelicopterSpreadsheet.csv`. Depending on the parameters, you may need a paper clip to stabilize the helicopter, hence the "paperclip" factor; the levels may be zero and one, or possibly one and two.

To illustrate the way sequential experimentation might proceed, we use a simple example from Chapter 7 of Hoerl and Snee (2012). A food company wants to create a hot chocolate drink that will compete favorably in the marketplace. Prior research had demonstrated that variables determine the quality of the hot chocolate drink are:

1. creaminess
2. chocolate amount
3. thickness

Other variables had also been considered, for example, temperature. A lukewarm hot chocolate is definitely inferior to a steaming hot chocolate. However, it is known that the optimal temperature for hot drinks is between $150^\circ and\ 170^\circ F$, so there was no need to experiment with this factor. The hunger of the consumer is another variable – hungry people like hot chocolate more than people who are not hungry. By ensuring that all the taste-testers were not hungry, this factor did not have to be varied. Prior experiments showed that there is an optimal ratio of cocoa to sugar, so only one of these variables has to be varied.

After some preliminary experimentation, ranges were determined for all three variables, f or example, a minimum amount of chocolate and a maximum amount of chocolate or a minimum amount of thickening agent and a maximum amount of thickening agent. Each range was scaled to be from 1 to 10, where 10 is the maximum. A good starting point for the first cup of hot chocolate was determined to be creaminess = 5, chocolatiness = 5, and thickness = 3.

Since there are three factors, a 2^3 design is in order. We'll use a center point for this design, which will be the abovementioned good starting point. We'll be aggressive for the first experiment, increasing and decreasing each factor by two units instead of one. For example, the hi and lo for creaminess will be seven and three. The taste-testers will taste each combination, and their scores will be combined into an overall score for the combination. Results for the first experiment are shown in Table 10.1. The runs are shown in standard order with the center point at the bottom, but of course the actual order of the experiments was randomized.

We don't need regression to analyze these data; we can do it by eyeball. The first experiment clearly shows that the taste-testers clearly preferred the creamiest, chocolatiest, and thickest combination, with a score of 78. We'll use this combination as the center point of the next experiment. While all the variables have an effect, cream seems to have the strongest effect. This result surprised the team, which has previously believed that the amount of chocolate would be the most important factor.

To see that cream is the most important factor, start at the maximum: 7,7,5, which has a score of 78. Changing cream while holding the others constant means going to the point 3,7,5, which has a score of 52, a 26-point drop. Changing chocolate while holding the others constant means going to the point 7,3,5 with a score of 68, only a 10-point drop. Changing thickness while holding the others constant means going to the point 7,7,1 with a score of 67, only an 11-point drop.

We're not going to be so aggressive for the second experiment. From the center point we'll increase and decrease each factor by one unit, as shown in Table 10.2. This time the most preferred combination is clearly cream = 8, chocolate = 6, and thickness = 4. We'll use this combination as the center point of the next experiment. Again, creaminess is the most important factor. Starting at the maximum (8,6,4), changing creaminess has a larger effect than changing the other factors. Make sure you can see this.

Table 10.1 Hot chocolate experiment I.

Run	Cream	Chocolate	Thickness	Score
1	3	3	1	21
2	7	3	1	55
3	3	7	1	34
4	7	7	1	67
5	3	3	5	38
6	7	3	5	68
7	3	7	5	52
8	7	7	5	78
9	5	5	3	53

Table 10.2 Hot chocolate experiment II.

Run	Cream	Chocolate	Thickness	Score
1	6	6	4	71
2	8	6	4	91
3	6	8	4	74
4	8	8	4	83
5	6	6	6	76
6	8	6	6	85
7	6	8	6	70
8	8	8	6	79
9	7	7	5	76

Table 10.3 Hot chocolate experiment III.

Run	Cream	Chocolate	Thickness	Score
1	7	5	3	77
2	9	5	3	84
3	7	7	3	73
4	9	7	3	85
5	7	5	5	75
6	9	5	5	84
7	7	7	5	81
8	9	7	5	89
9	8	6	4	93

Again there is no need to be aggressive with the factor levels, and each factor will be changed by one unit from the center point. The third experiment, results presented in Table 10.3, shows that the center point is the preferred combination – any deviation from these levels leads to a decrease in the score. Therefore, there is reason to believe we have found a global maximum.

More experiments could be conducted to verify the global maximum, but even as it is, an excellent formulation for hot chocolate has been achieved in only 27 experiments!

10.2 Approaches to Sequential Experimentation

Vining (2011) laments that books commonly teach experiments as "one-shot" deals. Given the nature of textbooks, this is understandable. The sequential nature can only be realized in an actual experiment. He gives several reasons for a second and third experiments:

- Move the experimental region.
- Add experimental points to an existing design.
- Change the levels of some or all of the factors.
- Drop some factors and add others.
- Replicate the entire experiment to increase the degrees of freedom in order to better see important effects.
- Perform a confirmatory experiment around the "best" set of conditions.

George Box began an article on the nature of sequential experimentation Box (1992b) thusly:

> Sir Ronald Fisher once said that "the best time to design an experiment is after you've done it." One manifestation of this seeming paradox is that after a preliminary experimental design has been run, questions are often raised about the results with an acuity of hindsight which is quite extraordinary. Questions like: "That factor doesn't seem to be doing anything. Wouldn't it have been better if you had included this other variable?"
> "You don't seem to have varied that factor over a wide enough range."
> "The experiments with high pressure and high temperature seem to give the best results; it's a pity you didn't experiment with these factors at even higher levels."
> And so on and so on.
> Such questions come up because the results from any experiment depend critically on decisions requiring *judgment* – Which factors should be studied? How should the response be measured? At which levels of a given factor should be experiments be run? Which experimental design should be measured? Human judgment is fallible, and these matters need to be reconsidered as you proceed and learn more about what is going on. If it can be avoided, therefore, it is best not to plan a large "all-encompassing" experiment at the outset. This is the time when you know *least* about the system.

It's not enough to know that you will conduct a sequence of experiments. You should have a methodology for conducting the sequence. In addition to offering much practical advice, Snee (2009) discusses four approaches to sequential experimentation. The overarching idea is to plan each step to gain maximal information for taking the next step. For example, there are times that, moving from one experiment to the next, factor ranges may change. A new experiment may be necessary to determine the extent of the range changes, and then the new ranges must be incorporated into subsequent experiments. The four approaches are as follows:

1) If you're familiar with the "hill-climbing" methods of optimization (and if you're not, skip this item), this idea is embodied in the "response surface methodology" to guide a sequence of experiments to a global extremum.
2) Fractional factorials can be used to establish important factors, and additional fractional factorials can be used to investigate possible interactions. Based on the fractional factorials, a full model can be estimated, though information learned from this may require additional experimentation.
3) Models with center points can be used to assess curvature. If curvature is detected, then multilevel models can be used to estimate quadratic effects and build a better model.
4) DuPont Corporation, over its many years of industrial experimentation, developed a methodology called "Screening, Characterization and Optimization" (SCO) that is described more fully in Snee's article.

Realizing that you probably have not a single experiment before you, but a series of experiments, it is important to plan ahead so that you don't exhaust your resources before the series of experiments is completed. Anderson-Cook and Lu have a pair of short articles on how to plan for a series of experiments (Anderson-Cook and Lu (2016a) and Anderson-Cook and Lu (2016b)).

References

R. L. Ackoff. *The Art of Problem Solving*. Wiley, 1978.

Andrew Althouse. Post-hoc power: Not empowering, just misleading. *Journal of Surgical Research*, 2020. https://doi.org/10.1016/j.jss.2019.10.049.

Eric T. Anderson and Duncan Simester. A step-by step guide to smart business experiments. *Harvard Business Review*, March: 98–105, 2011.

Christine Anderson-Cook and Lu Lu. Best bang for your buck – part 1. *Quality Progress*, 49(10): 45–48, 2016a.

Christine Anderson-Cook and Lu Lu. Best bang for your buck – part 2. *Quality Progress*, 49(11): 50–52, 2016b.

David R. Appleton, Joyce M. French, and Mark P. J. Vanderpump. Ignoring a covariate: An example of simpson's paradox. *The American Statistician*, 50 (4): 340–341, 1996.

Susan Athey and Guido Imbens. The econometrics of randomized experiments. In C. Rao and S. Sinharay, editors, *Handbook of Economic Field Experiments*, volume 1, pages 73–140. Elsevier, North-Holland, 2017.

P. C. Austin, M. M. Mamdani, D. N. Juurlinka, and J. E. Hux Testing multiple statistical hypotheses resulted in spurious associations: A study of astrological signs and health. *Journal of Clinical Epidemiology*, 59(9): 964–969, 2006.

Yanik J. Bababekov, Sahael M. Stapleton, Jessica L. Mueller, Zhi ven Fong, and David C. Chang. A proposal to mitigate the consequences of type 2 error in surgical science. *Annals of Surgery*, 267(4): 621–622, 2018.

Yanik J. Bababekov, Ya-Ching Hung, Yu-Tien Hsu, Brooks V. Udelsman, Jessica L. Mueller, Hsu-Ying Lin, Sahael M. Stapleton, and David C. Chang. Is the power threshold of 0.8 applicable to surgical science? – empowering the underpowered study. *Journal of Surgical Research*, 241(September): 235–239, 2019.

Larry B. Barrentine. *An Introduction to Design of Experiments: A Simplified Approach*. ASQ Quality Press, 1999.

S. Berman, L. DalleMule, M. Greene, and J. Lucker. Simpson's paradox: A cautionary tale in advanced analytics. *Significance*, September, 2012. https://www

Business Experiments with R, First Edition. B. D. McCullough.

Companion Website: www.wiley.com/go/mccullough/businessexperimentswithr

.significancemagazine.com/14-the-statistics-dictionary/106-simpson-s-paradox-a-cautionary-tale-in-advanced-analytics (accessed 19 June 2020).

P. J. Bickel, E. A. Hammel, and J. W. O'Connell. Sex bias in graduate admissions: Data from Berkeley. *Science*, 187(4175): 398–404, 1975.

Howard S. Bloom. Minimum detectable effects. *Evaluation Review*, 19(5): 547–556, 1995.

Roger Bohn. Chasing the noise in industrial a/b testing: What to do when all the low-hanging fruit have been picked?, 2018. http://andrewgelman.com/2018/06/16/chasing-noise-industrial-b-testing-low-hanging-fruit-picked/ (accessed 19 June 2020).

William M. Boldstad. *Introduction to Bayesian Statistics, 2e*. Wiley, New York, 2007.

Dennis Boos and J. M. Hughes-Oliver. How large does n have to be for z and t intervals? *The American Statistician*, 54(2): 121–128, 2000.

George Box. Do interactions matter? *Quality Engineering*, 2(3): 365–369, 1990.

George Box. George's column: Teaching engineers experimental design with a paper helicopter. *Quality Engineering*, 4(3): 453–459, 1992a.

George Box. Sequential experimentation and sequential assembly of designs. *Quality Engineering*, 5(2): 321–330, 1992b.

George Box. *Improving Almost Anything*. Wiley, 2006.

George Box, J. Stuart Hunter, and William G. Hunter. *Statistics for Experimenters: Design, Innovation, and Discovery, 2e*. Wiley, 2005.

Lawrence D. Brown and Xuefeng Li. Confidence intervals for two sample binomial distribution. *Journal of Statistical Planning and Inference*, 130: 359–375, 2005.

Lawrence D. Brown, T. Tony Cai, and Anirban DasGupta. Interval estimation for a binomial proportion. *Statistical Science*, 16(2): 101–133, 2005.

Richard H. Browne. On the use of a pilot sample for sample size determination. *Statistics in Medicine*, 14(17): 1933–1940, 1995.

K. A. Brownlee. Statistics of the 1954 polio vaccine trials. *Journal of the American Statistical Association*, 50(272): 1005–1013, 1955.

Richard Burnham. That voodoo we do – marketers are embracing statistical design of experiments. *The Direct Marketing Association*, 2004.

D. Campbell and J. Stanley. *Experimental and Quasi-Experimental Designs for Research*. Rand-McNally, Chicago, 1963.

Rolf Carlson. *Design and Optimization in Organic Synthesis*. Elsevier, Amsterdam, 1992.

Brian Christian. The a/b test: Inside the technology that's changing the rules of business. *Wired Magazine*, April, 2012.

W. S. Connor and Shirley Young. *Fractional Factorial Designs for Experiments with Factors at Two and Three Levels, AMS 58*. National Bureau of Standards, 1957.

Veronica Czitrom. One-factor-at-a-time versus designed experiments. *The American Statistician*, 53(2): 126–131, 1999.

Jerry Dallal. One sided tests. http://www.jerrydallal.com/LHSP/onesided.htm, 2012.

C. Daniel and F. S. Woods. *Fitting Equations to Data, 2e*. Wiley, 1980.

Thomas H. Davenport. How to design smart business experiments. *Harvard Business Review*, February: 68–76, 2009.

Angela Dean, Daniel Voss, and Danel Draguljic. *Design and Analysis of Experiments, 2e*. Springer, 2017.

Kevin Dunn. *Process Improvement Using Data*, 2019. manuscript. https://learnche.org/pid/ (accessed 19 June 2020).

Robert G. Easterling. *Fundamentals of Statistical Experimental Design and Analysis*. Wiley, New York, 2015.

Econsultancy. Conversion rate optimization report 2015. Technical report, 2015.

Gunnar Eliasson. *Advanced Public Procurement as Industrial Policy*. Springer, 2010.

Erik Barry Erhardt. Designing a better paper helicopter using response surface methodology. *Stats: The Magazine for Students of Statistics*, 48: 14–21, 2007.

L. Eriksson, E. Johansson, N. Kettaneh-Wold, C. Wikstrom, and S. Wold. *Design of Experiments: Principles and Applications*. Learnways AB, 2000.

Anna Errore, Bradley Jones, William Li, and Christopher Nachtsheim. Benefits and fast construction of efficient two-level foldover designs. *Technometrics*, 59 (1):48–57, 2017.

Elea Feit and Ron Berman. Test & roll: Profit maximizing a/b tests. *Management Science*, 38(6): 1038–1058, 2019.

Stephen Few. *Now You See It: Simple Visualization Techniques for Quantitative Analysis*. Analytics Press, Burlingame, CA, 2009.

Stephen Few. *Show Me the Numbers: Designing Tables and Graphs to Enlighten, 2e*. Analytics Press, Burlingame, CA, 2012.

R. A. Fisher. *The Design of Experiments, 9e*. Hafner Press, 1971.

Sara Fontdecaba, Pere Grima, and Xavier Tort-Martorell. Analyzing DOE with statistical software packages: Controversies and proposals. *The American Statistician*, 68(3): 205–211, 2014.

Charles H. Franklin. Efficient estimation in experiments. *The Political Methodologist*, 4(1): 13–15, 1991.

Laura J. Freeman, Anne G. Ryan, Jennifer L. K. Kensler, Rebecca M. Dickinson, and G. Geoffrey Vining. A tutorial on the planning of experiments. *Quality Engineering*, 25: 315–332, 2013.

Normand L. Frigon and David Mathews. *Practical Guide to Experimental Design*. Wiley, 1997.

Abhijit Ganguly and Jim Euchner. Conducting business experiments: Validating new business models. *Research Technology Management*, March-April: 27–35, 2018.

Andrew Gelman. Don't calculate post-hoc power using observed estimate of effect size. *Annals of Surgery*, 269(1): e9–e10, 2019.

Andrew Gelman and John Carlin. Beyond power calculations: Assessing Type S (Sign) and Type M (Magnitude) errors. *Psychological Science*, 9(6): 641–651, 2014.

Andrew Gelman and Erik Loken. The statistical crisis in science. *American Scientist*, 102: 460–465, 2014.

Andrew Gelman and Hal Stern. The difference between "significant" and "not significant" is not itself statistically significant. *The American Statistician*, 60: 328–331, 2006.

Alan S. Gerber and Donald P. Green. *Field Experiments: Design, Analysis and Interpretation*. W. W. Norton, New York, 2012.

Alexis Goncalves. *Innovation Hardwired*. Innovation Insight Network, 2008.

Luzia Goncalves, M. Rosario de Oliveira, Claudia Pascoal, and Ana Pires. Sample size for estimating a binomial proportion: Comparison of different methods. *Journal of Applied Statistics*, 39(11): 2453–2473, 2012.

S. N. Goodman and J. A. Berlin. The use of predicted confidence intervals when planning experiments and the misuse of power when interpreting results. *Annals of Internal Medicine*, 121: 200–206, 1994.

Peter Goos and Bradley Jones. *Optimal Design of Experiments: A Case Study Approach*. Wiley, New York, 2011.

Peter Goos and Herlinde Leemans. Teaching optimal design of experiments using a spreadsheet. *Journal of Statistics Education*, 12(3), 2004. DOI:10.1080/10691898.2004.11910631.

Jacques Goupy and Lee Creighton. *Introduction to Design of Experiments with JMP Examples, 3e*. SAS Press, 2007.

Joshua Green. The science behind those Obama campaign emails. *Bloomberg Businessweek*, 29 November, 2012.

Robert Greevy, Bo Lu, Jeffrey H. Silber, and Paul Rosenbaum. Optimal multivariate matching before randomization. *Biostatistics*, 5(2): 263–275, 2004.

Robert P. Hamlin. The rise and fall of the Latin square in marketing: A cautionary tale. *European Journal of Marketing*, 39(3/4): 328–350, 2005.

Frank Harrell. *Regression Modeling Strategies, 2e*. Springer, 2015.

Glenn W. Harrison and John A. List. Field experiments. *Journal of Economic Literature*, 42(4): 1009–1055, 2004.

Austin Bradford Hill. The environment and disease: Association or causation? *Proceedings of the Royal Society of Medicine*, 58(5): 295–300, 1965.

John M. Hoenig and Dennis M. Heisey. The abuse of power: The pervasive fallacy of power calculations for data analysis. *The American Statistician*, 55(1): 1–6, 2001.

Roger Hoerl and Ronald Snee. *Statistical Thinking, 2e*. Wiley, 2012.

Charles Holland and David Cochran. *Breakthrough Business Results with MVT*. Wiley, 2005.

Chetan Huded, Jill Rosno, and Vinay Prasad. When research evidence is misleading. *AMA Journal of Ethics*, 15: 29–33, 2013.

Kosuke Imai, Gary King, and Elizabeth A. Stuart. Misunderstandings between experimentalists and observationalists about causal inference. *Journal of the Royal Statistical Society A*, 171(2): 481–502, 2008.

ISIS-2. Collaborative group randomised trial of intravenous streptokinase, oral aspirin, both, or neither among 17 187 cases of suspected acute myocardial infarction. *The Lancet*, 332(8607): 349–360, 1988.

Ramesh Johari, Pete Koomen, Leonid Pekelis, and David Walsh. Peeking at a/b tests: Why it matters, and what to do about it. In *Proceedings of the 23rd ACM SIGKDD International Conference on Knowledge Discovery and Data Mining*, KDD '17, pages 1517–1525, New York, NY, USA, 2017. ACM.

Brian L. Joiner. Lurking variables: Some examples. *The American Statistician*, 35(4): 227–233, 1981.

Bradley Jones. 21st century screening experiments: What, why, and how. *Quality Engineering*, 28(1): 98–106, 2016.

Bradley Jones and Christopher J. Nachtsheim. Efficient designs with minimal aliasing. *Technometrics*, 53(1): 62–71, 2011.

Steven A. Julious. Sample size of 12 per group rule of thumb for a pilot study. *Pharmaceutical Statistics*, 4: 287–291, 2008.

Steven A. Julious and Mark A. Mullee. Confounding and Simpson's paradox. *BMJ*, 309(6967): 1480–1481, 1994.

Walter N. Kernan, Catherine M. Viscoli, Robert W. Makuch, Lawrence M. Brass, and Ralph I. Horwitz. Stratified randomization for clinical trials. *Journal of Clinical Epidemiology*, 52: 19–26, 1999.

C. N. Knaflic. *Storytelling with Data: A Data Visualization Guide for Business Professionals*. Wiley, 2015.

Ron Kohavi, Thomas Crook, Brian Frasca, and Roger Longbotham. Seven pitfalls to avoid when running controlled experiments on the web. In *Proceedings of the 15th ACM SIGKDD International Conference on Knowledge Discovery and Data Mining*, KDD '09, pages 1105–1114, New York, NY, USA, 2009a.

Ron Kohavi, Roger Longbotham, Dan Sommerfield, and Randal M. Henne. Controlled experiments on the web: Survey and practical guide. *Data Mining and Knowledge Discovery*, 18(1): 140–181, 2009b.

Ron Kohavi, Alex Deng, Brian Frasca, Roger Longbotham, Toby Walker, and Ya Xu. Trustworthy online controlled experiments: Five puzzling outcomes explained. In *Proceedings of the 18th ACM SIGKDD International Conference on Knowledge Discovery and Data Mining*, KDD '12, pages 786–794, New York, NY, USA, 2012.

Ron Kohavi, Alex Deng, Roger Longotham, and Ya Xu. Seven rules of thumb for web site experimenters. In *Proceedings of the 20th ACM SIGKDD International Conference on Knowledge Discovery and Data Mining*, KDD '14, page 1857–1866, 2014.

Warren F. Kuhfeld. Experimental design: Efficiency, coding, and choice designs. In *Marketing Research Methods in SAS*, chapter 3, pages 53–241. SAS Institute, Cary, NC, 2010.

M. H. Kutner, C. J. Nachtsheim, J. Neter, and W. Li. *Applied Linear Statistical Models, 5e*. McGraw Hill, 2005.

Russ Lenth. Some practical guidelines for effective sample size determination. *The American Statistician*, 55(3): 187–193, 2001.

Russ Lenth. Statistical power calculations. *Journal of Animal Science*, 85: E24–E29, 2007.

Xiang Li, Nandan Sudarsanam, and Daniel D. Frey. Regularities in data from factorial experiments. *Complexity*, 11(5): 32–45, 2006.

Morris M. Lightstone, David R. Dillingham, David D. Carpenter, and William M. Harral. The evolution of quality in the automotive industry. *Annual Quality Congress*, 47: 768–780, 1993.

Bo Lu, Robert Greevy, Xinyi Xu, and Cole Beck. Optimal nonbipartite matching and its statistical applications. *The American Statistician*, 65(1): 21–30, 2011.

Jim Manzi. *Uncontrolled: The Surprising Payoff of Trial-and-Error for Business, Politics, and Society*. Basic Books, New York, 2012.

Arfa Maqsood and Rafia Shafi. A soft-drink experiment using replicated full factorial (rff) design. *International Journal of Computer Applications*, 171(1): 25–30, 2017.

B. D. McCullough. The hazards of subgroup analysis in randomized business experiments and how to avoid them. In Kenneth Lawrence and Ron Klimberg, editors, *Contemporary Perspectives on Data Mining*, vol. 3, chapter 6, pages 79–92. Information Age Publishers, 2017.

Colin McFarland. *Experiment!: Website Conversion Rate Optimization with A/B and Multivariate Testing*. New Riders, 2012.

Robert Mee. *A Comprehensive Guide to Factorial Two-Level Experimentation*. Springer, 2009.

Paul Meier. The biggest public health experiment ever: The 1954 trial of the salk poliomyelitis vaccine. In Judith Tanur, editor, *Statistics: A Guide to the Unknown, 3e*, chapter 1, pages 3–14. Duxbury, 1989.

Douglas C. Montgomery. *Design and Analysis of Experiments, 9e*. Wiley, New York, 2017.

Douglas C. Montgomery and Cheryl L. Jennings. An overview of industrial screening experiments. In Angela Dean and Susan Lewis, editors, *Screening Methods for Experimentation in Industry, Drug Discovery, and Genetics*, chapter 1, pages 1–20. Springer, New York, 2006.

M. Moran. *Do It Wrong Quickly*. IBM Press, 2007.

Barry K. Moser and Gary R. Stevens. Homogeneity of variance in the two-sample means test. *The American Statistician*, 46(1): 19–21, 1992.

Lincoln E. Moses and Frederick Mosteller. Experimentation: Just do it! In Bruce D. Spencer, editor, *Statistics and Public Policy*, chapter 12, pages 212–232. Clarendon Press, Oxford, 1997.

J. A. Nelder. A reformulation of linear models. *Journal of the Royal Statistical Society*, 140(1): 48–77, 1977.

H. James Norton and George Divine. Simpson's paradox and how to avoid it. *Significance*, 12(4): 40–43, 2015.

Gary W. Oehlert. *A First Course in Design and Analysis of Experiments*. 2010. users.stat.umn.edu/gary/book/fcdae.pdf (accessed 19 June 2020)

Rebecca Olsen, John Lawson, Nathaniel Rohbock, and Brian Woodfield. Practical comparison of traditional and definitive screening designs in chemical process development. *International Journal of Experimental Design and Process Optimization*, 5(1–2): 2–33, 2016.

C. S. Peirce and J. Jastrow. On small differences in sensation. *Memoirs of the National Academy of Sciences*, 3: 73–83, 1885.

Robin Pemantle, Diana Mutz, and Philip Pham. The perils of balance testing in experimental design: Messy analyses of clean data. *The American Statistician*, 73(1): 32–42, 2019.

Richard Peto. Similarities and differences between adjuvant hormonal therapy in breast and prostate cancer. *Prostate Cancer Update*, 4(1): 9–11, 2005.

David B. Pillemer. One- versus two-tailed hypothesis tests in contemporary educational research. *Educational Researcher*, 20(9): 13–17, 1991.

Alex Reinhart. *Statistics Done Wrong*. No Starch Press, 2015.

Timothy J. Robinson and Christine M. Anderson-Cook. A closer look at d-optimality for screening designs. *Quality Engineering*, 23(1): 1–14, 2011.

Paul R. Rosenbaum. *Design of Observational Studies*. Springer, 2010.

Paul R. Rosenbaum. *Observation & Experiment: An Introduction to Causal Inference*. Harvard University Press, 2017.

Donald B. Rubin. Comment: The design and analysis of gold standard randomized experiments. *Journal of the American Statistical Association*, 103 (484): 1350–1356, 2008.

Graeme D. Ruxton. The unequal variance t-test is an underused alternative to student's t-test and the mann-whitney U test. *Behavioral Ecology*, 17(4): 688–690, 2006.

SalesForce.com. 2014 state of marketing report. Technical report, 2014.

Michael Schrage. *The Innovator's Hypothesis: How Cheap Experiments Are Worth More than Good Ideas*. MIT Press, Cambridge, MA, 2014.

Howard Seltman. *Experimental Design and Analysis*, 2018. http://www.stat.cmu.edu/~hseltman/309/Book/Book.pdf (accessed 19 June 2020).

Stephen J. Senn. Testing for baseline balance in clinical trials. *Statistics in Medicine*, 13: 1715–1726, 1994.

Stephen J. Senn. Power is indeed irrelevant in interpreting completed studies. *British Medical Journal*, 325(7535): 1304, 2002.

Stephen J. Senn. *Statistical Issues in Drug Development, 2e*. Wiley, West Sussex, England, 2007.

Stephen P. Senn. Francis galton and regression to the mean. *Significance*, 8(3): 124–126, 2011.

Stephen P. Senn. Tea for three: Of infusions and inferences and milk in first. *Significance*, 9(12): 30–33, 2012.

Walter A. Shewart. *Statistical Method from the Viewpoint of Quality Control*. Graduate School of the Department of Agriculture, 1939.

E. H. Simpson. The interpretation of interaction in contingency tables. *Journal of the Royal Statistical Society. Series B*, 13(2): 238–241, 1951.

James R. Simpson, Charles M. Listak, and Gergory T. Hutto. Guidelines for planning and evidence for assessing a well-designed experiment. *Quality Engineering*, 25: 333–355, 2013.

Dan Siroker and Pete Koomen. *A/B Testing: The Most Powerful Way to Turn Clicks Into Customers*. Wiley, 2013.

Peter Sleight. Debate: Subgroup analyses in clinical trials: Fun to look at – but don't believe them. *Current Controlled Trials in Cardiovascular Medicine*, 1 (1): 25–27, 2000.

Gary Smith. A fallacy that will not die. *The Journal of Investing*, 25(1): 7–15, 2016.

Ronald D. Snee. Computer-aided design of experiments – some practical experiences. *Journal of Quality Technology*, 17(4): 222–236, 1985.

Ronald D. Snee. Raise your batting average: Remember the importance of sequence in experimentation. *Quality Progress*, 42(12): 64–66, 2009.

Genichi Taguchi. *Introduction to Quality Engineering*. UNIPUB, Kraus International Publications: Asian Productivity Organization, 1986.

Ajit C. Tamhane. *Statistical Analysis of Designed Experiments: Theory and Applications*. Wiley, 2009.

Stefan Thomke. *Experimentation Works: The Surprising Power of Business Experiments*. Harvard Business Review Press, Cambridge, MA, 2020.

Stefan Thomke and Jim Manzi. *The discipline of business experimentation. Harvard Business Review*, December, 2014.

Brad Tuttle. The 5 big mistakes that led to ron johnson's ouster at jc penney, 2013. http://business.time.com/2013/04/09/the-5-big-mistakes-that-led-to-ron-johnsons-ouster-at-jc-penney/ (accessed 19 June 2020).

Gerald van Belle. *Statistical Rules of Thumb*. Wiley, 2008.

Penny Verhoeven and Victor Wakeling. Emphasizing one-sided interval estimation in introductory business statistics courses. *Proceedings of the 39th Annual Meeting of the Southeast Region of the DSI*, Charleston, SC, pages 988–992. 2009.

Geoff Vining. Technical advice: Design of experiments, response surface methodology, and sequential experimentation. *Quality Engineering*, 23(2): 217–220, 2011.

Geoff Vining. Technical advice: Experimental protocol and the basic principles of experimental design. *Quality Engineering*, 25: 307–311, 2013.

Thavatchai Vorapongsathorn, Sineenart Taejaroenkul, and Chukiat Viwatwongkasem. A comparison of type i error and power of bartlett's test, levene's test and cochran's test under violation of assumptions. *Songklanakarin J. Sci. Technol.*, 26(4): 537–547, 2004.

Stefan Wager and Susan Athey. Estimation and inference of heterogeneous treatment effects using random forests. *Journal of the American Statistical Association*, 113(523): 1228–1242, 2018.

Howard Wainer. *Truth or Truthiness: Distinguishing Fact from Fiction by Learning to Think Like a Data Scientist*. Cambridge University Press, 2016.

Howard Wainer and Lisa M. Brown. Three statistical paradoxes in the interpretation of group differences: Illustrated with medical school admission and licensing data. In C. Rao and S. Sinharay, editors, *Handbook of Statistics 26: Psychometrics*, pages 893–918. Elsevier, North-Holland, 2007.

Daniel Waisberg and Avinash Kaushik. Web analytics 2.0: Empowering customer centricity. *The Original Search Engine Marketing Journal*, 2(1): 5–11, 2009.

John A. Wass. First steps in experimental design: The screening experiment. *Journal of Validation Technology*, 16(2): 49–57, 2010.

John A. Wass. First steps in experimental design II: More on screening experiments. *Journal of Validation Technology*, 17(1): 12–20, 2011.

Ronald L. Wasserstein and Nicole A. Lazar. The ASA's statement on p-values: Context, process, and purpose. *The American Statistician*, 70(2): 129–133, 2016.

Kimberly Watson-Hemphill and Bill Kastle. Role for design of experiments in financial operations, 2012. https://www.isixsigma.com/tools-templates/design-of-experiments-doe/role-design-experiments-financial-operations/ (accessed 19 June 2020).

Donald J. Wheeler. *Screening Designs, 2e*. CRC Press, 1990.

Robert E. Wheeler. *Comments on Algorithmic Design*, 2009. https://cran.r-project.org/web/packages/AlgDesign/vignettes/AlgDesign.pdf (accessed 19 June 2020).

Dona M Wong. *The Wall Street Journal Guide to Information Graphics: The Dos and Don'ts of Presenting Data, Facts, and Figures*. WW Norton & Company, 2013.

C. F. Jeff Wu and Michael S. Hamada. *Experiments: Planning, Analysis and Optimization, 2e*. Wiley, 2009.

J. Clifton Young. Blocking, replication and randomization-the key to effective experimentation: A case study. *Quality Engineering*, 9(2): 269–277, 1996.

S. Stanley Young and Alan Karr. Deming, data and observational studies. *Significance*, September: 116–120, 2011.

Asad Zaman. Causal relations via econometrics. *International Econometric Review*, 2(1): 36–56, 2010.

Stephen T. Ziliak. Guinnessometrics: The economic foundation of "student's" t. *Journal of Economic Perspectives*, 22(4): 199–216, 2008.

Stephen T. Ziliak. Balanced versus randomized field experiments in economics: Why w. s. gosset aka "student" matters. *Review of Behavioral Economics*, 1: 167–208, 2014.

Index

Business Experiments with R, First Edition. B. D. McCullough.

Companion Website: www.wiley.com/go/mccullough/businessexperimentswithr

d

e

Printed and bound by CPI Group (UK) Ltd, Croydon, CR0 4YY

17/04/2025

14658854-0001